Building Machine Learning and Deep Learning Models on Google Cloud Platform

A Comprehensive Guide for Beginners

Ekaba Bisong

Apress®

Building Machine Learning and Deep Learning Models on Google Cloud Platform:
A Comprehensive Guide for Beginners

Ekaba Bisong
OTTAWA, ON, Canada

ISBN-13 (pbk): 978-1-4842-4469-2   ISBN-13 (electronic): 978-1-4842-4470-8
https://doi.org/10.1007/978-1-4842-4470-8

Managing Director, Apress Media LLC: Welmoed Spahr
Acquisitions Editor: Susan McDermott
Development Editor: Laura Berendson
Coordinating Editor: Rita Fernando

Cover designed by eStudioCalamar

Cover image designed by Freepik (www.freepik.com)

Distributed to the book trade worldwide by Springer Science+Business Media New York, 233 Spring Street, 6th Floor, New York, NY 10013. Phone 1-800-SPRINGER, fax (201) 348-4505, e-mail orders-ny@springer-sbm.com, or visit www.springeronline.com. Apress Media, LLC is a California LLC and the sole member (owner) is Springer Science + Business Media Finance Inc (SSBM Finance Inc). SSBM Finance Inc is a **Delaware** corporation.

For information on translations, please e-mail rights@apress.com, or visit http://www.apress.com/rights-permissions.

Apress titles may be purchased in bulk for academic, corporate, or promotional use. eBook versions and licenses are also available for most titles. For more information, reference our Print and eBook Bulk Sales web page at http://www.apress.com/bulk-sales.

Any source code or other supplementary material referenced by the author in this book is available to readers on GitHub via the book's product page, located at www.apress.com/9781484244692. For more detailed information, please visit http://www.apress.com/source-code.

Printed on acid-free paper

This book is dedicated to the Sovereign and Holy Triune God who created the Heavens and the Earth and is the source of all intelligence. To my parents Prof. and Prof. (Mrs.) Francis and Nonso Bisong, my mentors Prof. John Oommen and late Prof. Pius Adesanmi, and to Rasine, my best friend and companion.

Table of Contents

About the Author

Ekaba Bisong is a Data Science Lead at T4G. He previously worked as a Data Scientist/Data Engineer at Pythian. In addition, he maintains a relationship with the Intelligent Systems Lab at Carleton University with a research focus on learning systems (encompassing learning automata and reinforcement learning), machine learning, and deep learning. Ekaba is a Google Certified Professional Data Engineer and a Google Developer Expert in machine learning. Ekaba is from the Ejagham Nation.

About the Technical Reviewer

Vikram Tiwari is a co-founder of Omni Labs, Inc. where he handles all things tech. At Omni they are redesigning how we leverage Internet to get things done. He is also a Google Developer Expert for machine learning and Google Cloud Platform. He speaks at various conferences and runs hands-on workshops on cloud and machine learning topics. He loves working with start-ups and developers as mentor to help them navigate various challenges through their journey.

Other than work, he runs a community of developers at Google Developer Group Cloud, San Francisco. In his free time, he bikes around the city and hills of San Francisco.

Gonzalo Gasca Meza is a developer programs engineer working on the GCP Machine Learning platform. He currently works in the TensorFlow and Machine Learning infrastructure. Gonzalo holds a bachelor's degree in computer science and a master's degree in software engineering from the University of Oxford. Before joining Google, Gonzalo worked on Enterprise-focused products for voice and video communications. He is based in Sunnyvale, California.

Acknowledgments

I want to use this opportunity to appreciate the staff and faculty of Carleton University School of Computer Science; they made for a friendly and stimulating atmosphere during the writing of this book. I want to thank my friends at the graduate program, Abdolreza Shirvani, Omar Ghaleb, Anselm Ogbunugafor, Sania Hamid, Gurpreet Saran, Sean Benjamin, Steven Porretta, Kenniy Olorunnimbe, Moitry Das, Yajing Deng, Tansin Jahan, and Tahira Ghani. I also want to thank my dear friends at the undergraduate level during my time as a teaching assistant, Saranya Ravi and Geetika Sharma, as well as my friend and colleague at the Intelligent Systems Lab, Vojislav Radonjic. They all were most supportive with a kind and generous friendship that kept me going despite the obvious pressures at the time. I want to particularly thank Sania Hamid, who helped me type portions of the manuscript. Nonetheless, I take full responsibility for any typographical errors contained in this book.

I want to thank my friends Rasine Ukene, Yewande Marquis, Iyanu Obidele, Bukunmi Oyedeji, Deborah Braide, Akinola Odunlade, Damilola Adesinha, Chinenye Nwaneri, Chiamaka Chukwuemeka, Deji Marcus, Okoh Hogan, Somto Akaraiwe, Kingsley Munu and Ernest Onuiri who have offered words of encouragement along the way. Mr. Ernest taught a course in Artificial Intelligence during my undergraduate years at Babcock University that kickstarted my journey in this field. Somto provided valuable feedback that guided me in improving the arrangement of the chapters and parts of this book. In the same breadth, I thank my friends at the Carleton University Institute of African Studies, Femi Ajidahun and June Creighton Payne, with special thanks to a mentor and senior friend, who took me as a son and was most kind, supportive, and generous, (late) Prof. Pius Adesanmi, the then Director of the Institute. Prof. Adesanmi was sadly lost to the ill-fated Ethiopian Airlines Flight 302 (ET 302) crash shortly after takeoff from Addis Ababa on March 10, 2019. I want to appreciate Mrs. Muyiwa Adesanmi and Tise Adesanmi for their love, friendship, and strength. May they find comfort, now, and in the future.

ACKNOWLEDGMENTS

I would like to thank Emmanuel Okoi, Jesam Ujong, Adie Patrick, Redarokim Ikonga, and the Hope Waddell Old Students' Association (HWOSA) family; they provided community and fun to alleviate the mood during stressful periods. Special thanks to my roommates at the time, Jonathan Austin, Christina Austin, Margherita Ciccozzi, Thai Chin, and Chris Teal; I had a fantastic place to call home.

I am especially grateful to my former colleagues and friends at Pythian, Vanessa Simmons, Alex Gorbachev, and Paul Spiegelhalter, for their help and support along the way. I am thankful to the brethren of the House Fellowships at Ottawa, Toronto, Winnipeg, Calabar, and Owerri; they constitute my own company and are friends for life. I also want to thank the staff and crew at Standard Word Broadcasting Network (SWBN) for working together to keep the vision running as I engaged this project. Special thanks to Susan McDermott, Rita Fernando, and the publishing and editorial team at Apress for their support and belief in this project. Many thanks to Vikram Tiwari and Gonzalo Gasca Meza, who provided the technical review for this manuscript.

Finally, I conclude by giving special thanks to my family, starting with my loving Father and Mother, Francis Ebuta and Nonso Ngozika Bisong; they have been a rock in my life and have been there to counsel and encourage me, I am truly grateful for my loving parents. To my siblings, Osowo-Ayim, Chidera, and Ginika Bisong, and the extended Bisong family for the constant stream of love and support. Finally, a big shout-out to my aunty and friend, Joy Duncan, for her love and friendship; she is most dear to my heart. Also, my appreciation to Uncle Wilfred Achu and Aunty Blessing Bisong, they have been most kind to me and I am very thankful.

Introduction

Machine learning and deep learning technologies have impacted the world in profound ways, from how we interact with technological products and with one another. These technologies are disrupting how we relate, how we work, and how we engage life in general. Today, and in the foreseeable future, intelligent machines increasingly form the core upon which sociocultural and socioeconomic relationships rest. We are indeed already in the "age of intelligence."

What Are Machine Learning and Deep Learning?

Machine learning can be described as an assortment of tools and techniques for predicting or classifying a future event based on a set of interactions between variables (also referred to as features or attributes) in a particular dataset. Deep learning, on the other hand, extends a machine learning algorithm called neural network for learning complex tasks which are incredibly difficult for a computer to perform. Examples of these tasks may include recognizing faces and understanding languages in their varied contextual meanings.

The Role of Big Data

A key ingredient that is critical to the rise and future improved performance of machine learning and deep learning is data. Since the turn of the twenty-first century, there has been a steady exponential increase in the amount of data generated and stored. The rise of humongous data is partly due to the emergence of the Internet and the miniaturization of processors that have spurred the "Internet of Things (IoT)" technologies. These vast amounts of data have made it possible to train the computer to learn complex tasks where an explicit instruction set is infeasible.

The Computing Challenge

The increase in data available for training learning models throws up another kind of problem, and that is the availability of computational or processing power. Empirically, as data increases, the performance of learning models also goes up. However, due to the increasingly enormous size of datasets today, it is inconceivable to train sophisticated, state-of-the-art learning models on commodity machines.

Cloud Computing to the Rescue

Cloud is a term that is used to describe large sets of computers that are networked together in groups called data centers. These data centers are often distributed across multiple geographical locations. Big companies like Google, Microsoft, Amazon, and IBM own massive data centers where they manage computing infrastructure that is provisioned to the public (i.e., both enterprise and personal users) for use at a very reasonable cost.

Cloud technology/infrastructure is allowing individuals to leverage the computing resources of big business for machine learning/deep learning experimentation, design, and development. For example, by making use of cloud resources such as Google Cloud Platform (GCP), Amazon Web Services (AWS), or Microsoft Azure, we can run a suite of algorithms with multiple test grids for a fraction of time that it will take on a local machine.

Enter Google Cloud Platform (GCP)

One of the big competitors in the cloud computing space is Google, with their cloud resource offering termed "Google Cloud Platform," popularly referred to as GCP for short. Google is also one of the top technology leaders in the Internet space with a range of leading web products such as Gmail, YouTube, and Google Maps. These products generate, store, and process tons of terabytes of data each day from Internet users around the world.

To deal with this significant data, Google over the years has invested heavily in processing and storage infrastructure. As of today, Google boasts some of the most impressive data center design and technology in the world to support their computational demands and computing services. Through Google Cloud Platform, the public can leverage these powerful computational resources to design and develop cutting-edge machine learning and deep learning models.

The Aim of This Book

The goal of this book is to equip the reader from the ground up with the essential principles and tools for building learning models. Machine learning and deep learning are rapidly evolving, and often it is overwhelming and confusing for a beginner to engage the field. Many have no clue where to start. This book is a one-stop shop that takes the beginner on a journey to understanding the theoretical foundations and the practical steps for leveraging machine learning and deep learning techniques on problems of interest.

Book Organization

This book is divided into eight parts. Their breakdown is as follows:

- Part 1: Getting Started with Google Cloud Platform

- Part 2: Programming Foundations for Data Science

- Part 3: Introducing Machine Learning

- Part 4: Machine Learning in Practice

- Part 5: Introducing Deep Learning

- Part 6: Deep Learning in Practice

- Part 7: Advanced Analytics/Machine Learning on Google Cloud Platform

- Part 8: Productionalizing Machine Learning Solutions on GCP

It is best to go through the entire book in sequence. However, each part and its containing chapters are written in such a way that one can shop around and get out what is of primary interest. The code repository for this book is available at `https://github.com/Apress/building-ml-and-dl-models-on-gcp`. The reader can follow through the examples in this book by cloning the repository to Google Colab or GCP Deep Learning VM.

PART I

Getting Started with Google Cloud Platform

CHAPTER 1

What Is Cloud Computing?

Cloud computing is the practice where computing services such as storage options, processing units, and networking capabilities are exposed for consumption by users over the Internet (the cloud). These services range from free to pay-as-you-use billing.

The central idea behind cloud computing is to make aggregated computational power available for large-scale consumption. By doing so, the microeconomics principle of economies of scale kicks into effect where cost per unit output is minimized with increasing scale of operations.

In a cloud computing environment, enterprises or individuals can take advantage of the same speed and power of aggregated high-performance computing services and only pay for what they use and relinquish these compute resources when they are no longer needed.

The concept of cloud computing had existed as time-sharing systems from the early years of the modern computer where jobs submitted from different users were scheduled to execute on a mainframe. The idea of time-sharing machines fizzled away at the advent of the PC. Now, with the rise of enterprise data centers managed by big IT companies such as Google, Microsoft, Amazon, IBM, and Oracle, the cloud computing notion has resurfaced with the added twist of multi-tenancy as opposed to time-sharing. This computing model is set to disrupt the way we work and utilize software systems and services.

In addition to storage, networking, and processing services, cloud computing provides offer other product solutions such as databases, artificial intelligence, and data analytics capabilities and serverless infrastructures.

© Ekaba Bisong 2019
E. Bisong, *Building Machine Learning and Deep Learning Models on Google Cloud Platform*,
https://doi.org/10.1007/978-1-4842-4470-8_1

Categories of Cloud Solutions

The cloud is a terminology that describes large sets of computers that are networked together in groups called data centers. These clustered machines can be interacted with via dashboards, command-line interfaces, REST APIs, and client libraries. Data centers are often distributed across multiple geographical locations. The size of data centers is over 100,000 sq. ft. (and those are the smaller sizes!). Cloud computing solutions can be broadly categorized into three, namely, the public, private, and hybrid cloud. Let's briefly discuss them:

- Public cloud: Public clouds are the conventional cloud computing model, where cloud service providers make available their computing infrastructure and products for general use by other enterprises and individuals (see Figure 1-1). In public clouds, the cloud service provider is responsible for managing the hardware configuration and servicing.

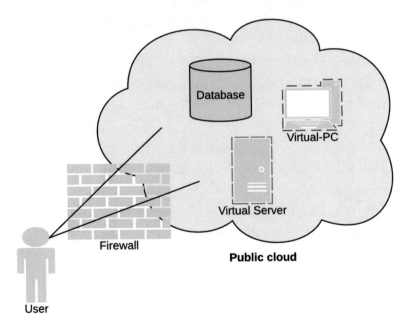

Figure 1-1. *The public cloud*

- Private cloud: In a private cloud, an organization is solely responsible for the management and servicing of its computing infrastructure. The machines in a private cloud can be located on-premises, or it can be hosted with a cloud service provider but routed on a private network.

- Hybrid cloud: The hybrid cloud is a compromise between the cost and efficiency of a public cloud and the data sovereignty and in-house security assurances of the private cloud. Many companies and institutions opt for a hybrid cloud and multi-cloud by using technology solutions to facilitate easy porting and sharing of data and applications between on-premise and cloud-based infrastructures.

Cloud Computing Models

Cloud computing is also categorized into three models of service delivery. They are illustrated as a pyramid as shown in Figure 1-2, where the layers of infrastructure abstraction increase as we approach the apex of the pyramid:

- Infrastructure as a Service (IaaS): This model is best suited for enterprises or individuals who want to manage the hardware infrastructure that hosts their data and applications. This level of fine-grained management requires the necessary system administration skills.

- Platform as a Service (PaaS): In the PaaS model, the hardware configuration is managed by the cloud service provider, as well as other system and development tools. This relieves the user to focus on the business logic for quick and easy deployment of application and database solutions. Another concept that comes up together with PaaS is the idea of **Serverless**, where the cloud service provider manages a scalable infrastructure that utilizes and relinquishes resources according to demand.

- Software as a Service (SaaS): The SaaS model is most recognizable by the general public, as a great deal of users interact with SaaS applications without knowing. The typical examples of SaaS applications are enterprise email suites such as Gmail, Outlook, and Yahoo! Mail. Others include storage platforms like Google Drive and Dropbox, photo software like Google Photos, and CRM e-suites like Salesforce and Oracle E-business Suite.

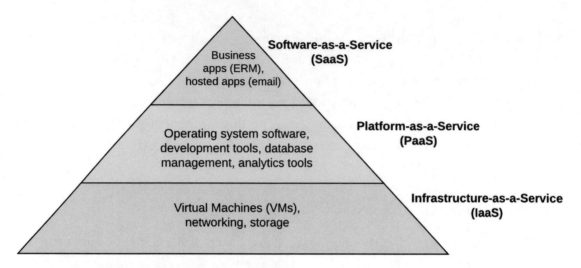

Figure 1-2. *Models of cloud computing*

In this chapter, we summarized the practice of cloud computing by explaining the different categories of cloud solutions and the models for service delivery over the cloud.

The next chapters in Part 1 will provide an introduction to Google Cloud Platform Infrastructure and Services and introduce JupyterLab Instances, and Google Colaboratory for prototyping machine learning models and doing data science and analytics tasks.

## CHAPTER 2

An Overview of Google Cloud Platform Services

Google Cloud Platform offers a wide range of services for securing, storing, serving, and analyzing data. These cloud services form a secure cloud perimeter for data, where different operations and transformations can be carried out on the data without it ever leaving the cloud ecosystem.

The services offered by Google Cloud include compute, storage, big data/analytics, artificial intelligence (AI), and other networking, developer, and management services. Let's briefly review some of the features of the Google Cloud ecosystem.

Cloud Compute

Google Compute offers a range of products shown in Figure 2-1 for catering to a wide range of computational needs. The compute products consist of the Compute Engine (virtual computing instances for custom processing), App Engine (a cloud-managed platform for developing web, mobile, and IoT app), Kubernetes Engine (orchestration manager for custom docker containers based on Kubernetes), Container Registry (private container storage), Serverless Cloud Functions (cloud-based functions to connect or extend cloud services), and Cloud Run (managed compute platform that automatically scales your stateless containers).

© Ekaba Bisong 2019

E. Bisong, *Building Machine Learning and Deep Learning Models on Google Cloud Platform*, https://doi.org/10.1007/978-1-4842-4470-8_2

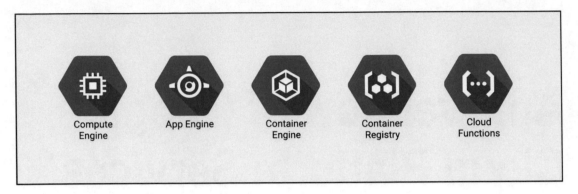

Figure 2-1. *Cloud compute services*

For our purposes of machine learning modeling, the cloud compute engine is what we will concentrate on. As later seen in Chapter 6, JupyterLab will provision a compute engine with all the relevant tools, packages, and frameworks for data analytics and modeling machine learning and deep learning solutions.

Cloud Storage

Google Cloud Storage options provide scalable and real-time storage access to live and archival data within the cloud perimeter. Cloud storage as an example is set up to cater for any conceivable storage demand. Data stored on cloud storage is available anytime and from any location around the world. What's more, this massive storage power comes at an almost negligible cost, taking into consideration the size and economic value of the stored data. Moreover, acknowledging the accessibility, security, and consistency provided by cloud storage, the cost is worth every penny.

The cloud storage products shown in Figure 2-2 include Cloud Storage (general-purpose storage platform), Cloud SQL (cloud-managed MySQL and PostgreSQL), Cloud Bigtable (NoSQL petabyte-sized storage), Cloud Spanner (scalable/high availability transactional storage), Cloud Datastore (transactional NoSQL database), and Persistent Disk (block storage for virtual machines).

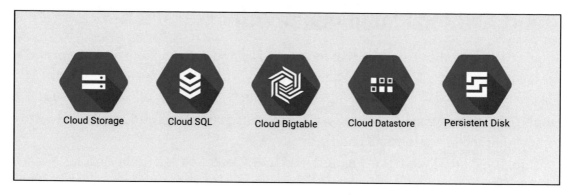

Figure 2-2. *Cloud storage products*

Big Data and Analytics

Google Cloud Platform offers a range of serverless big data and analytics solutions for data warehousing, stream, and batch analytics, cloud-managed Hadoop ecosystems, cloud-based messaging systems, and data exploration. These services provide multiple perspectives to mining/generating real-time intelligence from big data.

Examples of big data services shown in Figure 2-3 include Cloud BigQuery (serverless analytics/data warehousing platform), Cloud Dataproc (fully managed Hadoop/Apache Spark infrastructure), Cloud Dataflow (Batch/Stream data transformation/processing), Cloud Dataprep (serverless infrastructure for cleaning unstructured/structured data for analytics), Cloud Datastudio (data visualization/report dashboards), Cloud Datalab (managed Jupyter notebook for machine learning/data analytics), and Cloud Pub/Sub (serverless messaging infrastructure).

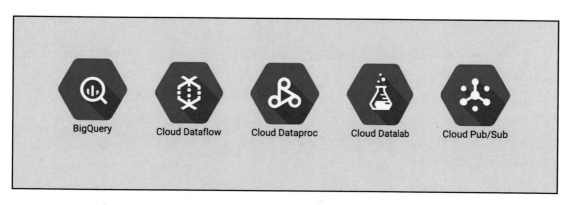

Figure 2-3. *Big data/analytics serverless platforms*

Cloud Artificial Intelligence (AI)

Google Cloud AI offers cloud services for businesses and individuals to leverage pre-trained models for custom artificial intelligence tasks through the use of REST APIs. It also exposes services for developing custom models for domain use cases such as AutoML Vision for image classification and object detection tasks and AutoML tables to deploy AI models on structured data.

Google Cloud AI services in Figure 2-4 include Cloud AutoML (train custom machine learning models leveraging transfer learning), Cloud Machine Learning Engine (for large-scale distributed training and deployment of machine learning models), Cloud TPU (to quickly train large-scale models), Video Intelligence (train custom video models), Cloud Natural Language API (extract/analyze text from documents), Cloud Speech API (transcribe audio to text), Cloud Vision API (classification/segmentation of images), Cloud Translate API (translate from one language to another), and Cloud Video Intelligence API (extract metadata from video files).

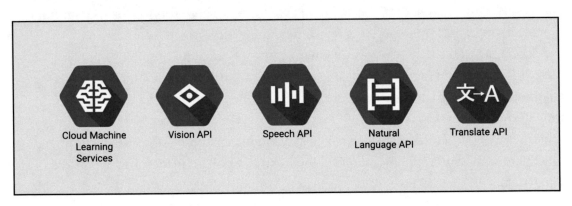

Figure 2-4. *Cloud AI services*

This chapter provides a high-level overview of the products and services offered on Google Cloud Platform.

The next chapter will introduce the Google Cloud software development kit (SDK) for interacting with cloud resources from the command line on the local machine and the cloud command-line interface (CLI) for doing the same via the cloud console interface on GCP.

CHAPTER 3

The Google Cloud SDK and Web CLI

GCP provides a command-line interface (CLI) for interacting with cloud products and services. GCP resources can be accessed via the web-based CLI on GCP or by installing the Google Cloud software development kit (SDK) on your local machine to interact with GCP via the local command-line terminal.

GCP contains shell commands for a wide range of GCP products such as the Compute Engine, Cloud Storage, Cloud ML Engine, BigQuery, and Datalab, to mention just a few. Major tools of the Cloud SDK include

- gcloud tool: Responsible for cloud authentication, configuration, and other interactions on GCP

- gsutil tool: Responsible for interacting with Google Cloud Storage buckets and objects

- bq tool: Used for interacting and managing Google BigQuery via the command line

- kubectl tool: Used for managing Kubernetes container clusters on GCP

The Google Cloud SDK also installs client libraries for developers to programmatically interact with GCP products and services through APIs.[1] As of this time of writing, the Go, Java, Node.js, Python, Ruby, PHP, and C# languages are covered. Many more are expected to be added to this list.

This chapter works through setting up an account on GCP, installing the Google Cloud SDK, and then exploring GCP commands using the CLI.

---

[1]APIs stands for application programming interfaces, which are packages and tools used in building software applications.

© Ekaba Bisong 2019
E. Bisong, *Building Machine Learning and Deep Learning Models on Google Cloud Platform*,
https://doi.org/10.1007/978-1-4842-4470-8_3

Setting Up an Account on Google Cloud Platform

This section shows how to set up an account on Google Cloud Platform. A GCP account gives access to all of the platform's products and services. For a new account, a $300 credit is awarded to be spent over a period of 12 months. This offer is great as it gives ample time to explore the different features and services of Google's cloud offering.

Note that a valid credit card is required to register an account to validate that it is an authentic user, as opposed to a robot. However, the credit card won't be charged after the trial ends, except Google is authorized to do so:

1. Go to `https://cloud.google.com/` to open an account (see Figure 3-1).

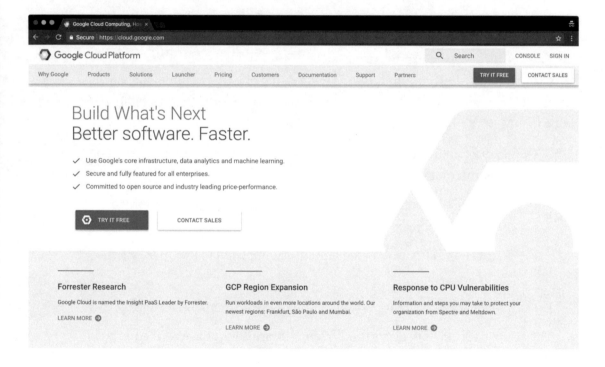

Figure 3-1. *Google Cloud Platform login page*

2. Fill in the necessary identity, address, and credit card details.

3. Wait a moment while an account is created on the platform (see Figure 3-2).

Figure 3-2. *Creating account*

4. After account creation, we're presented with the Welcome to GCP
    page (see Figure 3-3).

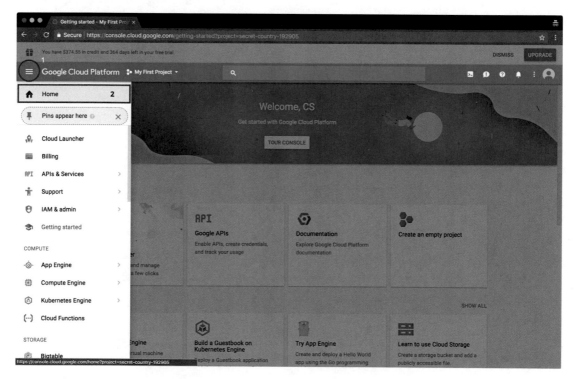

Figure 3-3. *Welcome to GCP*

5. Click the icon of three lines in the top-left corner of the page
    (marked with a circle in Figure 3-3), then click Home (marked
    with a rectangle in Figure 3-3) to open the Google Cloud Platform
    dashboard (Figure 3-4).

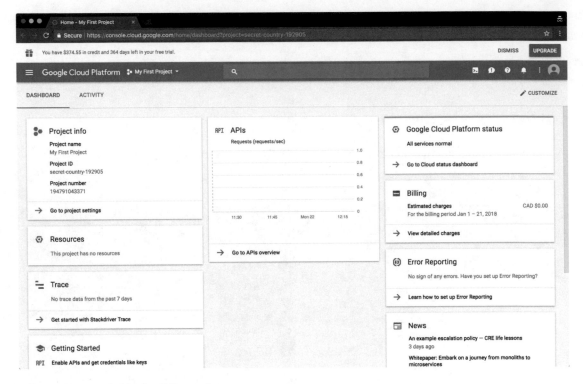

Figure 3-4. *GCP dashboard*

The Cloud dashboard provides a bird's-eye summary of the project such as the
current billing rate and other resource usage statistics. The activity tab to the right gives
a breakdown of the resource actions performed on the account. This feature is useful
when building an audit trail of events.

GCP Resources: Projects

All the services and features of the Google Cloud Platform are called resources. These
resources are arranged in a hierarchical order, with the top level being the project.
The project is like a container that houses all GCP resources. Billing on an account is
attached to a project. Multiple projects can be created for an account. A project must be
created before working with GCP.

To view the projects in the account in Figure 3-5, click the **scope picker in the cloud console** (marked with an oval in Figure 3-6).

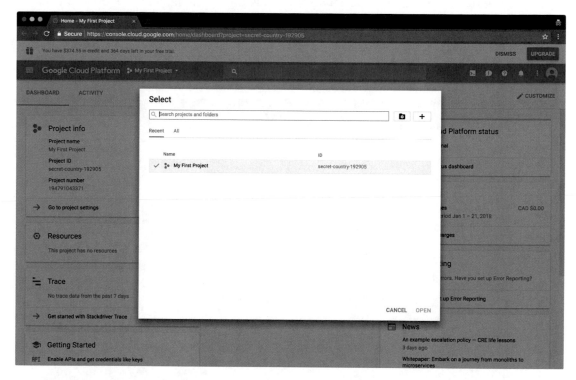

Figure 3-5. *Select projects*

Figure 3-6. *Scope picker to select projects*

Accessing Cloud Platform Services

To access the resources on the cloud platform, click the triple dash in the top-right corner of the window. Grouped service offerings are used to organize the resources. For example, in Figure 3-7, we can see the products under **STORAGE**: Bigtable, Datastore, Storage, SQL, and Spanner.

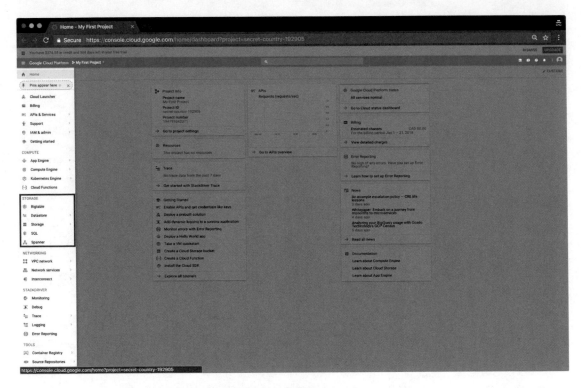

Figure 3-7. *Google Cloud Platform services*

Account Users and Permissions

GCP allows you to define security roles and permissions for every resource in a specific project. This feature is particularly useful when a project scales beyond one user. New roles and permissions are created for a user through the IAM & admin tab (see Figures 3-8 and 3-9).

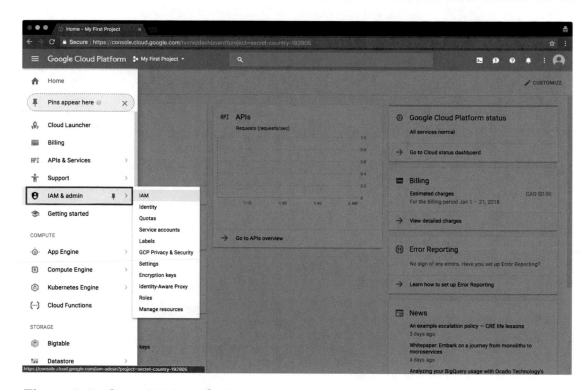

Figure 3-8. *Open IAM & admin*

The Cloud Shell

The Cloud Shell is a vital component for working with GCP resources. Cloud Shell provisions an ephemeral virtual machine with command-line tools installed for interacting with GCP resources. It gives the user cloud-based command-line access to manipulate resources directly from within the GCP perimeter without installing the Google Cloud SDK on a local machine.

The Cloud Shell is accessed by clicking the **prompt icon** in the top-left corner of the window. See Figures 3-9, 3-10, and 3-11.

Figure 3-9. *Activate Cloud Shell*

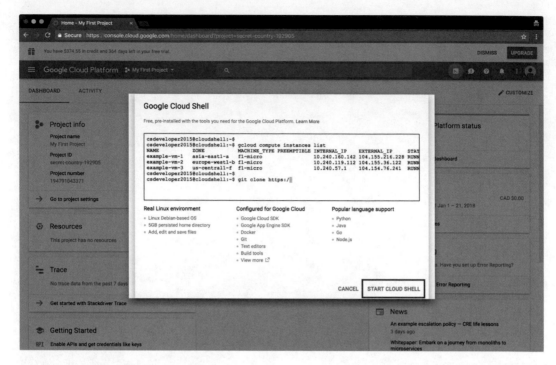

Figure 3-10. *Start Cloud Shell*

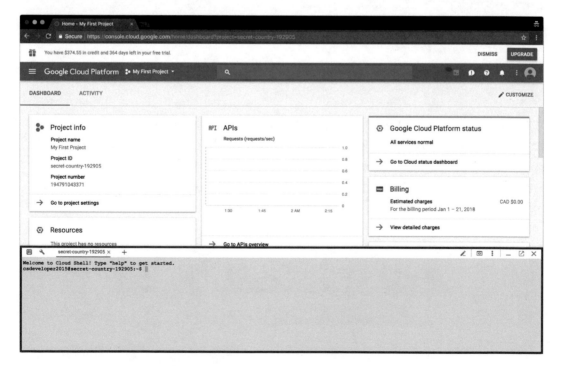

Figure 3-11. *Cloud Shell interface*

Google Cloud SDK

The Google Cloud SDK installs command-line tools for interacting with cloud resources from the terminal on the local machine:

1. Go to https://cloud.google.com/sdk/ to download and install the appropriate Cloud SDK for your machine type (see Figure 3-12).

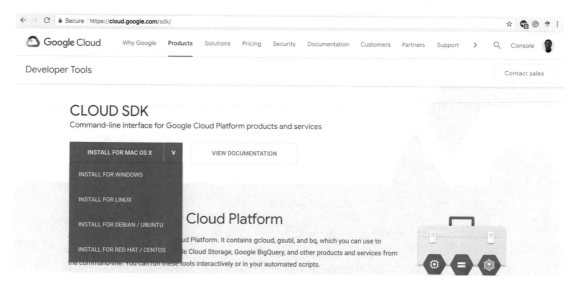

Figure 3-12. *Download Google Cloud SDK*

2. Follow the instructions for the operating system (OS) type to install the Google Cloud SDK. The installation installs the default Cloud SDK components.

3. Open the terminal application of your OS and run the command 'gcloud init' to begin authorization and configuration of the Cloud SDK.

```
gcloud init

Welcome! This command will take you through the configuration
of gcloud.

Pick configuration to use:
 [1] Create a new configuration
Please enter your numeric choice:  1
```

4. Select the name for your configuration. Here, it is set to the name 'your-email-id'.

   Enter configuration name. Names start with a lower case letter and contain only lower case letters a-z, digits 0-9, and hyphens '-': your-email-id

   Your current configuration has been set to: [your-email-id]

5. Select the Google account to use for the configuration. The browser will open to log in to the selected account (see Figures 3-13, 3-14, and 3-15). However, if a purely terminal initialization is desired, the user can run 'gcloud init --console-only'.

   Choose the account you would like to use to perform operations for this configuration:
    [1] Log in with a new account
   Please enter your numeric choice: 1

   Your browser has been opened to visit:

   https://accounts.google.com/o/oauth2/auth?redirect_
   uri=......=offline

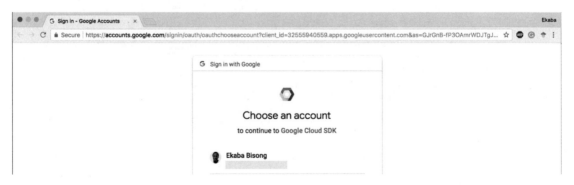

Figure 3-13. *Select Google account to authorize for Cloud SDK configuration*

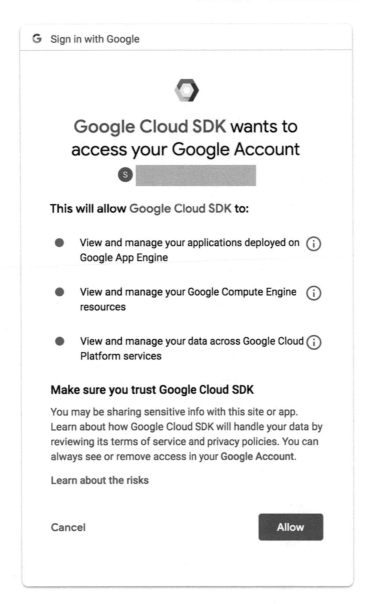

Figure 3-14. *Authenticate Cloud SDK to access Google account*

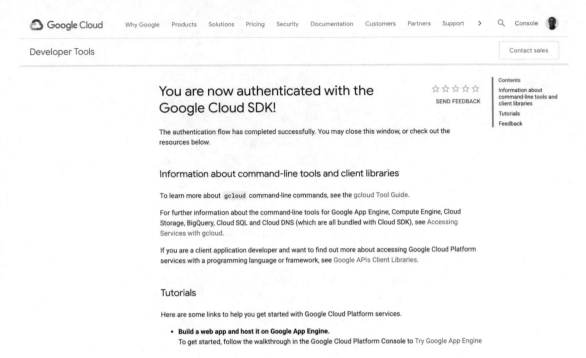

Figure 3-15. *Confirmation page for Cloud SDK authentication*

6. Select the cloud project to use after the browser-based authentication in a Google account.

```
You are logged in as: [your-email-id@gmail.com].
```

```
Pick cloud project to use:
 [1] secret-country-192905
 [2] Create a new project
Please enter numeric choice or text value (must exactly match list
item): 1
```

```
Your current project has been set to: [secret-country-192905].
```

```
Your Google Cloud SDK is configured and ready to use!
```

```
* Commands that require authentication will use your-email-id@
gmail.com by default
* Commands will reference project `secret-country-192905` by
default
```

```
Run `gcloud help config` to learn how to change individual
settings

This gcloud configuration is called [your-configuration-name].
You can create additional configurations if you work with multiple
accounts and/or projects.
Run `gcloud topic configurations` to learn more.

Some things to try next:

* Run `gcloud --help` to see the Cloud Platform services you can
interact with. And run `gcloud help COMMAND` to get help on any
gcloud command.
* Run `gcloud topic -h` to learn about advanced features of the
SDK like arg files and output formatting
```

The Google Cloud SDK is now configured and ready to use. The following are a few terminal commands for managing 'gcloud' configurations:

- 'gcloud auth list': Shows accounts with GCP credentials and indicates which account configuration is currently active.

```
gcloud auth list

                    Credentialed Accounts
ACTIVE   ACCOUNT
*        your-email-id@gmail.com

To set the active account, run:
    $ gcloud config set account `ACCOUNT`
```

- 'gcloud config configurations list': List existing Cloud SDK configurations.

```
gcloud config configurations list

NAME   IS_ACTIVE   ACCOUNT   PROJECT   DEFAULT_ZONE   DEFAULT_REGION
your-email-id  True  your-email-id@gmail.com     secret-
country-192905
```

- 'gcloud config configurations activate [CONFIGURATION_NAME]':
  Use this command to activate a configuration.

  ```
  gcloud config configurations activate your-email-id
  ```

  ```
  Activated [your-email-id].
  ```

- 'gcloud config configurations create [CONFIGURATION_NAME]':
  Use this command to create a new configuration.

This chapter covers how to set up command-line access for interacting with GCP resources. This includes working with the web-based Cloud Shell and installing the Cloud SDK to access GCP resources via the terminal on the local machine.

In the next chapter, we'll introduce Google Cloud Storage (GCS) for storing ubiquitous data assets on GCP.

CHAPTER 4

Google Cloud Storage (GCS)

Google Cloud Storage is a product for storing a wide range of diverse data objects. Cloud storage may be used to store both live and archival data. It has guarantees of scalability (can store increasingly large data objects), consistency (the most updated version is served on request), durability (data is redundantly placed in separate geographic locations to eliminate loss), and high availability (data is always available and accessible).

Let's take a brief tour through creating and deleting a storage bucket, as well as uploading and deleting files from a cloud storage bucket.

Create a Bucket

A bucket, as the name implies, is a container for storing data objects on GCP. A bucket is the base organizational structure on Cloud Storage. It is similar to the topmost directory on a file system. Buckets may have a hierarchy of sub-folders containing data assets.

To create a bucket,

1. Click 'Create bucket' on the cloud storage dashboard as shown in Figure 4-1.

2. Give the bucket a unique name (see Figure 4-2). Buckets in GCP must have a global unique name. That is to say, no two storage buckets on Google Cloud can have the same name. A common naming convention for buckets is to prefix with your organization's domain name.

© Ekaba Bisong 2019

E. Bisong, *Building Machine Learning and Deep Learning Models on Google Cloud Platform*, https://doi.org/10.1007/978-1-4842-4470-8_4

3. Select a storage class. A multi-region storage class is for buckets frequently accessed all over the world, whereas the cold-line storage is more or less for storing backup files. For now, the default selection is okay.

4. Click 'Create' to set up a bucket on Google Cloud Storage.

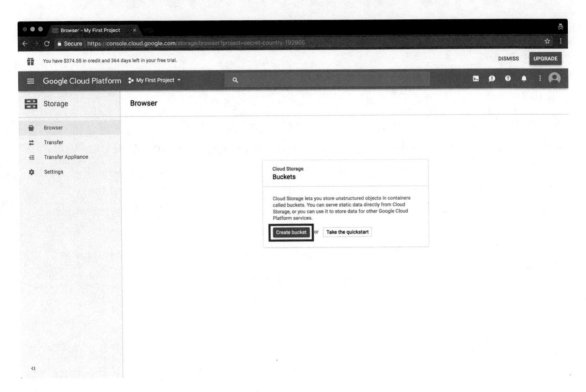

Figure 4-1. *Cloud Storage Console*

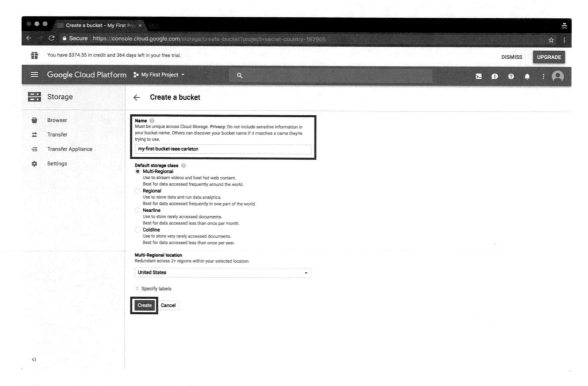

Figure 4-2.   *Create a bucket*

Uploading Data to a Bucket

Individual files or folders can be uploaded into a bucket on GCS. As an example, let's upload a file from the local machine.

To upload a file to a cloud storage bucket on GCP,

1. Click 'UPLOAD FILES' within the red highlight in Figure 4-3.

2. Select the file from the file upload window, and click 'Open' as shown in Figure 4-4.

3. Upon upload completion, the file is uploaded as an object in GCS bucket (see Figure 4-5).

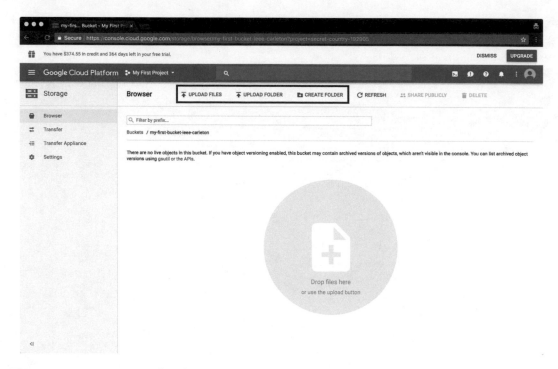

Figure 4-3. *An empty bucket*

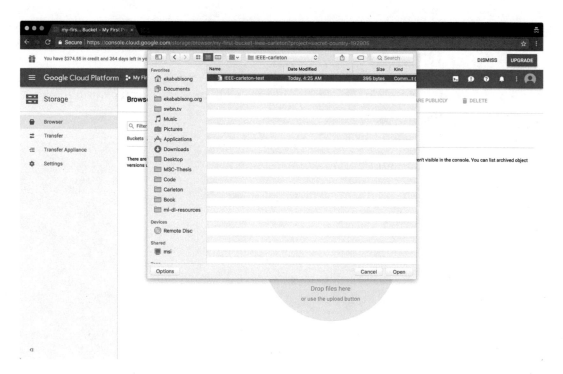

Figure 4-4. *Upload an object*

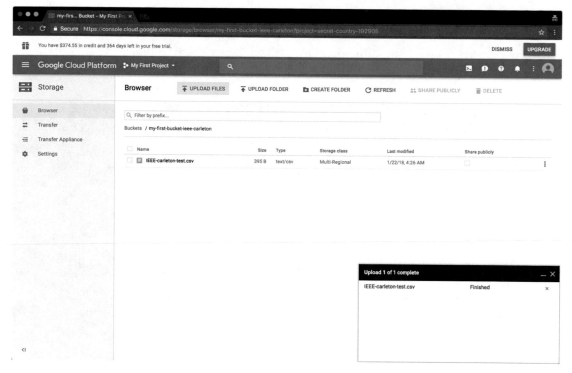

Figure 4-5. *Upload successful*

Delete Objects from a Bucket

Click the checkbox beside the file and click 'DELETE' as shown in Figure 4-6 to delete an object from a bucket.

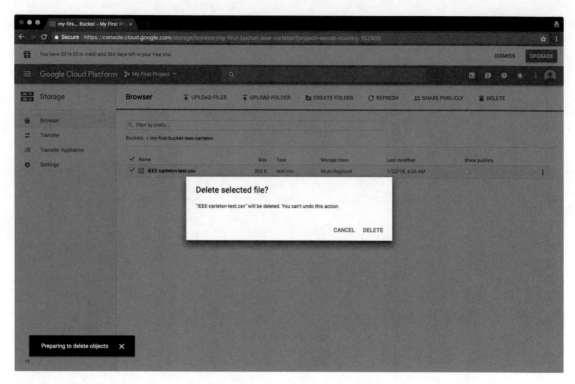

Figure 4-6. *Delete a file*

Free Up Storage Resource

To delete a bucket or free up a storage resource to prevent billing on a resource that is not used, click the checkbox beside the bucket in question, and click 'DELETE' to remove the bucket and its contents. This action is not recoverable. See Figures 4-7 and 4-8.

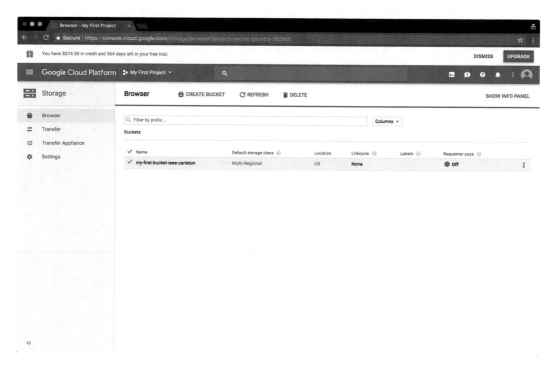

Figure 4-7. *Select bucket to delete*

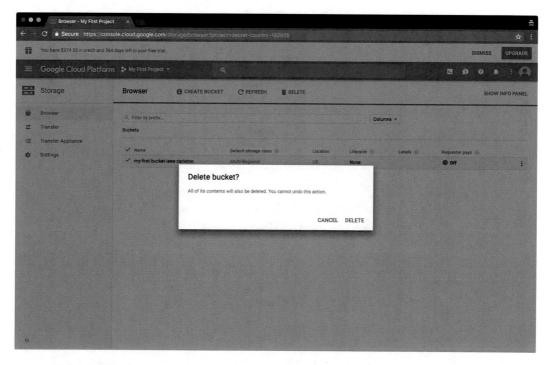

Figure 4-8. *Delete bucket*

Working with GCS from the Command Line

In this section, we'll carry out similar commands for creating and deleting buckets and objects on GCS from the command-line interface.

- Creating a bucket: To create a bucket, execute the command

  ```
  gsutil mb gs://[BUCKET_NAME]
  ```

  As an example, we'll create a bucket titled 'hwosa_09_docs'.

  ```
  gsutil mb gs://hwosa_09_docs

  Creating gs://hwosa_09_docs/...
  ```

  List buckets on GCP project.

  ```
  gsutil ls
  ```

  ```
  gs://hwosa_09_docs/
  gs://my-first-bucket-ieee-carleton/
  ```

- Uploading objects to cloud bucket: To transfer objects from a local directory to the cloud bucket, execute the command

  ```
  gsutil cp -r [LOCAL_DIR] gs://[DESTINATION BUCKET]
  ```

  Copy an image file from the desktop to a bucket on GCP.

  ```
  gsutil cp -r /Users/ekababisong/Desktop/Howad.jpeg
  gs://hwosa_09_docs/

  Copying file:///Users/ekababisong/Desktop/Howad.jpeg
  [Content-Type=image/jpeg]...
  - [1 files][ 49.8 KiB/ 49.8 KiB]
  Operation completed over 1 objects/49.8 KiB.
  ```

  List objects in bucket.

  ```
  gsutil ls gs://hwosa_09_docs
  ```

  ```
  gs://hwosa_09_docs/Howad.jpeg
  ```

- Deleting objects from the cloud bucket: To delete a specific file from the bucket, execute

```
gsutil rm -r gs://[SOURCE_BUCKET]/[FILE_NAME]
```

To delete all files from the bucket, execute

```
gsutil rm -a gs://[SOURCE_BUCKET]/**
```

As an example, let's delete the image file in the bucket 'gs://hwosa_09_docs'.

```
gsutil rm -r gs://hwosa_09_docs/Howad.jpeg
```

```
Removing gs://hwosa_09_docs/Howad.jpeg#1537539161893501...
/ [1 objects]
Operation completed over 1 objects.
```

- Deleting a bucket: When a bucket is deleted, all the files within that bucket are also deleted. This action is irreversible. To delete a bucket, execute the command

```
gsutil rm -r gs://[SOURCE_BUCKET]/
```

Delete the bucket 'gs://hwosa_09_docs'

```
gsutil rm -r gs://hwosa_09_docs/
```

```
Removing gs://hwosa_09_docs/...
```

This chapter works through uploading and deleting data from Google Cloud Storage using the Cloud GUI console and command-line tools.

In the next chapter, we will introduce Google Compute Engines, which are virtual machines running on Google's distributed data centers and are connected via state-of-the-art fiber optic network. These machines are provisioned to lower the cost and speed up the processing of computing workloads.

CHAPTER 5

Google Compute Engine (GCE)

Google Compute Engine (GCE) makes available to users virtual machines (VMs) that are running on Google's data centers around the world. These machines take advantage of Google's state-of-the-art fiber optic powered network capabilities to offer fast and high-performance machines that can scale based on usage and automatically deal with issues of load balancing.

GCE provides a variety of pre-defined machine types for use out of the box; also it has the option to create custom machines that are tailored to the specific needs of the user. Another major feature of GCE is the ability to use computing resources that are currently idle on Google infrastructure for a short period of time to enhance or speed up the processing capabilities of batch jobs or fault-tolerant workloads. These machines are called preemptible VMs and come at a huge cost-benefit to the user as they are about 80% cheaper than regular machines.

Again one of the major benefits of GCEs is that the user only pays for the time the machines are actually in operation. Also, when the machines are used for a long uninterrupted period of time, discounts are accrued to the prices.

In this chapter, we will go through a simple example of provisioning and tearing down a Linux machine on the cloud. The examples will cover using the Google Cloud web interface and the command-line interface for creating VMs on GCP.

Provisioning a VM Instance

To deploy a VM instance, click the triple dash in the top-left corner of the web page to pull out the GCP resources drawer. In the group named 'COMPUTE', click the arrow beside 'Compute Engine' and select 'VM instances' as shown in Figure 5-1.

© Ekaba Bisong 2019

E. Bisong, *Building Machine Learning and Deep Learning Models on Google Cloud Platform*, https://doi.org/10.1007/978-1-4842-4470-8_5

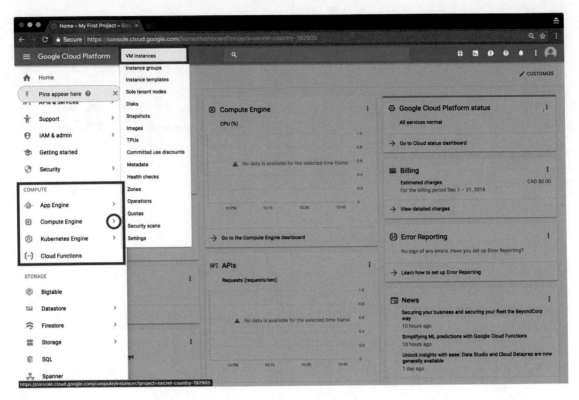

Figure 5-1. *Select VM instances*

Click 'Create' to begin the process of deploying a VM instance (see Figure 5-2).

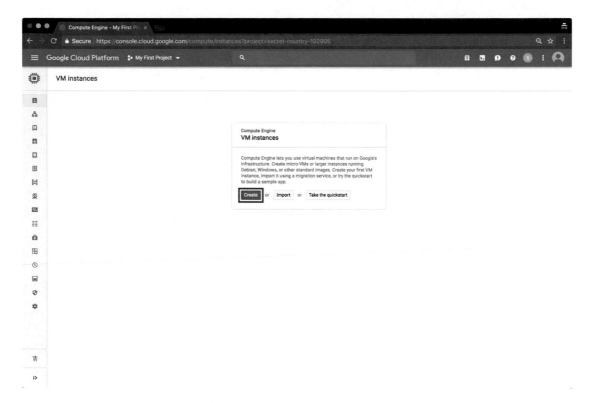

Figure 5-2. *Begin process of deploying a VM instance*

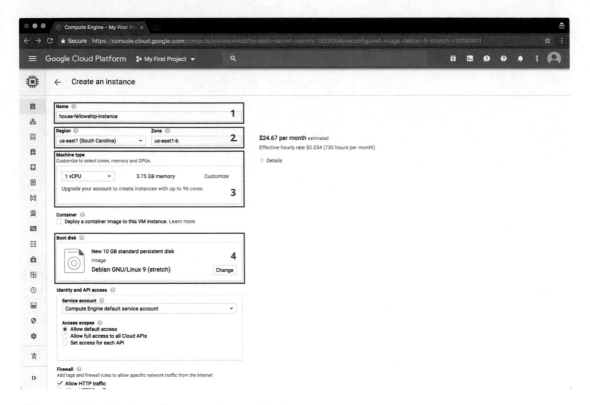

Figure 5-3.   *Options for creating an instance*

The labeled numbers in Figure 5-3 are explained here:

1. Choose the instance name. This name must start with a lowercase letter and can include numbers or hyphens, but should not end with a hyphen.

2. Select the instance region and zone. This is the geographical region where your computing instance is located, while the zone is a location within a region.

3. Select the machine type. This allows for customization of the cores, memory, and GPUs for the VM (see Figure 5-4).

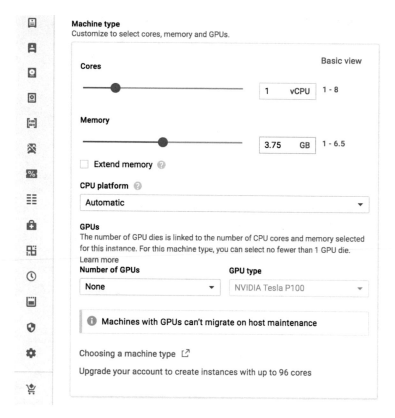

Figure 5-4. *Select machine type*

4. Select the boot disk. This option selects a disk to boot from. This disk could be created from an OS image, an application image, a custom image, or a snapshot of an image (see Figure 5-5).

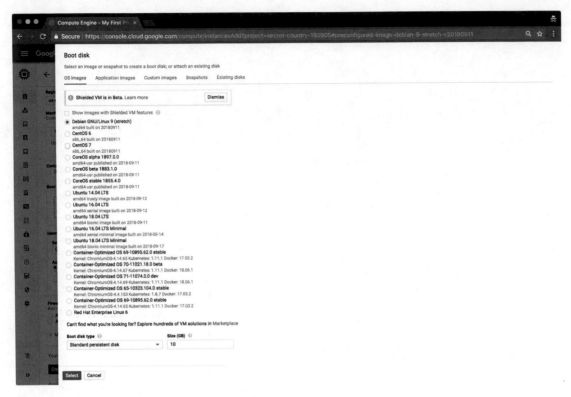

Figure 5-5. *Select boot disk*

5. Select 'Allow HTTP traffic' to allow network traffic from the
    Internet as shown in Figure 5-6.

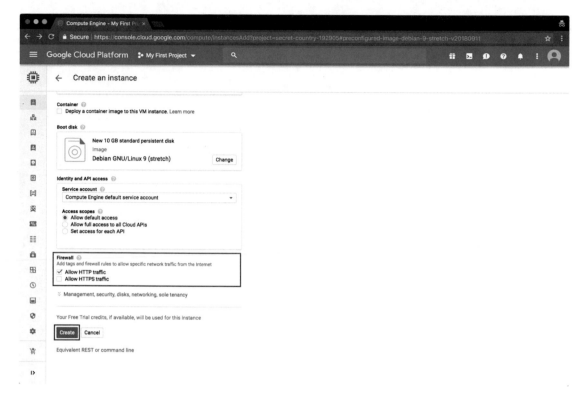

Figure 5-6. *Allow network traffic to VM*

> 6.   Click 'Create' in Figure 5-6 to deploy the VM instance.

Connecting to the VM Instance

In the VM instances page that lists the created VMs, click 'SSH' beside the created instance as shown in Figure 5-7. This launches a new window with terminal access to the created VM as shown in Figures 5-8 and 5-9.

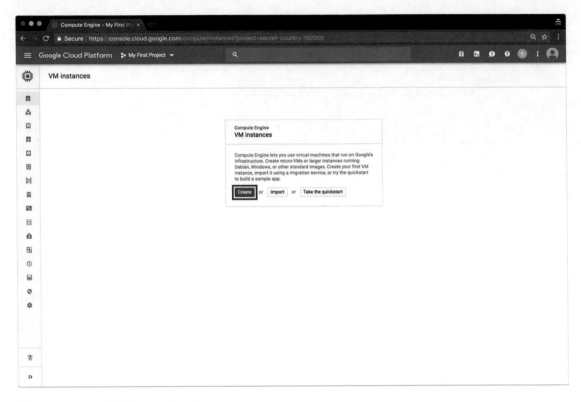

Figure 5-7. *SSH into VM instances*

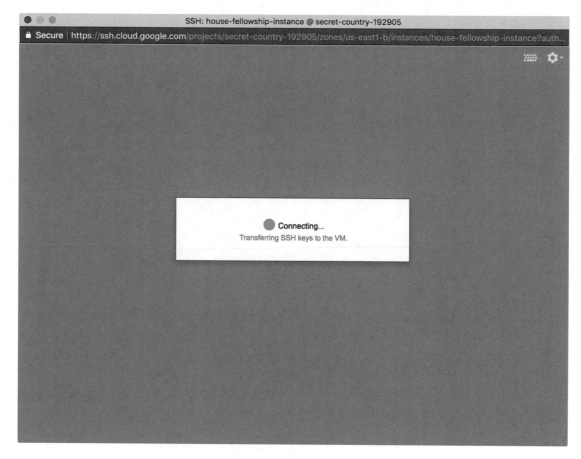

Figure 5-8. *Connecting to VM instances via SSH*

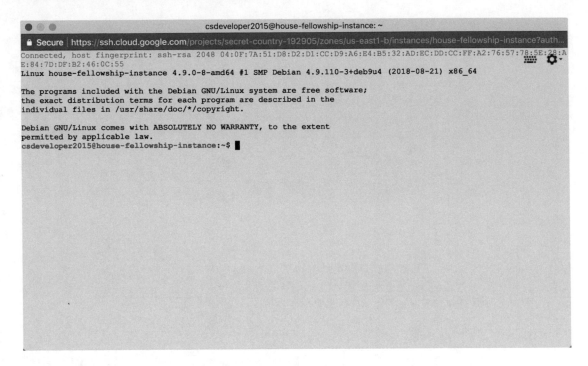

Figure 5-9. *Terminal window access to the instance*

Tearing Down the Instance

It is good practice to delete compute instances that are no longer in use to save cost for utilizing GCP resources. To delete a compute instance, on the 'VM instances' page, select the instance for deletion and click 'DELETE' (in red) as shown in Figure 5-10.

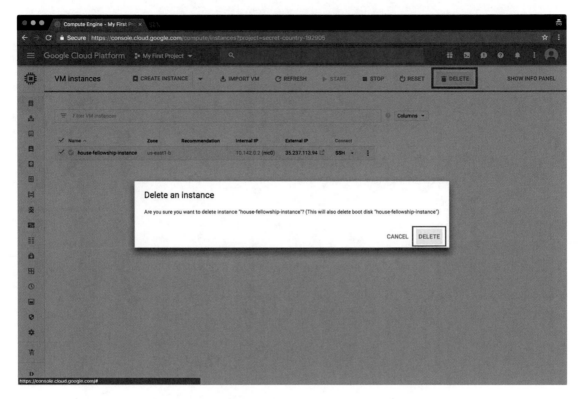

Figure 5-10. *Delete the VM instance*

Working with GCE from the Command Line

In this section, we'll sample the commands for creating and deleting a compute instance on GCP using the command-line interface. To create a compute instance using 'gcloud' from the command-line interface, there are a variety of options that can be added to the commands for different specifications of the machine. To learn more about a command, attach 'help' after the command:

- Provisioning a VM instance: To create a VM instance, use the code syntax

  ```
  gcloud compute instances create [INSTANCE_NAME]
  ```

  For example, let's create an instance named 'ebisong-howad-instance'

  ```
  gcloud compute instances create ebisong-howad-instance
  ```

Created [https://www.googleapis.com/compute/v1/projects/secret-
country-192905/zones/us-east1-b/instances/ebisong-howad-instance].
NAME   ZONE   MACHINE_TYPE   PREEMPTIBLE
INTERNAL_IP   EXTERNAL_IP   STATUS
ebisong-howad-instance   us-east1-b   n1-standard-1
10.142.0.2   35.196.17.39   RUNNING

To learn more of the options that can be included with the 'gcloud instance
create' command, run

gcloud compute instances create –help

NAME
    gcloud compute instances create - create Google Compute Engine
    virtual
        machine instances

SYNOPSIS
    gcloud compute instances create INSTANCE_NAMES [INSTANCE_
    NAMES ...]
        [--accelerator=[count=COUNT],[type=TYPE]] [--async]
        [--no-boot-disk-auto-delete]
        [--boot-disk-device-name=BOOT_DISK_DEVICE_NAME]
        [--boot-disk-size=BOOT_DISK_SIZE] [--boot-disk-type=BOOT_
            DISK_TYPE]
        [--can-ip-forward] [--create-disk=[PROPERTY=VALUE,...]]
        [--csek-key-file=FILE] [--deletion-protection]
        [--description=DESCRIPTION]
        [--disk=[auto-delete=AUTO-DELETE],
            [boot=BOOT],[device-name=DEVICE-NAME],[mode=MODE],
            [name=NAME]]
        [--labels=[KEY=VALUE,...]]
        [--local-ssd=[device-name=DEVICE-NAME],[interface=INTERFACE]]
        [--machine-type=MACHINE_TYPE] [--maintenance-
            policy=MAINTENANCE_POLICY]
        [--metadata=KEY=VALUE,[KEY=VALUE,...]]
        [--metadata-from-file=KEY=LOCAL_FILE_PATH,[...]]

```
[--min-cpu-platform=PLATFORM] [--network=NETWORK]
[--network-interface=[PROPERTY=VALUE,...]]
[--network-tier=NETWORK_TIER] [--preemptible]
[--private-network-ip=PRIVATE_NETWORK_IP]
```
:

To exit from the help page, type 'q' and then press the 'Enter' key on the keyboard.

To list the created instances, run

```
gcloud compute instances list
```

```
NAME                          ZONE        MACHINE_TYPE    PREEMPTIBLE
INTERNAL_IP   EXTERNAL_IP     STATUS
ebisong-howad-instance  us-east1-b   n1-standard-1
10.142.0.2    35.196.17.39   RUNNING
```

- Connecting to the instance: To connect to a created VM instance using SSH, run the command

```
gcloud compute ssh [INSTANCE_NAME]
```

For example, to connect to the 'ebisong-howad-instance' VM, run the command

```
gcloud compute ssh ebisong-howad-instance
```

```
Warning: Permanently added 'compute.84932566799990250176' (ECDSA)
to the list of known hosts.
Linux ebisong-howad-instance 4.9.0-8-amd64 #1 SMP Debian
4.9.110-3+deb9u4 (2018-08-21) x86_64
```

```
The programs included with the Debian GNU/Linux system are free
software;
the exact distribution terms for each program are described in the
individual files in /usr/share/doc/*/copyright.
```

```
Debian GNU/Linux comes with ABSOLUTELY NO WARRANTY, to the extent
permitted by applicable law.
ekababisong@ebisong-howad-instance:~$
```

- To leave the instance on the terminal, type 'exit' and then press the 'Enter' key on the keyboard.

  ```
  Debian GNU/Linux comes with ABSOLUTELY NO WARRANTY, to the extent
  permitted by applicable law.
  ekababisong@ebisong-howad-instance:~$ exit
  logout
  Connection to 35.196.17.39 closed.
  ```

- Tearing down the instance: To delete an instance, run the command

  ```
  gcloud compute instances delete [INSTANCE_NAME]
  ```

  Using our example, to delete the 'ebisong-howad-instance' VM, run the command

  ```
  gcloud compute instances delete ebisong-howad-instance
  ```

  ```
  The following instances will be deleted. Any attached disks
  configured to be auto-deleted will be deleted unless they are
  attached to any other instances or the `--keep-disks` flag
  is given and specifies them for keeping. Deleting a disk is
  irreversible and any data on the disk will be lost.
   - [ebisong-howad-instance] in [us-east1-b]
  ```

  ```
  Do you want to continue (Y/n)?  Y
  ```

  ```
  Deleted [https://www.googleapis.com/compute/v1/projects/secret-
  country-192905/zones/us-east1-b/instances/ebisong-howad-instance].
  ```

This chapter went through the step for launching a compute machine instance on GCP. It covered working with the web-based cloud console and using commands via the shell terminal.

In the next chapter, we'll discuss how to launch a Jupyter notebook instance on GCP called JupyterLab. A notebook provides an interactive environment for analytics, data science, and prototyping machine learning models.

CHAPTER 6

JupyterLab Notebooks

Google deep learning virtual machines (VMs) are a part of GCP AI Platform. It provisions a Compute Engine instance that comes pre-configured with the relevant software packages for carrying out analytics and modeling tasks. It also makes available high-performance computing TPU and GPU processing capabilities at a single click. These VMs expose a JupyterLab notebook environment for analyzing data and designing machine learning models.

In this chapter, we'll launch a JupyterLab notebook instance using the web-based console and the command line.

Provisioning a Notebook Instance

The following steps provide a walk-through for deploying a Notebook instance on a deep learning VM:

1. In the group named 'ARTIFICIAL INTELLIGENCE' on the GCP resources drawer, click the arrow beside 'AI Platform' and select 'Notebooks' as shown in Figure 6-1.

© Ekaba Bisong 2019
E. Bisong, *Building Machine Learning and Deep Learning Models on Google Cloud Platform*,
https://doi.org/10.1007/978-1-4842-4470-8_6

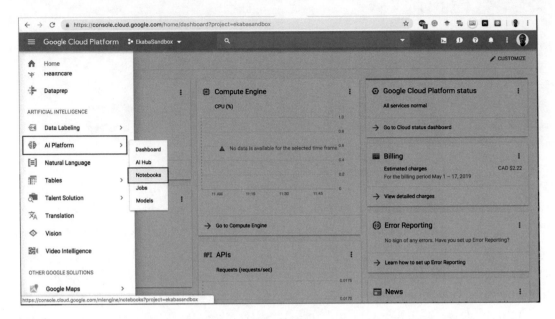

Figure 6-1. *Open Notebooks on GCP AI Platform*

2. Click 'NEW INSTANCE' to initiate a notebook instance as shown
    in Figure 6-2; there is an option to customize your instance or
    to use one of the pre-configured instances with TensorFlow,
    PyTorch, or RAPIDS XGBoost installed.

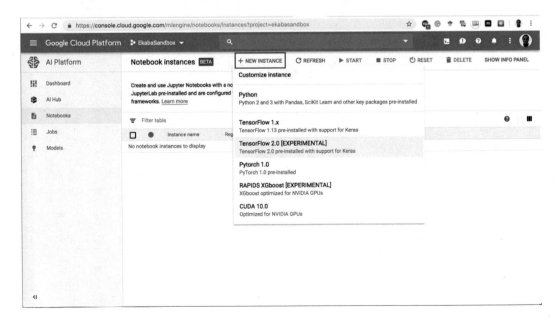

Figure 6-2. *Start a new Notebook instance*

3. For this example, we will create a Notebook instance pre-
    configured with TensorFlow 2.0 (see Figure 6-3).

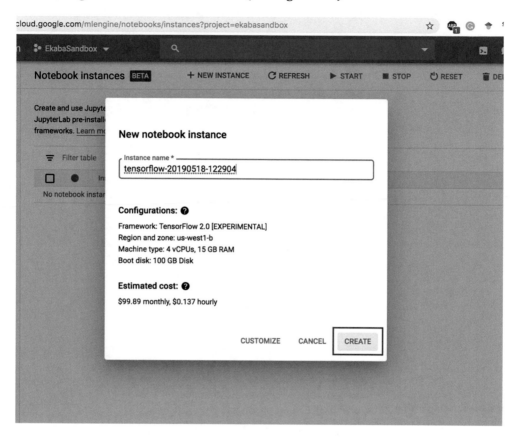

Figure 6-3. *Start a new Notebook instance*

4. Click 'OPEN JUPYTERLAB' to launch the JupyterLab notebook
    instance in a new window (see Figure 6-4).

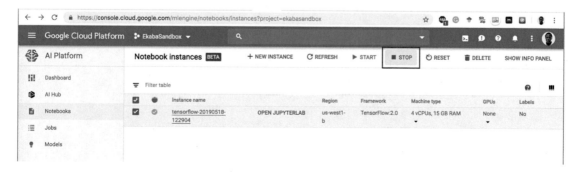

Figure 6-4. *Open JupyterLab*

5. From the JupyterLab Launcher in Figure 6-5, options exist to open a Python notebook, a Python interactive shell, a bash terminal, a text file, or a Tensorboard dashboard (more on Tensorboard in Part 6).

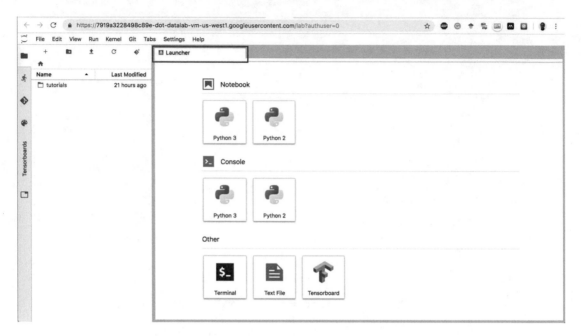

Figure 6-5. *JupyterLab Launcher*

6. Open a Python 3 Notebook (see Figure 6-6). We'll work with Python notebooks in later chapters to carry out data science tasks.

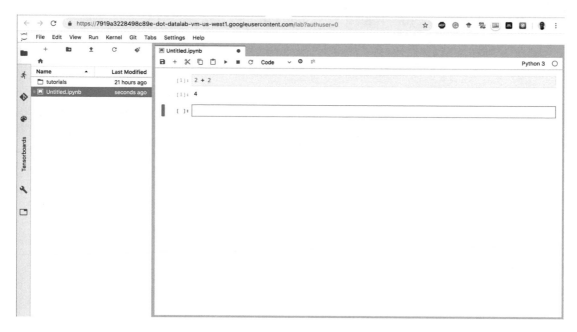

Figure 6-6. *Python 3 Notebook*

Shut Down/Delete a Notebook Instance

The following steps provide a walk-through for shutting down and deleting a Notebook instance:

1. From the 'Notebook instances' dashboard, click 'STOP' to shut down the instance when not in use so as to save compute costs on GCP (see Figure 6-7).

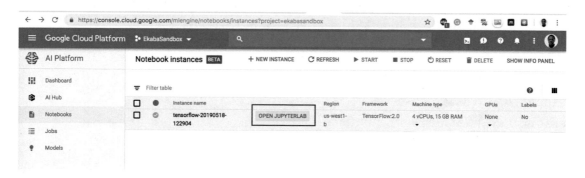

Figure 6-7. *Stop Notebook instance*

2. When the instance is no longer needed, click 'DELETE' to permanently remove the instance. Note that this option is non-recoverable (see Figure 6-8).

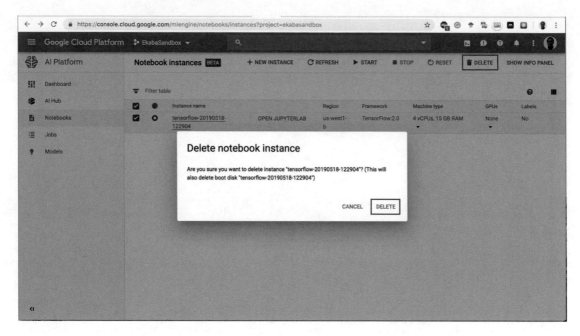

Figure 6-8. *Delete a Notebook instance*

Starting a Notebook Instance from the Command Line

In this section, we'll examine how the command line is used to launch and shut down a pre-configured deep learning VM integrated with JupyterLab.

Create a Datalab instance: To create a Notebook instance, execute the code

```
export IMAGE_FAMILY="tf-latest-cpu-experimental"
export ZONE="us-west1-b"
export INSTANCE_NAME="my-instance"
```

```
gcloud compute instances create $INSTANCE_NAME \
  --zone=$ZONE \
  --image-family=$IMAGE_FAMILY \
  --image-project=deeplearning-platform-release
```

where

- --image-family can be any of the available images supported by Google Deep Learning VM; "tf-latest-cpu-experimental" launches an image with TensorFlow 2.0 pre-configured.

- --image-project must be set to deeplearning-platform-release

Here's the output when the instance is created:

```
Created [https://www.googleapis.com/compute/v1/projects/ekabasandbox/zones/
us-west1-b/instances/my-instance].
NAME            ZONE          MACHINE_TYPE   PREEMPTIBLE   INTERNAL_IP
EXTERNAL_IP     STATUS
my-instance     us-west1-b    n1-standard-1                10.138.0.6
34.83.90.154    RUNNING
```

Connect to the instance: To connect to JupyterLab running on the instance, run the command

```
export INSTANCE_NAME="my-instance"
gcloud compute ssh $INSTANCE_NAME -- -L 8080:localhost:8080
```

Then on your local machine, visit http://localhost:8080 in your browser (see Figure 6-9).

Figure 6-9. *JupyterLab instance launched from terminal*

Stop the instance: To stop the instance, run the following command from your local terminal (not on the instance):

```
gcloud compute instances stop $INSTANCE_NAME
Stopping instance(s) my-instance...done.
Updated [https://www.googleapis.com/compute/v1/projects/ekabasandbox/zones/
us-west1-b/instances/my-instance].
```

Delete the instance: The Notebook instance is basically a Google Compute Engine. Hence, the instance is deleted the same way a Compute Engine VM is deleted.

```
gcloud compute instances delete $INSTANCE_NAME
The following instances will be deleted. Any attached disks configured
 to be auto-deleted will be deleted unless they are attached to any
other instances or the `--keep-disks` flag is given and specifies them
 for keeping. Deleting a disk is irreversible and any data on the disk
 will be lost.
 - [my-instance] in [us-west1-b]
```

```
Do you want to continue (Y/n)?   Y
```

Deleted [https://www.googleapis.com/compute/v1/projects/ekabasandbox/zones/
us-west1-b/instances/my-instance].

This chapter introduces Jupyter notebooks running on Google Deep Learning VMs for interactive programming of data science tasks and prototyping deep learning and machine learning models.

In the next chapter, we will introduce another product for programming and rapid prototyping of learning models called Google Colaboratory.

CHAPTER 7

Google Colaboratory

Google Colaboratory more commonly referred to as "Google Colab" or just simply "Colab" is a research project for prototyping machine learning models on powerful hardware options such as GPUs and TPUs. It provides a serverless Jupyter notebook environment for interactive development. Google Colab is free to use like other G Suite products.

Starting Out with Colab

The following steps provide a walk-through for launching a Notebook on Google Colab:

1. Go to `https://colab.research.google.com/` and log in using your existing Google account to access the Colab homepage (see Figure 7-1).

© Ekaba Bisong 2019
E. Bisong, *Building Machine Learning and Deep Learning Models on Google Cloud Platform*,
https://doi.org/10.1007/978-1-4842-4470-8_7

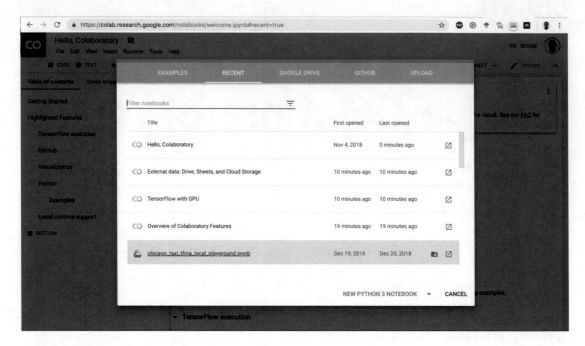

Figure 7-1. *Google Colab homepage*

2. Open a Python 3 Notebook (see Figure 7-2).

Figure 7-2. *Python 3 Notebook*

Change Runtime Settings

The following steps provide a walk-through for changing the Notebook runtime settings:

1. Go to Runtime ➤ Change runtime type (see Figure 7-3).

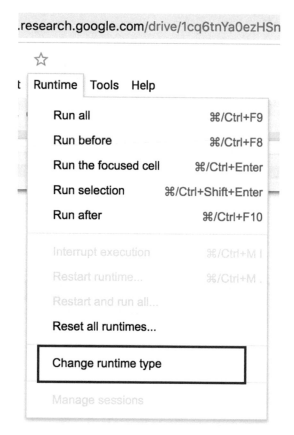

Figure 7-3. *Python 3 Notebook*

2. Here, the options exist to change the Python runtime and hardware accelerator to a GPU or TPU (see Figure 7-4).

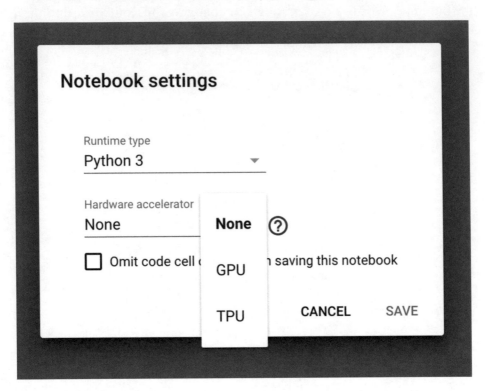

Figure 7-4. *Change runtime*

Storing Notebooks

Notebooks on Colab are stored on Google Drive. They can also be saved to GitHub or published as a GitHub Gist. They can also be downloaded to the local machine.

Figure 7-5 highlights the options for storing Jupyter notebooks running on Google Colab.

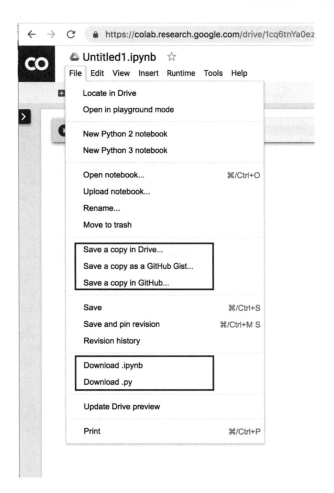

Figure 7-5. *Storing Notebooks*

Uploading Notebooks

Notebooks can be uploaded from Google Drive, GitHub, or the local machine (see Figure 7-6).

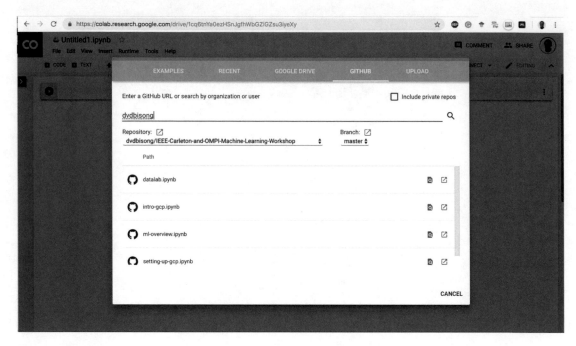

Figure 7-6. *Opening Notebooks*

This chapter introduces Google Colaboratory as an alternative platform to quickly spin up a high-performance computing infrastructure running Jupyter notebooks for rapid data science and data modeling tasks.

This is the last chapter in Part 1 on "Getting Started with Google Cloud Platform." In Part 2, containing Chapters 8–12, we will go over the fundamentals of "Programming for Data Science." The code samples in the ensuing chapters can be executed either using Jupyter notebooks running on Google Deep Learning VMs or running on Google Colab.

The advantage of working with Google Colab is that you do not need to log into the Google Cloud Console and it is free to use. When security and privacy are not a premium, Google Colab is a good option for modeling as it saves computing cost as far as data science and machine learning prototyping is concerned.

PART II

Programming Foundations for Data Science

## CHAPTER 8

What Is Data Science?

Data science encompasses the tools and techniques for extracting information from data. Data science techniques draw extensively from the field of mathematics, statistics, and computation. However, data science is now encapsulated into software packages and libraries, thus making them easily accessible and consumable by the software development and engineering communities. This is a major factor to the rise of intelligence capabilities now integrated as a major staple in software products across all sorts of domains.

This chapter will discuss broadly on the opportunities for data science and big data analytics integration as part of the transformation portfolio of businesses and institutions and give an overview on the data science process as a reusable template for fulfilling data science projects.

The Challenge of Big Data

Due to the expansion of data at the turn of the twenty-first century epitomized by the so-called 3Vs of big data, which are volume, velocity, and variety. Volume refers to the increasing size of data, velocity the speed at which data is acquired, and variety the diverse types of data that are available. For others, this becomes 5Vs with the inclusion of value and veracity to mean the usefulness of data and the truthfulness of data, respectively. We have observed data volume blowout from the megabyte (MB) to the terabyte (TB) scale and now exploding past the petabyte (PB). We have to find new and improved means of storing and processing this ever-increasing dataset. Initially, this challenge of storage and data processing was addressed by the Hadoop ecosystem and other supporting frameworks, but even these have become expensive to manage and scale, and this is why there is a pivot to cloud-managed, elastic, secure, and high-availability data storage and processing capabilities.

© Ekaba Bisong 2019
E. Bisong, *Building Machine Learning and Deep Learning Models on Google Cloud Platform*,
https://doi.org/10.1007/978-1-4842-4470-8_8

On the other hand, for most applications and business use cases, there is a need to carry out real-time analysis on data due to the vast amount of data created and available at a given moment. Previously, getting insights from data and unlocking value had been down to traditional analysis on batch data workloads using statistical tools such as Excel, Minitab, or SPSS. But in the era of big data, this is changing, as more and more businesses and institutions want to understand the information in their data at a real-time or at worst near real-time pace.

Another vertical to the big data conundrum is that of variety. Formerly, a pre-defined structure had to be imposed on data in order to easily store them as well as make it easy for data analysis. However, a wide diversity of datasets are now collected and stored such as spatial maps, image data, video data, audio data, text data from emails and other documents, and sensor data. As a matter of fact, a far larger amount of datasets in the wild are unstructured. This led to the development of unstructured or semi-structured databases such as Elasticsearch, Solr, HBase, Cassandra, and MongoDB, to mention just a few.

The Data Science Opportunity

In the new age, where data has inevitably and irreversibly become the new gold, the greatest needs of organizations are the skills required for data governance and analytics to unlock intelligence and value from data as well as the expertise to develop and productionize enterprise data products. This has led to new roles within the data science umbrella such as

- Data analysts/scientist who specialize in mining intelligence from data using statistical techniques and computational tools by understanding the business use case

- Data engineers/architects who specialize in architecting and managing the infrastructure for efficient big data pipelines by ensuring that the data platform is redundant, scalable, secure, and highly available

- Machine learning engineers who specialize in designing and developing machine learning algorithms as well as incorporating them into production systems for online or batch prediction services

The Data Science Process

The data science process involves components for data ingestion and serving of data models. However, we will discuss briefly on the steps for carrying out data analytics in lieu of data prediction modeling.

These steps consist of

1. Data summaries: The vital statistical summaries of the datasets' variables or features. This includes information such as the number of variables, their data types, the number of observations, and the count/percentage of missing data.

2. Data visualization: This involves employing univariate and multivariate data visualization methods to get a better intuition on the properties of the data variables and their relationship with each other. This includes metrics such as histograms, box and whisker plots, and correlation plots.

3. Data cleaning/preprocessing: This process involves sanitizing the data to make it amenable for modeling. Data rarely comes clean with each row representing an observation and each column an entity. In this phase of a data science effort, the tasks involved may include removing duplicate entries, choosing a strategy for dealing with missing data, as well as converting data features into numeric data types of encoded categories. This phase may also involve carrying out statistical transformation on the data features to normalize and/or standardize the data elements. Data features of wildly differing scales can lead to poor model results as they become more difficult for the learning algorithm to converge to the global minimum.

4. Feature engineering: This practice involves systematically pruning the data feature space to only select those features relevant to the modeling problem as part of the model task.
    Good feature engineering is often the difference between an average and high performant model.

5. Data modeling and evaluation: This phase involves passing the data through a learning algorithm to build a predictive model. This process is usually an iterative process that involves constant refinement in order to build a model that better minimizes the cost function on the hold-out validation set and the test set.

In this chapter, we provided a brief overview to the concept of data science, the challenge of big data, and its goal to unlock value from data. The next chapter will provide an introduction to programming with Python.

CHAPTER 9

Python

Python is one of the preferred languages for data science in the industry primarily because of its simple syntax and the number of reusable machine learning/deep learning packages. These packages make it easy to develop data science products without getting bogged down with the internals of a particular algorithm or method. They have been written, debugged, and tested by the best experts in the field, as well as by a large supporting community of developers that contribute their time and expertise to maintain and improve them.

In this section, we will go through the foundations of programming with Python 3. This section forms a framework for working with higher-level packages such as NumPy, Pandas, Matplotlib, TensorFlow, and Keras. The programming paradigm we will cover in this chapter can be easily adapted or applied to similar languages, such as R, which is also commonly used in the data science industry.

The best way to work through this chapter and the successive chapters in this part is to work through the code by executing them on Google Colab or GCP Deep Learning VMs.

Data and Operations

Fundamentally, programming involves storing data and operating on that data to generate information. Techniques for efficient data storage are studied in the field called data structures, while the techniques for operating on data are studied as algorithms.

Data is stored in a memory block on the computer. Think of a memory block as a container holding data (Figure 9-1). When data is operated upon, the newly processed data is also stored in memory. Data is operated by using arithmetic and boolean expressions and functions.

© Ekaba Bisong 2019   71
E. Bisong, *Building Machine Learning and Deep Learning Models on Google Cloud Platform*,
https://doi.org/10.1007/978-1-4842-4470-8_9

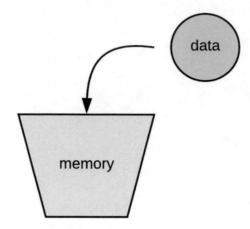

Figure 9-1. *An illustration of a memory cell holding data*

In programming, a memory location is called a variable. A **variable** is a container for storing the data that is assigned to it. A variable is usually given a unique name by the programmer to represent a particular memory cell. In python, variable names are programmer defined, but it must follow a valid naming condition of only alphanumeric lowercase characters with words separated by an underscore. Also, a variable name should have semantic meaning to the data that is stored in that variable. This helps to improve code readability later in the future.

The act of placing data to a variable is called **assignment**.

```
# assigning data to a variable
x = 1
user_name = 'Emmanuel Okoi'
```

Data Types

Python has the number and string data types in addition to other supported specialized datatypes. The number datatype, for instance, can be an int or a float. Strings are surrounded by quotes in Python.

```
# data types
type(3)
'Output': int
type(3.0)
'Output': float
```

```
type('Jesam Ujong')
'Output': str
```

Other fundamental data types in Python include the lists, tuple, and dictionary. These data types hold a group of items together in sequence. Sequences in Python are indexed from 0.

Tuples are an immutable ordered sequence of items. Immutable means the data cannot be changed after being assigned. Tuple can contain elements of different types. Tuples are surrounded by brackets (…).

```
my_tuple = (5, 4, 3, 2, 1, 'hello')
type(my_tuple)
'Output': tuple
my_tuple[5]            # return the sixth element (indexed from 0)
'Output': 'hello'
my_tuple[5] = 'hi'    # we cannot alter an immutable data type
Traceback (most recent call last):

  File "<ipython-input-49-f0e593f95bc7>", line 1, in <module>
    my_tuple[5] = 'hi'

TypeError: 'tuple' object does not support item assignment
```

Lists are very similar to tuples, only that they are mutable. This means that list elements can be changed after being assigned. Lists are surrounded by square brackets […].

```
my_list = [4, 8, 16, 32, 64]
print(my_list)    # print list items to console
'Output': [4, 8, 16, 32, 64]
my_list[3]        # return the fourth list element (indexed from 0)
'Output': 32
my_list[4] = 256
print(my_list)
'Output': [4, 8, 16, 32, 256]
```

Dictionaries contain a mapping from keys to values. A key/value pair is an item in a dictionary. The items in a dictionary are indexed by their keys. The keys in a dictionary can be any *hashable* datatype (hashing transforms a string of characters into a key to speed up search). Values can be of any datatype. In other languages, a dictionary is

analogous to a hash table or a map. Dictionaries are surrounded by a pair of braces {...}.
A dictionary is not ordered.

```
my_dict = {'name':'Rijami', 'age':42, 'height':72}
my_dict                    # dictionary items are un-ordered
'Output': {'age': 42, 'height': 72, 'name': 'Rijami'}
my_dict['age']          # get dictionary value by indexing on keys
'Output': 42
my_dict['age'] = 35    # change the value of a dictionary item
my_dict['age']
'Output': 35
```

More on Lists

As earlier mentioned, because list items are mutable, they can be changed, deleted, and
sliced to produce a new list.

```
my_list = [4, 8, 16, 32, 64]
my_list
'Output': [4, 8, 16, 32, 64]
my_list[1:3]        # slice the 2nd to 4th element (indexed from 0)
'Output': [8, 16]
my_list[2:]         # slice from the 3rd element (indexed from 0)
'Output': [16, 32, 64]
my_list[:4]         # slice till the 5th element (indexed from 0)
'Output': [4, 8, 16, 32]
my_list[-1]         # get the last element in the list
'Output': 64
min(my_list)        # get the minimum element in the list
'Output': 4
max(my_list)        # get the maximum element in the list
'Output': 64
sum(my_list)        # get the sum of elements in the list
'Output': 124
my_list.index(16) # index(k) - return the index of the first occurrence of
item k in the list
'Output': 2
```

When modifying a slice of elements in the list, the right-hand side can be of any length depending that the left-hand size is not a single index.

```
# modifying a list: extended index example
my_list[1:4] = [43, 59, 78, 21]
my_list
'Output': [4, 43, 59, 78, 21, 64]
my_list = [4, 8, 16, 32, 64]  # re-initialize list elements
my_list[1:4] = [43]
my_list
'Output': [4, 43, 64]

# modifying a list: single index example
my_list[0] = [1, 2, 3]       # this will give a list-on-list
my_list
'Output': [[1, 2, 3], 43, 64]
my_list[0:1] = [1, 2, 3]     # again - this is the proper way to extend lists
my_list
'Output': [1, 2, 3, 43, 64]
```

Some useful list methods include

```
my_list = [4, 8, 16, 32, 64]
len(my_list)              # get the length of the list
'Output': 5
my_list.insert(0,2)    # insert(i,k) - insert the element k at index i
my_list
'Output': [2, 4, 8, 16, 32, 64]
my_list.remove(8) # remove(k) - remove the first occurrence of element k in
                                the list
my_list
'Output': [2, 4, 16, 32, 64]
my_list.pop(3)    # pop(i) - return the value of the list at index i
'Output': 32
my_list.reverse() # reverse in-place the elements in the list
my_list
'Output': [64, 16, 4, 2]
```

```
my_list.sort()     # sort in-place the elements in the list
my_list
'Output': [2, 4, 16, 64]
my_list.clear()    # clear all elements from the list
my_list
'Output': []
```

The append() method adds an item (could be a list, string, or number) to the end of a list. If the item is a list, the list as a whole is appended to the end of the current list.

```
my_list = [4, 8, 16, 32, 64]    # initial list
my_list.append(2)                    # append a number to the end of list
my_list.append('wonder')             # append a string to the end of list
my_list.append([256, 512])           # append a list to the end of list
my_list
'Output': [4, 8, 16, 32, 64, 2, 'wonder', [256, 512]]
```

The extend() method extends the list by adding items from an iterable. An iterable in Python are objects that have special methods that enable you to access elements from that object sequentially. Lists and strings are iterable objects. So extend() appends all the elements of the iterable to the end of the list.

```
my_list = [4, 8, 16, 32, 64]
my_list.extend(2)                    # a number is not an iterable
Traceback (most recent call last):

  File "<ipython-input-24-092b23c845b9>", line 1, in <module>
    my_list.extend(2)

TypeError: 'int' object is not iterable

my_list.extend('wonder')        # append a string to the end of list
my_list.extend([256, 512])      # append a list to the end of list
my_list
'Output': [4, 8, 16, 32, 64, 'w', 'o', 'n', 'd', 'e', 'r', 256, 512]
```

We can combine a list **with another list** by overloading the operator +.

```
my_list = [4, 8, 16, 32, 64]
my_list + [256, 512]
'Output': [4, 8, 16, 32, 64, 256, 512]
```

Strings

Strings in Python are enclosed by a pair of single quotes (' ... '). Strings are immutable. This means they cannot be altered when assigned or when a string variable is created. Strings can be indexed like a list as well as sliced to create new lists.

```
my_string = 'Schatz'
my_string[0]        # get first index of string
'Output': 'S'
my_string[1:4]      # slice the string from the 2nd to the 5th element
                        (indexed from 0)
'Output': 'cha'
len(my_string)      # get the length of the string
'Output': 6
my_string[-1]       # get last element of the string
'Output': 'z'
```

We can operate on string values with the boolean operators.

```
't' in my_string
'Output': True
't' not in my_string
'Output': False
't' is my_string
'Output': False
't' is not my_string
'Output': True
't' == my_string
'Output': False
't' != my_string
'Output': True
```

We can concatenate two strings to create a new string using the overloaded operator +.

```
a = 'I'
b = 'Love'
c = 'You'
```

```
a + b + c
'Output': 'ILoveYou'

# let's add some space
a + ' ' + b + ' ' + c
```

Arithmetic and Boolean Operations

This section introduces operators for programming arithmetic and logical constructs.

Arithmetic Operations

In Python, we can operate on data using familiar algebra operations such as addition +, subtraction -, multiplication *, division /, and exponentiation **.

```
2 + 2       # addition
'Output': 4
5 - 3       # subtraction
'Output': 2
4 * 4       # multiplication
'Output': 16
10 / 2      # division
'Output': 5.0
2**4 / (5 + 3)      # use brackets to enforce precedence
'Output': 2.0
```

Boolean Operations

Boolean operations evaluate to True or False. Boolean operators include the comparison and logical operators. The comparison operators include less than or equal to <=, less than <, greater than or equal to >=, greater than >, not equal to !=, and equal to ==.

```
2 < 5
'Output': True
2 <= 5
'Output': True
```

```
2 > 5
'Output': False
2 >= 5
'Output': False
2 != 5
'Output': True
2 == 5
'Output': False
```

The logical operators include Boolean NOT (not), Boolean AND (and), and Boolean OR (or). We can also carry out identity and membership tests using

- is, is not (identity)

- in, not in (membership)

```
a = [1, 2, 3]
2 in a
'Output': True
2 not in a
'Output': False
2 is a
'Output': False
2 is not a
'Output': True
```

The print() Statement

The print() statement is a simple way to show the output of data values to the console. Variables can be concatenated using the comma. Space is implicitly added after the comma.

```
a = 'I'
b = 'Love'
c = 'You'
print(a, b, c)
'Output': I Love You
```

Using the Formatter

Formatters add a placeholder for inputting a data value into a string output using the curly brace {}. The format method from the str class is invoked to receive the value as a parameter. The number of parameters in the format method should match the number of placeholders in the string representation. Other format specifiers can be added with the placeholder curly brackets.

```
print("{} {} {}".format(a, b, c))
'Output': I Love You
# re-ordering the output
print("{2} {1} {0}".format(a, b, c))
'Output': You Love I
```

Control Structures

Programs need to make decisions which result in executing a particular set of instructions or a specific block of code repeatedly. With control structures, we would have the ability to write programs that can make logical decisions and execute an instruction set until a terminating condition occurs.

The if/elif (else-if) Statements

The if/elif (else-if) statement executes a set of instructions if the tested condition evaluates to true. The else statement specifies the code that should execute if none of the previous conditions evaluate to true. It can be visualized by the flowchart in Figure 9-2.

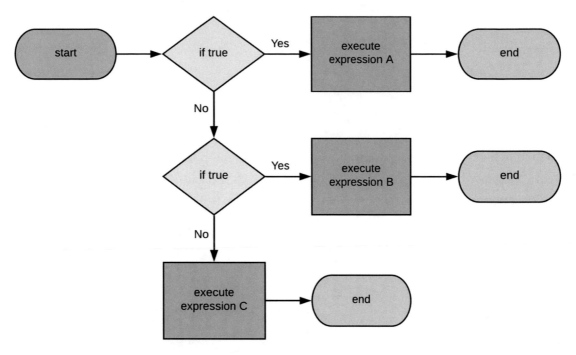

Figure 9-2. *Flowchart of the if statement*

The syntax for the if/elif statement is given as follows:

```
if expressionA:
    statementA
elif expressionB:
    statementB
...
...
else:
    statementC
```

Here is a program example:

```
a = 8
if type(a) is int:
    print('Number is an integer')
elif a > 0:
    print('Number is positive')
```

```
else:
    print('The number is negative and not an integer')

'Output': Number is an integer
```

The while Loop

The while loop evaluates a condition, which, if true, repeatedly executes the set of instructions within the while block. It does so until the condition evaluates to false. The while statement is visualized by the flowchart in Figure 9-3.

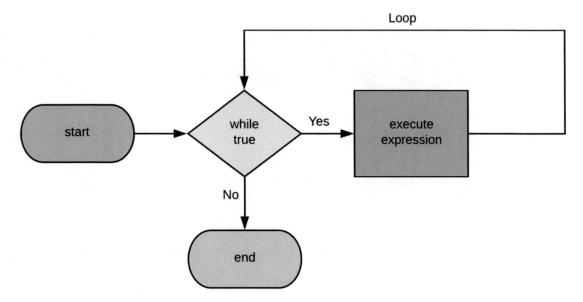

Figure 9-3. *Flowchart of the while loop*

Here is a program example:

```
a = 8
while a > 0:
    print('Number is', a)

    # decrement a
    a -= 1
```

```
'Output': Number is 8
         Number is 7
         Number is 6
         Number is 5
         Number is 4
         Number is 3
         Number is 2
         Number is 1
```

The for Loop

The for loop repeats the statements within its code block until a terminating condition is reached. It is different from the while loop in that it knows exactly how many times the iteration should occur. The for loop is controlled by an iterable expression (i.e., expressions in which elements can be accessed sequentially). The for statement is visualized by the flowchart in Figure 9-4.

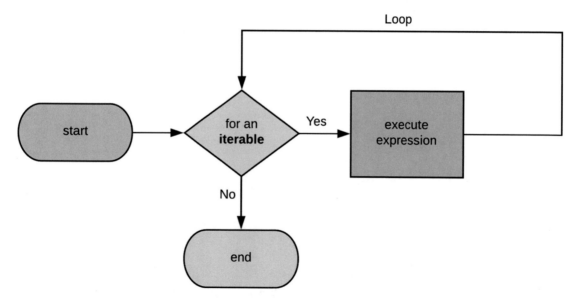

Figure 9-4. *Flowchart of the for loop*

The syntax for the for loop is as follows:

```
for item in iterable:
    statement
```

Note that in the for loop syntax is not the same as the membership logical operator earlier discussed.

Here is a program example:

```
a = [2, 4, 6, 8, 10]
for elem in a:
    print(elem**2)

'Output': 4
    16
    36
    64
    100
```

To loop for a specific number of time, use the range() function.

```
for idx in range(5):
    print('The index is', idx)

'Output': The index is 0
    The index is 1
    The index is 2
    The index is 3
    The index is 4
```

List Comprehensions

Using list comprehension, we can succinctly rewrite a for loop that iteratively builds a new list using an elegant syntax. Assuming we want to build a new list using a for loop, we will write it as

```
new_list = []
for item in iterable:
    new_list.append(expression)
```

We can rewrite this as

```
[expression for item in iterable]
```

Let's have some program examples.

```
squares = []
for elem in range(0,5):
    squares.append((elem+1)**2)

squares
'Output': [1, 4, 9, 16, 25]
```

The preceding code can be concisely written as

```
[(elem+1)**2 for elem in range(0,5)]
'Output': [1, 4, 9, 16, 25]
```

This is even more elegant in the presence of nested control structures.

```
evens = []
for elem in range(0,20):
    if elem % 2 == 0 and elem != 0:
        evens.append(elem)

evens
'Output': [2, 4, 6, 8, 10, 12, 14, 16, 18]
```

With list comprehension, we can code this as

```
[elem for elem in range(0,20) if elem % 2 == 0 and elem != 0]
'Output': [2, 4, 6, 8, 10, 12, 14, 16, 18]
```

The break and continue Statements

The break statement terminates the execution of the nearest enclosing loop (for, while loops) in which it appears.

```
for val in range(0,10):
    print("The variable val is:", val)
    if val > 5:
```

```
    print("Break out of for loop")
    break
```

```
'Output': The variable val is: 0
    The variable val is: 1
    The variable val is: 2
    The variable val is: 3
    The variable val is: 4
    The variable val is: 5
    The variable val is: 6
    Break out of for loop
```

The continue statement skips the next iteration of the loop to which it belongs, ignoring any code after it.

```
a = 6
while a > 0:
    if a != 3:
        print("The variable a is:", a)
    # decrement a
    a = a - 1
    if a == 3:
        print("Skip the iteration when a is", a)
        continue
```

```
'Output': The variable a is: 6
    The variable a is: 5
    The variable a is: 4
    Skip the iteration when a is 3
    The variable a is: 2
    The variable a is: 1
```

Functions

A function is a code block that carries out a particular action (Figure 9-5). Functions are called by the programmer when needed by making a **function call**. Python comes pre-packaged with lots of useful functions to simplify programming. The programmer can also write custom functions.

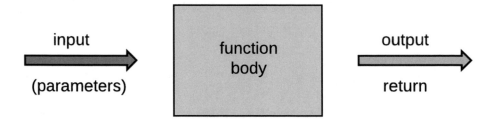

Figure 9-5. Functions

A function receives data into its parameter list during a function call. The inputed data is used to complete the function execution. At the end of its execution, a function always returns a result – this result could be 'None' or a specific data value.

Functions are treated as first-class objects in Python. That means a function can be passed as data into another function, the result of a function execution can also be a function, and a function can also be stored as a variable.

Functions are visualized as a black box that receives a set of objects as input, executes some code, and returns another set of objects as output.

User-Defined Functions

A function is defined using the def keyword. The syntax for creating a function is as follows:

```
def function-name(parameters):
    statement(s)
```

Let's create a simple function:

```
def squares(number):
    return number**2

squares(2)
'Output': 4
```

Here's another function example:

```
def _mean_(*number):
    avg = sum(number)/len(number)
    return avg

_mean_(1,2,3,4,5,6,7,8,9)
'Output': 5.0
```

The * before the parameter number indicates that the variable can receive any number of values, which is implicitly bound to a tuple.

Lambda Expressions

Lambda expressions provide a concise and succinct way to write simple functions that contain just a single line. Lambdas now and again can be very useful, but in general, working with **def** may be more readable. The syntax for lambdas are as follows:

```
lambda parameters: expression
```

Let's see an example:

```
square = lambda x: x**2
square(2)
'Output': 4
```

Packages and Modules

A module is simply a Python source file, and packages are a collection of modules. Modules written by other programmers can be incorporated into your source code by using **import** and **from** statements.

import Statement

The **import** statement allows you to load any Python module into your source file. It has the following syntax:

```
import module_name [as user_defined_name][,...]
```

where the following is optional:

```
[as user_defined_name]
```

Let us take an example by importing a very important package called **numpy** that is used for numerical processing in Python and very critical for machine learning.

```
import numpy as np

np.abs(-10)    # the absolute value of -10
'Output': 10
```

from Statement

The **from** statement allows you to import a specific feature from a module into your source file. The syntax is as follows:

```
from module_name import module_feature [as user_defined_name][,...]
```

Let's see an example:

```
from numpy import mean
```

```
mean([2,4,6,8])
'Output': 5.0
```

This chapter provides the fundamentals for programming with Python. Programming is a very active endeavor, and competency is gained by experience and repetition. What is presented in this chapter provides just enough to be dangerous.

In the next chapter, we'll introduce NumPy, a Python package for numerical computing.

CHAPTER 10

NumPy

NumPy is a Python library optimized for numerical computing. It bears close semblance with MATLAB and is equally as powerful when used in conjunction with other packages such as SciPy for various scientific functions, Matplotlib for visualization, and Pandas for data analysis. NumPy is short for numerical python.

NumPy's core strength lies in its ability to create and manipulate n-dimensional arrays. This is particularly critical for building machine learning and deep learning models. Data is often represented in a matrix-like grid of rows and columns, where each row represents an observation and each column a variable or feature. Hence, NumPy's 2-D array is a natural fit for storing and manipulating datasets.

This tutorial will cover the basics of NumPy to get you very comfortable working with the package and also get you to appreciate the thinking behind how NumPy works. This understanding forms a foundation from which one can extend and seek solutions from the NumPy reference documentation when a specific functionality is needed.

To begin using NumPy, we'll start by importing the NumPy module:

```
import numpy as np
```

NumPy 1-D Array

Let's create a simple 1-D NumPy array:

```
my_array = np.array([2,4,6,8,10])
my_array
'Output': array([ 2,  4,  6,  8, 10])
# the data-type of a NumPy array is the ndarray
type(my_array)
'Output': numpy.ndarray
```

© Ekaba Bisong 2019
E. Bisong, *Building Machine Learning and Deep Learning Models on Google Cloud Platform*,
https://doi.org/10.1007/978-1-4842-4470-8_10

```
# a NumPy 1-D array can also be seen a vector with 1 dimension
my_array.ndim
'Output': 1
# check the shape to get the number of rows and columns in the array \
# read as (rows, columns)
my_array.shape
'Output': (5,)
```

We can also create an array from a Python list.

```
my_list = [9, 5, 2, 7]
type(my_list)
'Output': list
# convert a list to a numpy array
list_to_array = np.array(my_list) # or np.asarray(my_list)
type(list_to_array)
'Output': numpy.ndarray
```

Let's explore other useful methods often employed for creating arrays.

```
# create an array from a range of numbers
np.arange(10)
'Output': [0 1 2 3 4 5 6 7 8 9]
# create an array from start to end (exclusive) via a step size - (start,
stop, step)
np.arange(2, 10, 2)
'Output': [2 4 6 8]
# create a range of points between two numbers
np.linspace(2, 10, 5)
'Output': array([ 2.,    4.,    6.,    8.,   10.])
# create an array of ones
np.ones(5)
'Output': array([ 1.,   1.,   1.,   1.,   1.])
# create an array of zeros
np.zeros(5)
'Output': array([ 0.,   0.,   0.,   0.,   0.])
```

NumPy Datatypes

NumPy boasts a broad range of numerical datatypes in comparison with vanilla Python. This extended datatype support is useful for dealing with different kinds of signed and unsigned integer and floating-point numbers as well as booleans and complex numbers for scientific computation. NumPy datatypes include the **bool_**, **int**(8,16,32,64), **uint**(8,16,32,64), **float**(16,32,64), **complex**(64,128) as well as the **int_**, **float_**, and **complex_**, to mention just a few.

The datatypes with a _ appended are base Python datatypes converted to NumPy datatypes. The parameter **dtype** is used to assign a datatype to a NumPy function. The default NumPy type is **float_**. Also, NumPy infers contiguous arrays of the same type.

Let's explore a bit with NumPy datatypes:

```
# ints
my_ints = np.array([3, 7, 9, 11])
my_ints.dtype
'Output': dtype('int64')

# floats
my_floats = np.array([3., 7., 9., 11.])
my_floats.dtype
'Output': dtype('float64')

# non-contiguous types - default: float
my_array = np.array([3., 7., 9, 11])
my_array.dtype
'Output': dtype('float64')

# manually assigning datatypes
my_array = np.array([3, 7, 9, 11], dtype="float64")
my_array.dtype
'Output': dtype('float64')
```

Indexing + Fancy Indexing (1-D)

We can index a single element of a NumPy 1-D array similar to how we index a Python list.

```
# create a random numpy 1-D array
my_array = np.random.rand(10)
my_array
'Output': array([ 0.7736445 ,  0.28671796,  0.61980802,  0.42110553,
                  0.86091567,  0.93953255,  0.300224  ,  0.56579416,
                  0.58890282,  0.97219289])
# index the first element
my_array[0]
'Output': 0.77364449999999996
# index the last element
my_array[-1]
'Output': 0.97219288999999998
```

Fancy indexing in NumPy is an advanced mechanism for indexing array elements based on integers or boolean. This technique is also called *masking*.

Boolean Mask

Let's index all the even integers in the array using a boolean mask.

```
# create 10 random integers between 1 and 20
my_array = np.random.randint(1, 20, 10)
my_array
'Output': array([14,  9,  3, 19, 16,  1, 16,  5, 13,  3])
# index all even integers in the array using a boolean mask
my_array[my_array % 2 == 0]
'Output': array([14, 16, 16])
```

Observe that the code my_array % 2 == 0 outputs an array of booleans.

```
my_array % 2 == 0
'Output': array([ True, False, False, False,  True, False,  True, False,
False, False], dtype=bool)
```

Integer Mask

Let's select all elements with even indices in the array.

```
# create 10 random integers between 1 and 20
my_array = np.random.randint(1, 20, 10)
my_array
'Output': array([ 1, 18,  8, 12, 10,  2, 17,  4, 17, 17])
my_array[np.arange(1,10,2)]
'Output': array([18, 12,  2,  4, 17])
```

Remember that array indices are indexed from 0. So the second element, 18, is in index 1.

```
np.arange(1,10,2)
'Output': array([1, 3, 5, 7, 9])
```

Slicing a 1-D Array

Slicing a NumPy array is also similar to slicing a Python list.

```
my_array = np.array([14,  9,  3, 19, 16,  1, 16,  5, 13,  3])
my_array
'Output': array([14,  9,  3, 19, 16,  1, 16,  5, 13,  3])
# slice the first 2 elements
my_array[:2]
'Output': array([14,  9])
# slice the last 3 elements
my_array[-3:]
'Output': array([ 5, 13,  3])
```

Basic Math Operations on Arrays: Universal Functions

The core power of NumPy is in its highly optimized vectorized functions for various mathematical, arithmetic, and string operations. In NumPy these functions are called universal functions. We'll explore a couple of basic arithmetic with NumPy 1-D arrays.

```
# create an array of even numbers between 2 and 10
my_array = np.arange(2,11,2)
'Output': array([ 2,  4,  6,  8, 10])
# sum of array elements
np.sum(my_array) # or my_array.sum()
'Output': 30
# square root
np.sqrt(my_array)
'Output': array([ 1.41421356,  2.        ,  2.44948974,  2.82842712,
                  3.16227766])
# log
np.log(my_array)
'Output': array([ 0.69314718,  1.38629436,  1.79175947,  2.07944154,
                  2.30258509])
# exponent
np.exp(my_array)
'Output': array([  7.38905610e+00,   5.45981500e+01,   4.03428793e+02,
                  2.98095799e+03,   2.20264658e+04])
```

Higher-Dimensional Arrays

As we've seen earlier, the strength of NumPy is its ability to construct and manipulate n-dimensional arrays with highly optimized (i.e., vectorized) operations. Previously, we covered the creation of 1-D arrays (or vectors) in NumPy to get a feel of how NumPy works.

This section will now consider working with 2-D and 3-D arrays. 2-D arrays are ideal for storing data for analysis. Structured data is usually represented in a grid of rows and columns. And even when data is not necessarily represented in this format, it is often transformed into a tabular form before doing any data analytics or machine learning. Each column represents a feature or attribute and each row an observation.

Also, other data forms like images are adequately represented using 3-D arrays. A colored image is composed of $n \times n$ pixel intensity values with a color depth of three for the red, green, and blue (RGB) color profiles.

Creating 2-D Arrays (Matrices)

Let us construct a simple 2-D array.

```
# construct a 2-D array
my_2D = np.array([[2,4,6],
                  [8,10,12]])
my_2D
'Output':
array([[ 2,  4,  6],
       [ 8, 10, 12]])
# check the number of dimensions
my_2D.ndim
'Output': 2
# get the shape of the 2-D array - this example has 2 rows and
3 columns: (r, c)
my_2D.shape
'Output': (2, 3)
```

Let's explore common methods in practice for creating 2-D NumPy arrays, **which are also matrices**.

```
# create a 3x3 array of ones
np.ones([3,3])
'Output':
array([[ 1.,  1.,  1.],
       [ 1.,  1.,  1.],
       [ 1.,  1.,  1.]])
# create a 3x3 array of zeros
np.zeros([3,3])
'Output':
array([[ 0.,  0.,  0.],
       [ 0.,  0.,  0.],
       [ 0.,  0.,  0.]])
# create a 3x3 array of a particular scalar - full(shape, fill_value)
np.full([3,3], 2)
```

```
'Output':
array([[2, 2, 2],
       [2, 2, 2],
       [2, 2, 2]])
# create a 3x3, empty uninitialized array
np.empty([3,3])
'Output':
array([[ -2.00000000e+000,   -2.00000000e+000,    2.47032823e-323],
       [  0.00000000e+000,    0.00000000e+000,    0.00000000e+000],
       [ -2.00000000e+000,   -1.73060571e-077,   -2.00000000e+000]])
# create a 4x4 identity matrix - i.e., a matrix with 1's on its diagonal
np.eye(4) # or np.identity(4)
'Output':
array([[ 1.,   0.,   0.,   0.],
       [ 0.,   1.,   0.,   0.],
       [ 0.,   0.,   1.,   0.],
       [ 0.,   0.,   0.,   1.]])
```

Creating 3-D Arrays

Let's construct a basic 3-D array.

```
# construct a 3-D array
my_3D = np.array([[
                    [2,4,6],
                    [8,10,12]
                  ],[
                    [1,2,3],
                    [7,9,11]
                  ]])
my_3D
'Output':
array([[[ 2,  4,  6],
        [ 8, 10, 12]],

       [[ 1,  2,  3],
        [ 7,  9, 11]]])
```

```
# check the number of dimensions
my_3D.ndim
'Output': 3
# get the shape of the 3-D array - this example has 2 pages, 2 rows and 3
columns: (p, r, c)
my_3D.shape
'Output': (2, 2, 3)
```

We can also create 3-D arrays with methods such as **ones**, **zeros**, **full**, and **empty** by passing the configuration for [page, row, columns] into the **shape** parameter of the methods. For example:

```
# create a 2-page, 3x3 array of ones
np.ones([2,3,3])
'Output':
array([[[ 1.,   1.,   1.],
        [ 1.,   1.,   1.],
        [ 1.,   1.,   1.]],

       [[ 1.,   1.,   1.],
        [ 1.,   1.,   1.],
        [ 1.,   1.,   1.]]])
# create a 2-page, 3x3 array of zeros
np.zeros([2,3,3])
'Output':
array([[[ 0.,   0.,   0.],
        [ 0.,   0.,   0.],
        [ 0.,   0.,   0.]],

       [[ 0.,   0.,   0.],
        [ 0.,   0.,   0.],
        [ 0.,   0.,   0.]]])
```

Indexing/Slicing of Matrices

Let's see some examples of indexing and slicing 2-D arrays. The concept extends nicely from doing the same with 1-D arrays.

```
# create a 3x3 array contain random normal numbers
my_3D = np.random.randn(3,3)
'Output':
array([[ 0.99709882, -0.41960273,  0.12544161],
       [-0.21474247,  0.99555079,  0.62395035],
       [-0.32453132,  0.3119651 , -0.35781825]])
# select a particular cell (or element) from a 2-D array.
my_3D[1,1]    # In this case, the cell at the 2nd row and column
'Output': 0.99555079000000002
# slice the last 3 columns
my_3D[:,1:3]
'Output':
array([[-0.41960273,  0.12544161],
       [ 0.99555079,  0.62395035],
       [ 0.3119651 , -0.35781825]])
# slice the first 2 rows and columns
my_3D[0:2, 0:2]
'Output':
array([[ 0.99709882, -0.41960273],
       [-0.21474247,  0.99555079]])
```

Matrix Operations: Linear Algebra

Linear algebra is a convenient and powerful system for manipulating a set of data features and is one of the strong points of NumPy. Linear algebra is a crucial component of machine learning and deep learning research and implementation of learning algorithms. NumPy has vectorized routines for various matrix operations. Let's go through a few of them.

Matrix Multiplication (Dot Product)

First let's create random integers using the method **np.random.randint(low, high=None, size=None,)** which returns random integers from low (inclusive) to high (exclusive).

```
# create a 3x3 matrix of random integers in the range of 1 to 50
A = np.random.randint(1, 50, size=[3,3])
B = np.random.randint(1, 50, size=[3,3])
# print the arrays
A
'Output':
array([[15, 29, 24],
       [ 5, 23, 26],
       [30, 14, 44]])
B
'Output':
array([[38, 32, 22],
       [32, 30, 46],
       [33, 47, 24]])
```

We can use the following routines for matrix multiplication, **np.matmul(a,b)** or **a @ b** if using Python 3.6. Using **a @ b** is preferred. Remember that when multiplying matrices, the inner matrix dimensions must agree. For example, if A is an $m \times n$ matrix and B is an $n \times p$ matrix, the product of the matrices will be an $m \times p$ matrix with the inner dimensions of the respective matrices n agreeing (see Figure 10-1).

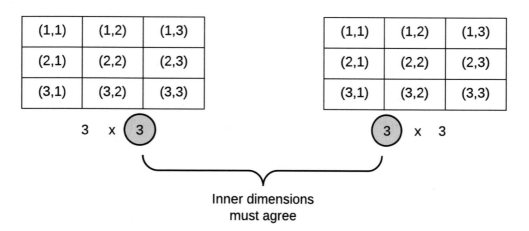

Figure 10-1. *Matrix multiplication*

```
# multiply the two matrices A and B (dot product)
A @ B     # or np.matmul(A,B)
```

'Output':
array([[2290, 2478, 2240],
        [1784, 2072, 1792],
        [3040, 3448, 2360]])

Element-Wise Operations

Element-wise matrix operations involve matrices operating on themselves in an element-wise fashion. The action can be an addition, subtraction, division, or multiplication (which is commonly called the Hadamard product). The matrices must be of the same shape. **Please note** that while a matrix is of shape $n \times n$, a vector is of shape $n \times 1$. These concepts easily apply to vectors as well. See Figure 10-2.

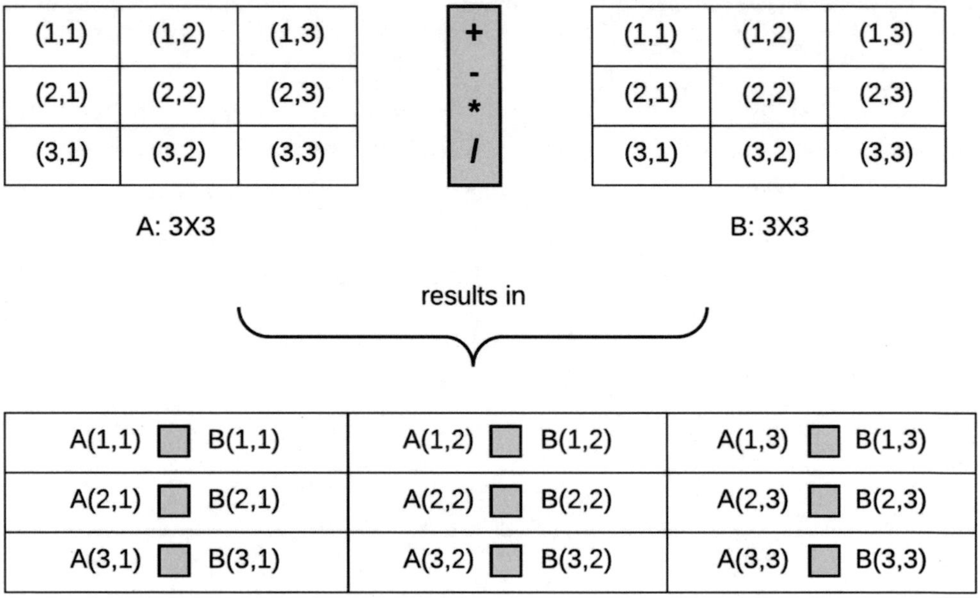

(1,1)	(1,2)	(1,3)
(2,1)	(2,2)	(2,3)
(3,1)	(3,2)	(3,3)

A: 3X3

+
-
*
/

(1,1)	(1,2)	(1,3)
(2,1)	(2,2)	(2,3)
(3,1)	(3,2)	(3,3)

B: 3X3

results in

A(1,1) ☐ B(1,1)	A(1,2) ☐ B(1,2)	A(1,3) ☐ B(1,3)
A(2,1) ☐ B(2,1)	A(2,2) ☐ B(2,2)	A(2,3) ☐ B(2,3)
A(3,1) ☐ B(3,1)	A(3,2) ☐ B(3,2)	A(3,3) ☐ B(3,3)

Figure 10-2. *Element-wise matrix operations*

Let's have some examples.

```
# Hadamard multiplication of A and B
A * B
'Output':
array([[ 570,  928,  528],
       [ 160,  690, 1196],
       [ 990,  658, 1056]])
```

```
# add A and B
A + B
'Output':
array([[53, 61, 46],
       [37, 53, 72],
       [63, 61, 68]])
# subtract A from B
B - A
'Output':
array([[ 23,   3,  -2],
       [ 27,   7,  20],
       [  3,  33, -20]])
# divide A with B
A / B
'Output':
array([[ 0.39473684,  0.90625   ,  1.09090909],
       [ 0.15625   ,  0.76666667,  0.56521739],
       [ 0.90909091,  0.29787234,  1.83333333]])
```

Scalar Operation

A matrix can be acted upon by a scalar (i.e., a single numeric entity) in the same way element-wise fashion. This time the scalar operates upon each element of the matrix or vector. See Figure 10-3.

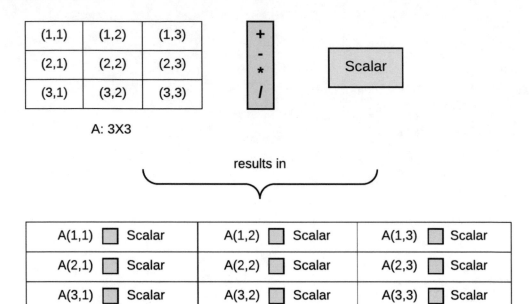

Figure 10-3. *Scalar operations*

Let's look at some examples.

```
# Hadamard multiplication of A and a scalar, 0.5
A * 0.5
'Output':
array([[  7.5,  14.5,  12. ],
       [  2.5,  11.5,  13. ],
       [ 15. ,   7. ,  22. ]])
# add A and a scalar, 0.5
A + 0.5
'Output':
array([[ 15.5,  29.5,  24.5],
       [  5.5,  23.5,  26.5],
       [ 30.5,  14.5,  44.5]])
# subtract a scalar 0.5 from B
B - 0.5
'Output':
array([[ 37.5,  31.5,  21.5],
       [ 31.5,  29.5,  45.5],
       [ 32.5,  46.5,  23.5]])
```

```
# divide A and a scalar, 0.5
A / 0.5
'Output':
array([[ 30.,   58.,   48.],
       [ 10.,   46.,   52.],
       [ 60.,   28.,   88.]])
```

Matrix Transposition

Transposition is a vital matrix operation that reverses the rows and columns of a matrix by flipping the row and column indices. The transpose of a matrix is denoted as A^T. Observe that the diagonal elements remain unchanged. See Figure 10-4.

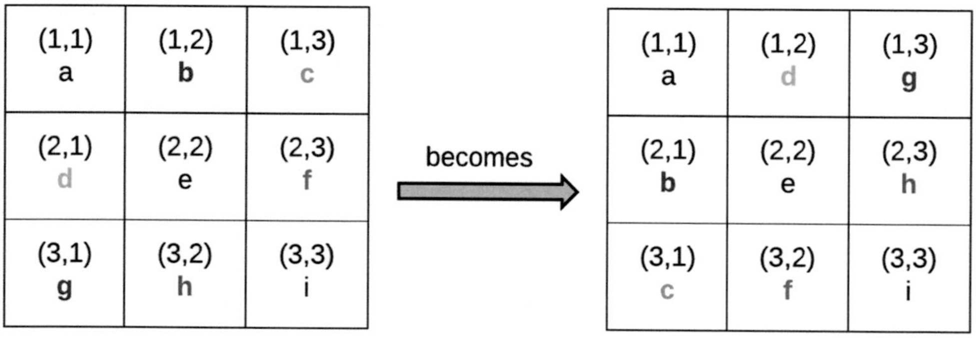

(1,1) a	(1,2) b	(1,3) c
(2,1) d	(2,2) e	(2,3) f
(3,1) g	(3,2) h	(3,3) i

becomes →

(1,1) a	(1,2) d	(1,3) g
(2,1) b	(2,2) e	(2,3) h
(3,1) c	(3,2) f	(3,3) i

Figure 10-4. *Matrix transpose*

Let's see an example.

```
A = np.array([[15, 29, 24],
              [ 5, 23, 26],
              [30, 14, 44]])
# transpose A
A.T   # or A.transpose()
'Output':
array([[15,  5, 30],
       [29, 23, 14],
       [24, 26, 44]])
```

The Inverse of a Matrix

A $m \times m$ matrix A (also called a square matrix) has an inverse if A times another matrix B results in the identity matrix I also of shape $m \times m$. This matrix B is called the inverse of A and is denoted as A^{-1}. This relationship is formally written as

$$AA^{-1} = A^{-1}A = I$$

However, not all matrices have an inverse. A matrix with an inverse is called a *nonsingular* or *invertible* matrix, while those without an inverse are known as *singular* or *degenerate*.

Note   A square matrix is a matrix that has the same number of rows and columns.

Let's use NumPy to get the inverse of a matrix. Some linear algebra modules are found in a sub-module of NumPy called **linalg**.

```
A = np.array([[15, 29, 24],
              [ 5, 23, 26],
              [30, 14, 44]])
# find the inverse of A
np.linalg.inv(A)
'Output':
array([[ 0.05848375, -0.08483755,  0.01823105],
       [ 0.05054152, -0.00541516, -0.02436823],
       [-0.05595668,  0.05956679,  0.01805054]])
```

NumPy also implements the *Moore-Penrose pseudo inverse*, which gives an inverse derivation for degenerate matrices. Here, we use the **pinv** method to find the inverses of invertible matrices.

```
# using pinv()
np.linalg.pinv(A)
'Output':
array([[ 0.05848375, -0.08483755,  0.01823105],
       [ 0.05054152, -0.00541516, -0.02436823],
       [-0.05595668,  0.05956679,  0.01805054]])
```

Reshaping

A NumPy array can be restructured to take on a different shape. Let's convert a 1-D array to a $m \times n$ matrix.

```
# make 20 elements evenly spaced between 0 and 5
a = np.linspace(0,5,20)
a
'Output':
array([ 0.        ,  0.26315789,  0.52631579,  0.78947368,  1.05263158,
        1.31578947,  1.57894737,  1.84210526,  2.10526316,  2.36842105,
        2.63157895,  2.89473684,  3.15789474,  3.42105263,  3.68421053,
        3.94736842,  4.21052632,  4.47368421,  4.73684211,  5.        ])
# observe that a is a 1-D array
a.shape
'Output': (20,)
# reshape into a 5 x 4 matrix
A = a.reshape(5, 4)
A
'Output':
array([[ 0.        ,  0.26315789,  0.52631579,  0.78947368],
       [ 1.05263158,  1.31578947,  1.57894737,  1.84210526],
       [ 2.10526316,  2.36842105,  2.63157895,  2.89473684],
       [ 3.15789474,  3.42105263,  3.68421053,  3.94736842],
       [ 4.21052632,  4.47368421,  4.73684211,  5.        ]])
# The vector a has been reshaped into a 5 by 4 matrix A
A.shape
'Output': (5, 4)
```

Reshape vs. Resize Method

NumPy has the **np.reshape** and **np.resize** methods. The reshape method returns an ndarray with a modified shape without changing the original array, whereas the resize method changes the original array. Let's see an example.

```
# generate 9 elements evenly spaced between 0 and 5
a = np.linspace(0,5,9)
a
'Output':  array([ 0.   ,  0.625,  1.25 ,  1.875,  2.5  ,  3.125,  3.75 ,
4.375,  5.   ])
# the original shape
a.shape
'Output':  (9,)
# call the reshape method
a.reshape(3,3)
'Output':
array([[ 0.   ,  0.625,  1.25 ],
       [ 1.875,  2.5  ,  3.125],
       [ 3.75 ,  4.375,  5.   ]])
# the original array maintained its shape
a.shape
'Output':  (9,)
# call the resize method - resize does not return an array
a.resize(3,3)
# the resize method has changed the shape of the original array
a.shape
'Output':  (3, 3)
```

Stacking Arrays

NumPy has methods for concatenating arrays – also called stacking. The methods hstack and vstack are used to stack several arrays along the horizontal and vertical axis, respectively.

```
# create a 2x2 matrix of random integers in the range of 1 to 20
A = np.random.randint(1, 50, size=[3,3])
B = np.random.randint(1, 50, size=[3,3])
# print out the arrays
A
```

```
'Output':
array([[19, 40, 31],
       [ 5, 16, 38],
       [22, 49,  9]])

B
'Output':
array([[15, 22, 16],
       [49, 26,  9],
       [42, 13, 39]])
```

Let's stack **A** and **B** horizontally using **hstack**. To use **hstack**, the arrays must have the same number of rows. Also, the arrays to be stacked are passed as a tuple to the **hstack** method.

```
# arrays are passed as tuple to hstack
np.hstack((A,B))
'Output':
array([[19, 40, 31, 15, 22, 16],
       [ 5, 16, 38, 49, 26,  9],
       [22, 49,  9, 42, 13, 39]])
```

To stack **A** and **B** vertically using **vstack**, the arrays must have the same number of columns. The arrays to be stacked are also passed as a tuple to the **vstack** method.

```
# arrays are passed as tuple to hstack
np.vstack((A,B))
'Output':
array([[19, 40, 31],
       [ 5, 16, 38],
       [22, 49,  9],
       [15, 22, 16],
       [49, 26,  9],
       [42, 13, 39]])
```

Broadcasting

NumPy has an elegant mechanism for arithmetic operation on arrays with different dimensions or shapes. As an example, when a scalar is added to a vector (or 1-D array). The scalar value is conceptually broadcasted or stretched across the rows of the array and added element-wise. See Figure 10-5.

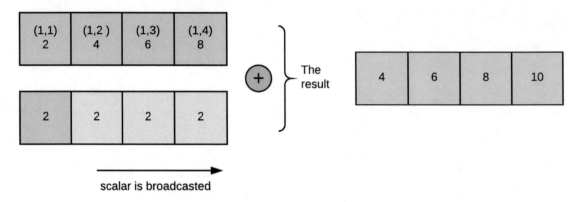

Figure 10-5. *Broadcasting example of adding a scalar to a vector (or 1-D array)*

Matrices with different shapes can be broadcasted to perform arithmetic operations by stretching the dimension of the smaller array. Broadcasting is another vectorized operation for speeding up matrix processing. However, not all arrays with different shapes can be broadcasted. For broadcasting to occur, the trailing axes for the arrays must be the same size or 1.

In the example that follows, the matrices **A** and **B** have the same rows, but the column of matrix **B** is 1. Hence, an arithmetic operation can be performed on them by broadcasting and adding the cells element-wise.

```
A      (2d array):  4 x 3       + <perform addition>
B      (2d array):  4 x 1
Result (2d array):  4 x 3
```

See Figure 10-6 for more illustration.

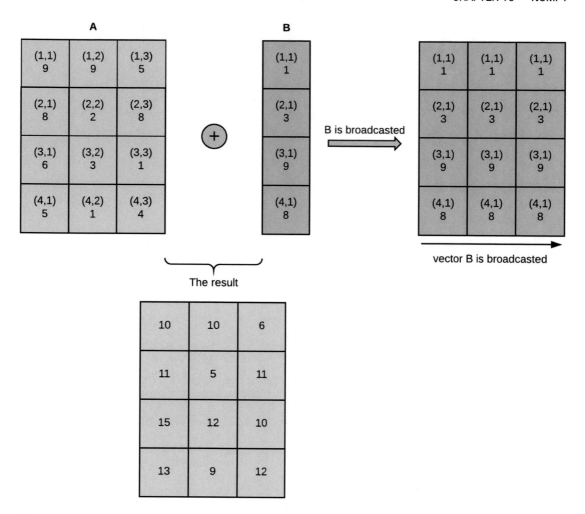

Figure 10-6. *Matrix broadcasting example*

Let's see this in code:

```
# create a 4 X 3 matrix of random integers between 1 and 10
A = np.random.randint(1, 10, [4, 3])
A
'Output':
array([[9, 9, 5],
       [8, 2, 8],
       [6, 3, 1],
       [5, 1, 4]])
```

```
# create a 4 X 1 matrix of random integers between 1 and 10
B = np.random.randint(1, 10, [4, 1])
B
'Output':
array([[1],
       [3],
       [9],
       [8]])
# add A and B
A + B
'Output':
array([[10, 10,  6],
       [11,  5, 11],
       [15, 12, 10],
       [13,  9, 12]])
```

The example that follows cannot be broadcasted and will result in a *ValueError: operands could not be broadcasted together with shapes (4,3) (4,2)* because the matrices **A** and **B** have different columns and do not fit with the aforementioned rules of broadcasting that the trailing axes for the arrays must be the same size or 1.

```
A       (2d array):   4 x 3
B       (2d array):   4 x 2
The dimensions do not match - they must be either the same or 1
```

When we try to add the preceding example in Python, we get an error.

```
A = np.random.randint(1, 10, [4, 3])
B = np.random.randint(1, 10, [4, 2])
A + B
'Output':
Traceback (most recent call last):

  File "<ipython-input-145-624e41e41a31>", line 1, in <module>
    A + B

ValueError: operands could not be broadcast together with shapes (4,3) (4,2)
```

Loading Data

Loading data is an important process in the data analysis/machine learning pipeline. Data usually comes in **.csv** format. **csv** files can be loaded into Python by using the **loadtxt** method. The parameter **skiprows** skips the first row of the dataset – it is usually the header row of the data.

```
np.loadtxt(open("the_file_name.csv", "rb"), delimiter=",", skiprows=1)
```

Pandas is a preferred package for loading data in Python.

We will learn more about Pandas for data manipulation in the next chapter.

CHAPTER 11

Pandas

Pandas is a specialized Python library for data analysis, especially on humongous datasets. It boasts easy-to-use functionality for reading and writing data, dealing with missing data, reshaping the dataset, and massaging the data by slicing, indexing, inserting, and deleting data variables and records. Pandas also has an important **groupBy** functionality for aggregating data for defined conditions – useful for plotting and computing data summaries for exploration.

Another key strength of Pandas is in re-ordering and cleaning time series data for time series analysis. In short, Pandas is the go-to tool for data cleaning and data exploration.

To use Pandas, first import the Pandas module:

```
import pandas as pd
```

Pandas Data Structures

Just like NumPy, Pandas can store and manipulate a multi-dimensional array of data. To handle this, Pandas has the **Series** and **DataFrame** data structures.

Series

The **Series** data structure is for storing a 1-D array (or vector) of data elements. A series data structure also provides labels to the data items in the form of an **index**. The user can specify this label via the **index** parameter in the **Series** function, but if the **index** parameter is left unspecified, a default label of 0 to one minus the size of the data elements is assigned.

© Ekaba Bisong 2019
E. Bisong, *Building Machine Learning and Deep Learning Models on Google Cloud Platform*,
https://doi.org/10.1007/978-1-4842-4470-8_11

Let us consider an example of creating a **Series** data structure.

```
# create a Series object
my_series = pd.Series([2,4,6,8], index=['e1','e2','e3','e4'])
# print out data in Series data structure
my_series
'Output':
e1    2
e2    4
e3    6
e4    8
dtype: int64
# check the data type of the variable
type(my_series)
'Output': pandas.core.series.Series
# return the elements of the Series data structure
my_series.values
'Output': array([2, 4, 6, 8])
# retrieve elements from Series data structure based on their assigned
indices
my_series['e1']
'Output': 2
# return all indices of the Series data structure
my_series.index
'Output': Index(['e1', 'e2', 'e3', 'e4'], dtype='object')
```

Elements in a Series data structure can be assigned the same indices.

```
# create a Series object with elements sharing indices
my_series = pd.Series([2,4,6,8], index=['e1','e2','e1','e2'])
# note the same index assigned to various elements
my_series
'Output':
e1    2
e2    4
e1    6
e2    8
```

```
dtype: int64
# get elements using their index
my_series['e1']
'Output':
e1    2
e1    6
dtype: int64
```

DataFrames

A DataFrame is a Pandas data structure for storing and manipulating 2-D arrays. A 2-D array is a table-like structure that is similar to an Excel spreadsheet or a relational database table. A DataFrame is a very natural form for storing structured datasets.

A DataFrame consists of rows and columns for storing records of information (in rows) across heterogeneous variables (in columns).

Let's see examples of working with DataFrames.

```
# create a data frame
my_DF = pd.DataFrame({'age': [15,17,21,29,25], \
          'state_of_origin':['Lagos', 'Cross River', 'Kano', 'Abia',
          'Benue']})
my_DF
'Output':
   age state_of_origin
0   15           Lagos
1   17     Cross River
2   21            Kano
3   29            Abia
4   25           Benue
```

We will observe from the preceding example that a DataFrame is constructed from a dictionary of records where each value is a **Series** data structure. Also note that each row has an **index** that can be assigned when creating the DataFrame, else the default from 0 to one off the number of records in the DataFrame is used. Creating an index manually is usually not feasible except when working with small dummy datasets.

NumPy is frequently used together with Pandas. Let's import the NumPy library and use some of its functions to demonstrate other ways of creating a quick DataFrame.

```
import numpy as np

# create a 3x3 dataframe of numbers from the normal distribution
my_DF = pd.DataFrame(np.random.randn(3,3),\
            columns=['First','Second','Third'])
my_DF
'Output':
      First     Second      Third
0 -0.211218 -0.499870 -0.609792
1 -0.295363  0.388722  0.316661
2  1.397300 -0.894861  1.127306
# check the dimensions
my_DF.shape
'Output': (3, 3)
```

Let's examine some other operations with DataFrames.

```
# create a python dictionary
my_dict = {'State':['Adamawa', 'Akwa-Ibom', 'Yobe', 'Rivers', 'Taraba'], \
            'Capital':['Yola','Uyo','Damaturu','Port-Harcourt','Jalingo'], \
            'Population':[3178950, 5450758, 2321339, 5198716, 2294800]}
my_dict
'Output':
{'Capital': ['Yola', 'Uyo', 'Damaturu', 'Port-Harcourt', 'Jalingo'],
  'Population': [3178950, 5450758, 2321339, 5198716, 2294800],
  'State': ['Adamawa', 'Akwa-Ibom', 'Yobe', 'Rivers', 'Taraba']}
# confirm dictionary type
type(my_dict)
'Output': dict
# create DataFrame from dictionary
my_DF = pd.DataFrame(my_dict)
my_DF
```

```
'Output':
          Capital    Population        State
0            Yola       3178950     Adamawa
1             Uyo       5450758   Akwa-Ibom
2        Damaturu       2321339        Yobe
3   Port-Harcourt       5198716      Rivers
4         Jalingo       2294800      Taraba
# check DataFrame type
type(my_DF)
'Output': pandas.core.frame.DataFrame
# retrieve column names of the DataFrame
my_DF.columns
'Output': Index(['Capital', 'Population', 'State'], dtype='object')
# the data type of `DF.columns` method is an Index
type(my_DF.columns)
'Output': pandas.core.indexes.base.Index
# retrieve the DataFrame values as a NumPy ndarray
my_DF.values
'Output':
array([['Yola', 3178950, 'Adamawa'],
       ['Uyo', 5450758, 'Akwa-Ibom'],
       ['Damaturu', 2321339, 'Yobe'],
       ['Port-Harcourt', 5198716, 'Rivers'],
       ['Jalingo', 2294800, 'Taraba']], dtype=object)
# the data type of  `DF.values` method is an numpy ndarray
type(my_DF.values)
'Output': numpy.ndarray
```

In summary, a DataFrame is a tabular structure for storing a structured dataset where each column contains a **Series** data structure of records. Here's an illustration (Figure 11-1).

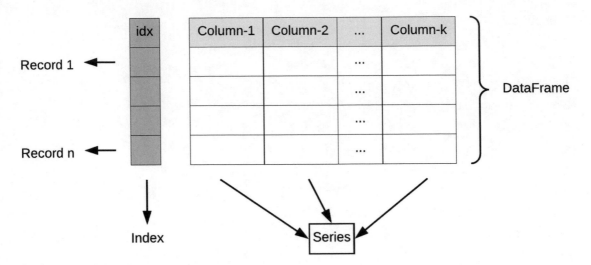

Figure 11-1. *Pandas data structure*

Let's check the data type of each column in the DataFrame.

```
my_DF.dtypes
'Output':
Capital        object
Population      int64
State          object
dtype: object
```

An **object** data type in Pandas represents **Strings**.

Data Indexing (Selection/Subsets)

Similar to NumPy, Pandas objects can index or subset the dataset to retrieve a specific sub-record of the larger dataset. Note that data indexing returns a new **DataFrame** or **Series** if a 2-D or 1-D array is retrieved. They do not, however, alter the original dataset. Let's go through some examples of indexing a Pandas DataFrame.

First let's create a dataframe. Observe the default integer indices assigned.

```
# create the dataframe
my_DF = pd.DataFrame({'age': [15,17,21,29,25], \
         'state_of_origin':['Lagos', 'Cross River', 'Kano', 'Abia',
         'Benue']})
```

```
my_DF
'Output':
   age state_of_origin
0   15           Lagos
1   17     Cross River
2   21            Kano
3   29            Abia
4   25           Benue
```

Selecting a Column from a DataFrame

Remember that the data type of a DataFrame column is a **Series** because it is a vector or 1-D array.

```
my_DF['age']
'Output':
0    15
1    17
2    21
3    29
4    25
Name: age, dtype: int64
# check data type
type(my_DF['age'])
'Output':  pandas.core.series.Series
```

To select multiple columns, enclose the column names as **strings** with the double square brackets **[[ ]]**. The following code is an example:

```
my_DF[['age','state_of_origin']]
'Output':
   age state_of_origin
0   15           Lagos
1   17     Cross River
2   21            Kano
3   29            Abia
4   25           Benue
```

Selecting a Row from a DataFrame

Pandas makes use of two unique wrapper attributes for indexing rows from a **DataFrame** or a cell from a **Series** data structure. These attributes are the **iloc** and **loc** – they are also known as indexers. The **iloc** attribute allows you to select or slice row(s) of a DataFrame using the intrinsic Python index format, whereas the **loc** attribute uses the explicit indices assigned to the DataFrame. If no explicit index is found, **loc** returns the same value as **iloc**.

Remember that the data type of a DataFrame row is a **Series** because it is a vector or 1-D array.

Let's select the first row from the DataFrame.

```
# using explicit indexing
my_DF.loc[0]
'Output':
age                     15
state_of_origin     Lagos
Name: 0, dtype: object
# using implicit indexing
my_DF.iloc[0]
'Output':
age                     15
state_of_origin     Lagos
Name: 0, dtype: object
# let's see the data type
type(my_DF.loc[0])
'Output':  pandas.core.series.Series
```

Now let's create a DataFrame with explicit indexing and test out the **iloc** and **loc** methods. Pandas will return an error if **iloc** is used for explicit indexing or if **loc** is used for implicit Python indexing.

```
my_DF = pd.DataFrame({'age': [15,17,21,29,25], \
            'state_of_origin':['Lagos', 'Cross River', 'Kano', 'Abia', \
            'Benue']},\
            index=['a','a','b','b','c'])
# observe the string indices
```

```
my_DF
'Output':
   age state_of_origin
a   15             Lagos
a   17       Cross River
b   21              Kano
b   29              Abia
c   25             Benue
# select using explicit indexing
my_DF.loc['a']
Out[196]:
   age state_of_origin
a   15             Lagos
a   17       Cross River
# let's try to use loc for implicit indexing
my_DF.loc[0]
'Output':
    Traceback (most recent call last):
    TypeError: cannot do label indexing on <class 'pandas.core.indexes.
    base.Index'>
        with these indexers [0] of <class 'int'>
```

Selecting Multiple Rows and Columns from a DataFrame

Let's use the **loc** method to select multiple rows and columns from a Pandas DataFrame.

```
# select rows with age greater than 20
my_DF.loc[my_DF.age > 20]
'Output':
   age state_of_origin
2   21              Kano
3   29              Abia
4   25             Benue
# find states of origin with age greater than or equal to 25
my_DF.loc[my_DF.age >= 25, 'state_of_origin']
```

```
'Output':
Out[29]:
3     Abia
4     Benue
```

Slice Cells by Row and Column from a DataFrame

First let's create a DataFrame. Remember, we use **iloc** when no explicit index or row labels are assigned.

```
my_DF = pd.DataFrame({'age': [15,17,21,29,25], \
            'state_of_origin':['Lagos', 'Cross River', 'Kano', 'Abia',
            'Benue']})
my_DF
'Output':
   age state_of_origin
0   15           Lagos
1   17     Cross River
2   21            Kano
3   29            Abia
4   25           Benue
# select the third row and second column
my_DF.iloc[2,1]
'Output': 'Kano'
# slice the first 2 rows - indexed from zero, excluding the final index
my_DF.iloc[:2,]
'Output':
   age state_of_origin
0   15           Lagos
1   17     Cross River
# slice the last three rows from the last column
my_DF.iloc[-3:,-1]
'Output':
2     Kano
3     Abia
4     Benue
Name: state_of_origin, dtype: object
```

124

DataFrame Manipulation

Let's go through some common tasks for manipulating a DataFrame.

Removing a Row/Column

In many cases during the data cleaning process, there may be a need to drop unwanted rows or data variables (i.e., columns). We typically do this using the **drop** function. The **drop** function has a parameter **axis** whose default is 0. If **axis** is set to 1, it drops columns in a dataset, but if left at the default, rows are dropped from the dataset.

Note that when a column or row is dropped, a new **DataFrame** or **Series** is returned without altering the original data structure. However, when the attribute **inplace** is set to **True**, the original DataFrame or Series is modified. Let's see some examples.

```
# the data frame
my_DF = pd.DataFrame({'age': [15,17,21,29,25], \
            'state_of_origin':['Lagos', 'Cross River', 'Kano', 'Abia',
            'Benue']})

my_DF
'Output':
   age state_of_origin
0   15          Lagos
1   17    Cross River
2   21           Kano
3   29           Abia
4   25          Benue
# drop the 3rd and 4th column
my_DF.drop([2,4])
'Output':
   age state_of_origin
0   15          Lagos
1   17    Cross River
3   29           Abia
# drop the `age` column
my_DF.drop('age', axis=1)
```

```
'Output':
   state_of_origin
0            Lagos
1      Cross River
2             Kano
3             Abia
4            Benue
```

```python
# original DataFrame is unchanged
my_DF
```

```
'Output':
   age state_of_origin
0   15           Lagos
1   17     Cross River
2   21            Kano
3   29            Abia
4   25           Benue
```

```python
# drop using 'inplace' - to modify the original DataFrame
my_DF.drop('age', axis=1, inplace=True)
# original DataFrame altered
my_DF
```

```
'Output':
   state_of_origin
0            Lagos
1      Cross River
2             Kano
3             Abia
4            Benue
```

Let's see examples of removing a row given a condition.

```python
my_DF = pd.DataFrame({'age': [15,17,21,29,25], \
            'state_of_origin':['Lagos', 'Cross River', 'Kano', 'Abia', \
            'Benue']})
my_DF
```

```
'Output':
    age state_of_origin
0   15            Lagos
1   17      Cross River
2   21             Kano
3   29             Abia
4   25            Benue

# drop all rows less than 20
my_DF.drop(my_DF[my_DF['age'] < 20].index, inplace=True)
my_DF

'Output':
    age state_of_origin
2   21             Kano
3   29             Abia
4   25            Benue
```

Adding a Row/Column

We can add a new column to a Pandas DataFrame by using the **assign** method.

```
# show dataframe
my_DF = pd.DataFrame({'age': [15,17,21,29,25], \
          'state_of_origin':['Lagos', 'Cross River', 'Kano', 'Abia', \
          'Benue']})
my_DF
'Output':
    age state_of_origin
0   15            Lagos
1   17      Cross River
2   21             Kano
3   29             Abia
4   25            Benue
# add column to data frame
```

```
my_DF = my_DF.assign(capital_city = pd.Series(['Ikeja', 'Calabar', \
                                    'Kano', 'Umuahia',
                                    'Makurdi']))

my_DF
'Output':
   age state_of_origin capital_city
0   15           Lagos        Ikeja
1   17     Cross River      Calabar
2   21            Kano         Kano
3   29            Abia      Umuahia
4   25           Benue      Makurdi
```

We can also add a new DataFrame column by computing some function on another column. Let's take an example by adding a column computing the absolute difference of the ages from their mean.

```
mean_of_age = my_DF['age'].mean()
my_DF['diff_age'] = my_DF['age'].map(lambda x: abs(x-mean_of_age))
my_DF
'Output':
   age state_of_origin  diff_age
0   15           Lagos       6.4
1   17     Cross River       4.4
2   21            Kano       0.4
3   29            Abia       7.6
4   25           Benue       3.6
```

Typically in practice, a fully formed dataset is converted into Pandas for cleaning and data analysis, which does not ideally involve adding a new observation to the dataset. But in the event that this is desired, we can use the **append()** method to achieve this. However, it may not be a computationally efficient action. Let's see an example.

```
# show dataframe
my_DF = pd.DataFrame({'age': [15,17,21,29,25], \
          'state_of_origin':['Lagos', 'Cross River', 'Kano', 'Abia',
          'Benue']})

my_DF
```

```
'Output':
   age state_of_origin
0   15              Lagos
1   17        Cross River
2   21               Kano
3   29               Abia
4   25              Benue
# add a row to data frame
my_DF = my_DF.append(pd.Series([30 , 'Osun'], index=my_DF.columns), \
                                               ignore_index=True)

my_DF
'Output':
   age state_of_origin
0   15              Lagos
1   17        Cross River
2   21               Kano
3   29               Abia
4   25              Benue
5   30               Osun
```

We observe that adding a new row involves passing to the **append** method, a **Series** object with the **index** attribute set to the columns of the main DataFrame. Since typically, in given datasets, the index is nothing more than the assigned defaults, we set the attribute **ignore_index** to create a new set of default index values with the new row(s).

Data Alignment

Pandas utilizes data alignment to align indices when performing some binary arithmetic operation on DataFrames. If two or more DataFrames in an arithmetic operation do not share a common index, a **NaN** is introduced denoting missing data. Let's see examples of this.

```
# create a 3x3 dataframe - remember randint(low, high, size)
df_A = pd.DataFrame(np.random.randint(1,10,[3,3]),\
          columns=['First','Second','Third'])
df_A
```

```
'Output':
   First  Second  Third
0      2       3      9
1      8       7      7
2      8       6      4
# create a 4x3 dataframe
df_B = pd.DataFrame(np.random.randint(1,10,[4,3]),\
           columns=['First','Second','Third'])
df_B
'Output':
   First  Second  Third
0      3       6      3
1      2       2      1
2      9       3      8
3      2       9      2
# add df_A and df_B together
df_A + df_B
'Output':
    First  Second  Third
0     5.0     9.0   12.0
1    10.0     9.0    8.0
2    17.0     9.0   12.0
3     NaN     NaN    NaN
# divide both dataframes
df_A / df_B
'Output':
        First  Second  Third
0    0.666667     0.5    3.0
1    4.000000     3.5    7.0
2    0.888889     2.0    0.5
3         NaN     NaN    NaN
```

If we do not want a **NaN** signifying missing values to be imputed, we can use the **fill_value** attribute to substitute with a default value. However, to take advantage of the **fill_value** attribute, we have to use the Pandas arithmetic methods: **add(), sub(), mul(),**

div(), **floordiv()**, **mod()**, and **pow()** for addition, subtraction, multiplication, integer division, numeric division, remainder division, and exponentiation. Let's see examples.

```
df_A.add(df_B, fill_value=10)
'Output':
   First  Second  Third
0    5.0     9.0   12.0
1   10.0     9.0    8.0
2   17.0     9.0   12.0
3   12.0    19.0   12.0
```

Combining Datasets

We may need to combine two or more datasets together; Pandas provides methods for such operations. We would consider the simple case of combining data frames with shared column names using the **concat** method.

```
# combine two dataframes column-wise
pd.concat([df_A, df_B])
'Output':
   First  Second  Third
0      2       3      9
1      8       7      7
2      8       6      4
0      3       6      3
1      2       2      1
2      9       3      8
3      2       9      2
```

Observe that the **concat** method preserves indices by default. We can also concatenate or combine two dataframes by rows (or horizontally). This is done by setting the **axis** parameter to 1.

```
# combine two dataframes horizontally
pd.concat([df_A, df_B], axis=1)
```

```
'Output':
Out[246]:
   First  Second  Third  First  Second  Third
0    2.0     3.0    9.0      3       6      3
1    8.0     7.0    7.0      2       2      1
2    8.0     6.0    4.0      9       3      8
3    NaN     NaN    NaN      2       9      2
```

Handling Missing Data

Dealing with missing data is an integral part of the data cleaning/data analysis process. Moreover, some machine learning algorithms will not work in the presence of missing data. Let's see some simple Pandas methods for identifying and removing missing data, as well as imputing values into missing data.

Identifying Missing Data

In this section, we'll use the **isnull()** method to check if missing cells exist in a DataFrame.

```
# let's create a data frame with missing data
my_DF = pd.DataFrame({'age': [15,17,np.nan,29,25], \
          'state_of_origin':['Lagos', 'Cross River', 'Kano',
          'Abia', np.nan]})
my_DF
'Output':
    age state_of_origin
0  15.0            Lagos
1  17.0      Cross River
2   NaN             Kano
3  29.0             Abia
4  25.0              NaN
```

Let's check for missing data in this data frame. The **isnull()** method will return **True** where there is a missing data, whereas the **notnull()** function returns **False**.

```
my_DF.isnull()
'Output':
    age   state_of_origin
0  False             False
1  False             False
2   True             False
3  False             False
4  False              True
```

However, if we want a single answer (i.e., either **True** or **False**) to report if there is a missing data in the data frame, we will first convert the DataFrame to a NumPy array and use the function **any()**.

The **any** function returns **True** when at least one of the elements in the dataset is **True**. In this case, **isnull()** returns a DataFrame of booleans where **True** designates a cell with a missing value.

Let's see how that works.

```
my_DF.isnull().values.any()
'Output':  True
```

Removing Missing Data

Pandas has a function **dropna()** which is used to filter or remove missing data from a DataFrame. **dropna()** returns a new DataFrame without missing data. Let's see examples of how this works.

```
# let's see our dataframe with missing data
my_DF = pd.DataFrame({'age': [15,17,np.nan,29,25], \
        'state_of_origin':['Lagos', 'Cross River', 'Kano',
        'Abia', np.nan]})
my_DF
'Output':
    age state_of_origin
0  15.0           Lagos
1  17.0     Cross River
2   NaN            Kano
3  29.0            Abia
4  25.0             NaN
```

```
# let's run dropna() to remove all rows with missing values
my_DF.dropna()
'Output':
    age state_of_origin
0   15.0            Lagos
1   17.0      Cross River
3   29.0             Abia
```

As we will observe from the preceding code block, **dropna()** drops all rows that contain a missing value. But we may not want that. We may rather, for example, want to drop columns with missing data or drop rows where all the observations are missing or better still remove consequent on the number of observations present in a particular row.

Let's see examples of this option. First let's expand our example dataset.

```
my_DF = pd.DataFrame({'Capital': ['Yola', np.nan, np.nan, 'Port-Harcourt',
                      'Jalingo'],
  'Population': [3178950, np.nan, 2321339, np.nan, 2294800],
  'State': ['Adamawa', np.nan, 'Yobe', np.nan, 'Taraba'],
  'LGAs': [22, np.nan, 17, 23, 16]})
my_DF
'Output':
         Capital  LGAs  Population     State
0          Yola   22.0   3178950.0   Adamawa
1           NaN   NaN         NaN       NaN
2           NaN   17.0   2321339.0      Yobe
3  Port-Harcourt  23.0         NaN       NaN
4       Jalingo   16.0   2294800.0    Taraba
```

Drop columns with **NaN**. This option is not often used in practice.

```
my_DF.dropna(axis=1)
'Output':
Empty DataFrame
Columns: []
Index: [0, 1, 2, 3, 4]
```

Drop rows where all the observations are missing.

```
my_DF.dropna(how='all')
'Output':
          Capital  LGAs  Population      State
0            Yola  22.0   3178950.0  Adamawa
2             NaN  17.0   2321339.0     Yobe
3  Port-Harcourt  23.0         NaN      NaN
4         Jalingo  16.0   2294800.0    Taraba
```

Drop rows based on an observation threshold. By adjusting the **thresh** attribute, we can drop rows where the number of observations in the row is less than the **thresh** value.

```
# drop rows where number of NaN is less than 3
my_DF.dropna(thresh=3)
'Output':
   Capital  LGAs  Population     State
0     Yola  22.0   3178950.0  Adamawa
2      NaN  17.0   2321339.0     Yobe
4  Jalingo  16.0   2294800.0    Taraba
```

Imputing Values into Missing Data

Imputing values as substitutes for missing data is a standard practice in preparing data for machine learning. Pandas has a **fillna()** function for this purpose. A simple approach is to fill **NaNs** with zeros.

```
my_DF.fillna(0) # we can also run my_DF.replace(np.nan, 0)
'Output':
          Capital  LGAs  Population      State
0            Yola  22.0   3178950.0  Adamawa
1               0   0.0         0.0        0
2               0  17.0   2321339.0     Yobe
3  Port-Harcourt  23.0         0.0        0
4         Jalingo  16.0   2294800.0    Taraba
```

Another tactic is to fill missing values with the mean of the column value.

```
my_DF.fillna(my_DF.mean())
'Output':
          Capital  LGAs   Population    State
0            Yola  22.0    3178950.0  Adamawa
1             NaN  19.5    2598363.0      NaN
2             NaN  17.0    2321339.0     Yobe
3   Port-Harcourt  23.0    2598363.0      NaN
4         Jalingo  16.0    2294800.0   Taraba
```

Data Aggregation (Grouping)

We will touch briefly on a common practice in data science, and that is grouping a set of data attributes, either for retrieving some group statistics or applying a particular set of functions to the group. Grouping is commonly used for data exploration and plotting graphs to understand more about the dataset. Missing data are automatically excluded in a grouping operation.

Let's see examples of how this works.

```
# create a data frame
my_DF = pd.DataFrame({'Sex': ['M', 'F', 'M', 'F','M', 'F','M', 'F'],
  'Age': np.random.randint(15,60,8),
  'Salary': np.random.rand(8)*10000})
my_DF
'Output':
   Age      Salary Sex
0   54  6092.596170   M
1   57  3148.886141   F
2   37  5960.916038   M
3   23  6713.133849   F
4   34  5208.240349   M
5   25  2469.118934   F
6   50  1277.511182   M
7   54  3529.201109   F
```

Let's find the mean age and salary for observations in our dataset grouped by **Sex**.

```
my_DF.groupby('Sex').mean()
'Output':
        Age       Salary
Sex
F     39.75   3965.085008
M     43.75   4634.815935
```

We can group by more than one variable. In this case for each Sex group, also group the age and find the mean of the other numeric variables.

```
my_DF.groupby([my_DF['Sex'], my_DF['Age']]).mean()
'Output':
              Salary
Sex Age
F   23    6713.133849
    25    2469.118934
    54    3529.201109
    57    3148.886141
M   34    5208.240349
    37    5960.916038
    50    1277.511182
    54    6092.596170
```

Also, we can use a variable as a group key to run a group function on another variable or sets of variables.

```
my_DF['Age'].groupby(my_DF['Salary']).mean()
'Output':
Salary
1277.511182    50
2469.118934    25
3148.886141    57
3529.201109    54
5208.240349    34
5960.916038    37
```

```
6092.596170     54
6713.133849     23
Name: Age, dtype: int64
```

Statistical Summaries

Descriptive statistics is an essential component of the data science pipeline. By investigating the properties of the dataset, we can gain a better understanding of the data and the relationship between the variables. This information is useful in making decisions about the type of data transformations to carry out or the types of learning algorithms to spot check. Let's see some examples of simple statistical functions in Pandas.

First, we'll create a Pandas dataframe.

```
my_DF = pd.DataFrame(np.random.randint(10,80,[7,4]),\
          columns=['First','Second','Third', 'Fourth'])
'Output':
```

	First	Second	Third	Fourth
0	47	32	66	52
1	37	66	16	22
2	24	16	63	36
3	70	47	62	12
4	74	61	44	18
5	65	73	21	37
6	44	47	23	13

Use the **describe** function to obtain summary statistics of a dataset. Eight statistical measures are displayed. They are count, mean, standard deviation, minimum value, 25th percentile, 50th percentile or median, 75th percentile, and the maximum value.

```
my_DF.describe()
'Output':
```

	First	Second	Third	Fourth
count	7.000000	7.000000	7.000000	7.000000
mean	51.571429	48.857143	42.142857	27.142857
std	18.590832	19.978560	21.980511	14.904458
min	24.000000	16.000000	16.000000	12.000000

25%	40.500000	39.500000	22.000000	15.500000
50%	47.000000	47.000000	44.000000	22.000000
75%	67.500000	63.500000	62.500000	36.500000
max	74.000000	73.000000	66.000000	52.000000

Correlation

Correlation shows how much relationship exists between two variables. Parametric machine learning methods such as logistic and linear regression can take a performance hit when variables are highly correlated. The correlation values range from –1 to 1, with 0 indicating no correlation at all. –1 signifies that the variables are strongly negatively correlated, while 1 shows that the variables are strongly positively correlated. In practice, it is safe to eliminate variables that have a correlation value greater than –0.7 or 0.7. A common correlation estimate in use is the Pearson's correlation coefficient.

```
my_DF.corr(method='pearson')
'Output':
```

	First	Second	Third	Fourth
First	1.000000	0.587645	-0.014100	-0.317333
Second	0.587645	1.000000	-0.768495	-0.345265
Third	-0.014100	-0.768495	1.000000	0.334169
Fourth	-0.317333	-0.345265	0.334169	1.000000

Skewness

Another important statistical metric is the skewness of the dataset. Skewness is when a bell-shaped or normal distribution is shifted toward the right or the left. Pandas offers a convenient function called **skew()** to check the skewness of each variable. Values close to 0 are more normally distributed with less skew.

```
my_DF.skew()
'Output':
First    -0.167782
Second   -0.566914
Third    -0.084490
Fourth    0.691332
dtype: float64
```

Importing Data

Again, getting data into the programming environment for analysis is a fundamental and first step for any data analytics or machine learning task. In practice, data usually comes in a comma-separated value, **csv,** format.

```
my_DF = pd.read_csv('link_to_file/csv_file', sep=',', header = None)
```

To export a DataFrame back to **csv**

```
my_DF.to_csv('file_name.csv')
```

For the next example, the dataset 'states.csv' is found in the chapter folder of the code repository of this book.

```
my_DF = pd.read_csv('states.csv', sep=',', header = 0)

# read the top 5 rows
my_DF.head()

# save DataFrame to csv
my_DF.to_csv('save_states.csv')
```

Timeseries with Pandas

One of the core strengths of Pandas is its powerful set of functions for manipulating timeseries datasets. A couple of these functions are covered in this material.

Importing a Dataset with a DateTime Column

When importing a dataset that has a column containing datetime entries, Pandas has an attribute in the **read_csv** method called **parse_dates** that converts the datetime column from strings into Pandas **date** datatype. The attribute **index_col** uses the column of datetimes as an index to the DataFrame.

The method **head()** prints out the first five rows of the DataFrame, while the method **tail()** prints out the last five rows of the DataFrame. This function is very useful for taking a peek at a large DataFrame without having to bear the computational cost of printing it out entirely.

```
# load the data
data = pd.read_csv('crypto-markets.csv', parse_dates=['date'], index_
    col='date')
data.head()
'Output':
  slug date  symbol name  ranknow      open     high      low     close
volume    market    close_ratio  spread
2013-04-28  bitcoin BTC Bitcoin 1    135.30   135.98   132.10   134.21
    0    1500520000       0.5438    3.88
2013-04-29  bitcoin BTC Bitcoin 1    134.44   147.49   134.00   144.54
    0    1491160000       0.7813    13.49
2013-04-30  bitcoin BTC Bitcoin 1    144.00   146.93   134.05   139.00
    0    1597780000       0.3843    12.88
2013-05-01  bitcoin BTC Bitcoin 1    139.00   139.89   107.72   116.99
    0    1542820000       0.2882    32.17
2013-05-02  bitcoin BTC Bitcoin 1    116.38   125.60   92.28    105.21
    0    1292190000       0.3881    33.32
```

Let's examine the index of the imported data. Notice that they are the datetime entries.

```
# get the row indices
data.index
'Output':
DatetimeIndex(['2013-04-28', '2013-04-29', '2013-04-30', '2013-05-01',
               '2013-05-02', '2013-05-03', '2013-05-04', '2013-05-05',
               '2013-05-06', '2013-05-07',
               ...
               '2018-01-01', '2018-01-02', '2018-01-03', '2018-01-04',
               '2018-01-05', '2018-01-06', '2018-01-07', '2018-01-08',
               '2018-01-09', '2018-01-10'],
              dtype='datetime64[ns]', name='date', length=659373,
              freq=None)
```

Selection Using DatetimeIndex

The **DatetimeIndex** can be used to select the observations of the dataset in various interesting ways. For example, we can select the observation of an exact day or the observations belonging to a particular month or year. The selected observation can be subsetted by columns and grouped to give more insight in understanding the dataset.

Let's see some examples.

Select a Particular Date

Let's select a particular date from a DataFrame.

```
# select a particular date
data['2018-01-05'].head()
'Output':
```

	slug	symbol	name	ranknow	open	high \
date						
2018-01-05	bitcoin	BTC	Bitcoin	1	15477.20	17705.20
2018-01-05	ethereum	ETH	Ethereum	2	975.75	1075.39
2018-01-05	ripple	XRP	Ripple	3	3.30	3.56
2018-01-05	bitcoin-cash	BCH	Bitcoin Cash	4	2400.74	2648.32
2018-01-05	cardano	ADA	Cardano	5	1.17	1.25

	low	close	volume	market \
date				
2018-01-05	15202.800000	17429.500000	23840900000	259748000000
2018-01-05	956.330000	997.720000	6683150000	94423900000
2018-01-05	2.830000	3.050000	6288500000	127870000000
2018-01-05	2370.590000	2584.480000	2115710000	40557600000
2018-01-05	0.903503	0.999559	508100000	30364400000

	close_ratio	spread
date		
2018-01-05	0.8898	2502.40
2018-01-05	0.3476	119.06
2018-01-05	0.3014	0.73
2018-01-05	0.7701	277.73
2018-01-05	0.2772	0.35

```
# select a range of dates
data['2018-01-05':'2018-01-06'].head()
'Output':
```

	slug	symbol	name	ranknow	open	high	low \
date							
2018-01-05	bitcoin	BTC	Bitcoin	1	15477.20	17705.20	15202.80
2018-01-06	bitcoin	BTC	Bitcoin	1	17462.10	17712.40	16764.60
2018-01-05	ethereum	ETH	Ethereum	2	975.75	1075.39	956.33
2018-01-06	ethereum	ETH	Ethereum	2	995.15	1060.71	994.62
2018-01-05	ripple	XRP	Ripple	3	3.30	3.56	2.83

	close	volume	market	close_ratio	spread
date					
2018-01-05	17429.50	23840900000	259748000000	0.8898	2502.40
2018-01-06	17527.00	18314600000	293091000000	0.8044	947.80
2018-01-05	997.72	6683150000	94423900000	0.3476	119.06
2018-01-06	1041.68	4662220000	96326500000	0.7121	66.09
2018-01-05	3.05	6288500000	127870000000	0.3014	0.73

Select a Month

Let's select a particular month from a DataFrame.

```
# select a particular month
data['2018-01'].head()
'Output':
```

	slug	symbol	name	ranknow	open	high	low \
date							
2018-01-01	bitcoin	BTC	Bitcoin	1	14112.2	14112.2	13154.7
2018-01-02	bitcoin	BTC	Bitcoin	1	13625.0	15444.6	13163.6
2018-01-03	bitcoin	BTC	Bitcoin	1	14978.2	15572.8	14844.5
2018-01-04	bitcoin	BTC	Bitcoin	1	15270.7	15739.7	14522.2
2018-01-05	bitcoin	BTC	Bitcoin	1	15477.2	17705.2	15202.8

	close	volume	market	close_ratio	spread
date					
2018-01-01	13657.2	10291200000	236725000000	0.5248	957.5
2018-01-02	14982.1	16846600000	228579000000	0.7972	2281.0
2018-01-03	15201.0	16871900000	251312000000	0.4895	728.3
2018-01-04	15599.2	21783200000	256250000000	0.8846	1217.5
2018-01-05	17429.5	23840900000	259748000000	0.8898	2502.4

Select a Year

Let's select a particular year from a DataFrame.

```
# select a particular year
data['2018'].head()
'Output':
```

	slug	symbol	name	ranknow	open	high	low \
date							
2018-01-01	bitcoin	BTC	Bitcoin	1	14112.2	14112.2	13154.7
2018-01-02	bitcoin	BTC	Bitcoin	1	13625.0	15444.6	13163.6
2018-01-03	bitcoin	BTC	Bitcoin	1	14978.2	15572.8	14844.5
2018-01-04	bitcoin	BTC	Bitcoin	1	15270.7	15739.7	14522.2
2018-01-05	bitcoin	BTC	Bitcoin	1	15477.2	17705.2	15202.8

	close	volume	market	close_ratio	spread
date					
2018-01-01	13657.2	10291200000	236725000000	0.5248	957.5
2018-01-02	14982.1	16846600000	228579000000	0.7972	2281.0
2018-01-03	15201.0	16871900000	251312000000	0.4895	728.3
2018-01-04	15599.2	21783200000	256250000000	0.8846	1217.5
2018-01-05	17429.5	23840900000	259748000000	0.8898	2502.4

Subset Data Columns and Find Summaries

Get the closing prices of Bitcoin stocks for the month of January.

```
data.loc[data.slug == 'bitcoin', 'close']['2018-01']
'Output':
date
2018-01-01      13657.2
2018-01-02      14982.1
2018-01-03      15201.0
2018-01-04      15599.2
2018-01-05      17429.5
2018-01-06      17527.0
2018-01-07      16477.6
2018-01-08      15170.1
2018-01-09      14595.4
2018-01-10      14973.3
```

Find the mean market value of Ethereum for the month of January.

```
data.loc[data.slug == 'ethereum', 'market']['2018-01'].mean()
'Output':
96739480000.0
```

Resampling Datetime Objects

A Pandas DataFrame with an index of **DatetimeIndex**, **PeriodIndex**, or **TimedeltaIndex** can be resampled to any of the date time frequencies from seconds, to minutes, to months. Let's see some examples.

Let's get the average monthly closing values for Litecoin.

```
data.loc[data.slug == 'bitcoin', 'close'].resample('M').mean().head()
'Output':
date
2013-04-30      139.250000
2013-05-31      119.993226
2013-06-30      107.761333
2013-07-31       90.512258
2013-08-31      113.905161
Freq: M, Name: close, dtype: float64
```

Get the average weekly market value of Bitcoin Cash.

```
data.loc[data.symbol == 'BCH', 'market'].resample('W').mean().head()
'Output':
date
2017-07-23    0.000000e+00
2017-07-30    0.000000e+00
2017-08-06    3.852961e+09
2017-08-13    4.982661e+09
2017-08-20    7.355117e+09
Freq: W-SUN, Name: market, dtype: float64
```

Convert to Datetime Datatype Using 'to_datetime'

Pandas uses the **to_datetime** method to convert strings to Pandas datetime datatype. The **to_datetime** method is smart enough to infer a **datetime** representation from a string of dates passed with different formats. The default output format of **to_datetime** is in the following order: **year, month, day, minute, second, millisecond, microsecond, nanosecond**.

The input to **to_datetime** is recognized as **month, day, year**. Although, it can easily be modified by setting the attributes **dayfirst** or **yearfirst** to **True**.

For example, if **dayfirst** is set to **True**, the input is recognized as **day, month, year**.

Let's see an example of this.

```
# create list of dates
my_dates = ['Friday, May 11, 2018', '11/5/2018', '11-5-2018', '5/11/2018',
            '2018.5.11']
pd.to_datetime(my_dates)
'Output':
DatetimeIndex(['2018-05-11', '2018-11-05', '2018-11-05', '2018-05-11',
               '2018-05-11'],
              dtype='datetime64[ns]', freq=None)
```

Let's set dayfirst to True. Observe that the first input in the string is treated as a day in the output.

```
# set dayfirst to True
pd.to_datetime('5-11-2018', dayfirst = True)
'Output':
Timestamp('2018-11-05 00:00:00')
```

The shift() Method

A typical step in a timeseries use case is to convert the timeseries dataset into a supervised learning framework for predicting the outcome for a given time instant. The **shift()** method is used to adjust a Pandas DataFrame column by shifting the observations forward or backward. If the observations are pulled backward (or lagged), **NaNs** are attached at the tail of the column. But if the values are pushed forward, the head of the column will contain **NaNs**. This step is important for adjusting the **target** variable of a dataset to predict outcomes n-days or steps or instances into the future. Let's see some examples.

Subset columns for the observations related to Bitcoin Cash.

```
# subset a few columns
data_subset_BCH = data.loc[data.symbol == 'BCH',
['open','high','low','close']]
data_subset_BCH.head()
'Output':
                open     high      low    close
date
2017-07-23    555.89   578.97   411.78   413.06
2017-07-24    412.58   578.89   409.21   440.70
2017-07-25    441.35   541.66   338.09   406.90
2017-07-26    407.08   486.16   321.79   365.82
2017-07-27    417.10   460.97   367.78   385.48
```

Now let's create a target variable that contains the closing rates 3 days into the future.

```
data_subset_BCH['close_4_ahead'] = data_subset_BCH['close'].shift(-4)
data_subset_BCH.head()
```

```
'Output':
              open      high      low     close   close_4_ahead
date
2017-07-23   555.89   578.97   411.78   413.06            385.48
2017-07-24   412.58   578.89   409.21   440.70            406.05
2017-07-25   441.35   541.66   338.09   406.90            384.77
2017-07-26   407.08   486.16   321.79   365.82            345.66
2017-07-27   417.10   460.97   367.78   385.48            294.46
```

Observe that the tail of the column **close_4_head** contains **NaNs**.

```
data_subset_BCH.tail()
'Output':
               open       high        low      close   close_4_ahead
date
2018-01-06   2583.71   2829.69   2481.36   2786.65         2895.38
2018-01-07   2784.68   3071.16   2730.31   2786.88             NaN
2018-01-08   2786.60   2810.32   2275.07   2421.47             NaN
2018-01-09   2412.36   2502.87   2346.68   2391.56             NaN
2018-01-10   2390.02   2961.20   2332.48   2895.38             NaN
```

Rolling Windows

Pandas provides a function called **rolling()** to find the rolling or moving statistics of values in a column over a specified window. The window is the "number of observations used in calculating the statistic." So we can find the rolling sums or rolling means of a variable. These statistics are vital when working with timeseries datasets. Let's see some examples.

Let's find the rolling means for the closing variable over a 30-day window.

```
# find the rolling means for Bitcoin cash
rolling_means = data_subset_BCH['close'].rolling(window=30).mean()
```

The first few values of the **rolling_means** variable contain **NaNs** because the method computes the rolling statistic from the earliest time to the latest time in the dataset. Let's print out the first five values using the **head** method.

```
rolling_means.head()
Out[75]:
date
2017-07-23    NaN
2017-07-24    NaN
2017-07-25    NaN
2017-07-26    NaN
2017-07-27    NaN
```

Now let's observe the last five values using the **tail** method.

```
rolling_means.tail()
'Output':
date
2018-01-06    2403.932000
2018-01-07    2448.023667
2018-01-08    2481.737333
2018-01-09    2517.353667
2018-01-10    2566.420333
Name: close, dtype: float64
```

Let's do a quick plot of the rolling means using the Pandas plotting function. The output of the plot is shown in Figure 11-2.

```
# plot the rolling means for Bitcoin cash
data_subset_BCH['close'].rolling(window=30).mean().plot(label='Rolling
Average over 30 days')
```

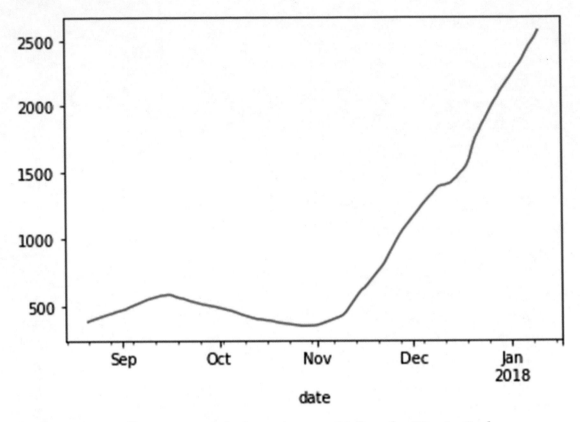

Figure 11-2. *Rolling average closing price over 30 days for Bitcoin Cash*

More on plotting in the next chapter.

CHAPTER 12

Matplotlib and Seaborn

It is critical to be able to plot the observations and variables of a dataset before subjecting the dataset to some machine learning algorithm or another. Data visualization is essential to understand your data and to glean insights into the underlying structure of the dataset. These insights help the scientist in deciding with statistical analysis or which learning algorithm is more appropriate for the given dataset. Also, the scientist can get ideas on suitable transformations to apply to the dataset.

In general, visualization in data science can conveniently be split into **univariate** and **multivariate** data visualizations. Univariate data visualization involves plotting a single variable to understand more about its distribution and structure, while multivariate plots expose the relationship and structure between two or more variables.

Matplotlib and Seaborn

Matplotlib is a graphics package for data visualization in Python. Matplotlib has arisen as a key component in the Python data science stack and is well integrated with NumPy and Pandas. The **pyplot** module mirrors the MATLAB plotting commands closely. Hence, MATLAB users can easily transit to plotting with Python.

Seaborn, on the other hand, extends the Matplotlib library for creating beautiful graphics with Python using a more straightforward set of methods. Seaborn is more integrated for working with Pandas DataFrames. We will go through creating simple essential plots with Matplotlib and seaborn.

Pandas Plotting Methods

Pandas also has a robust set of plotting functions which we will also use for visualizing our dataset. The reader will observe how we can easily convert datasets from NumPy to Pandas and vice versa to take advantage of one functionality or the other. The plotting features of Pandas are found in the **plotting** module.

© Ekaba Bisong 2019
E. Bisong, *Building Machine Learning and Deep Learning Models on Google Cloud Platform*,
https://doi.org/10.1007/978-1-4842-4470-8_12

There are many options and properties for working with **matplotlib**, **seaborn**, and **pandas.plotting** functions for data visualization, but as is the theme of this material, the goal is to keep it simple and give the reader just enough to be dangerous. Deep competency comes with experience and continuous usage. These cannot really be taught.

To begin, we will load Matplotlib by importing the **pyplot** module from the **matplotlib** package and the **seaborn** package.

```
import matplotlib.pyplot as plt
import seaborn as sns
```

We'll also import the **numpy** and **pandas** packages to create our datasets.

```
import pandas as pd
import numpy as np
```

Univariate Plots

Some common and essential univariate plots are line plots, bar plots, histograms and density plots, and the box and whisker plot, to mention just a few.

Line Plot

Let's plot a sine graph of 100 points from the negative to positive **exponential** range. The **plot** method allows us to plot lines or markers to the figure. The outputs of the sine and cosine line plot are shown in Figure 12-1 and Figure 12-2, respectively.

```
data = np.linspace(-np.e, np.e, 100, endpoint=True)
# plot a line plot of the sine wave
plt.plot(np.sin(data))
plt.show()
# plot a red cosine wave with dash and dot markers
plt.plot(np.cos(data), 'r-.')
plt.show()
```

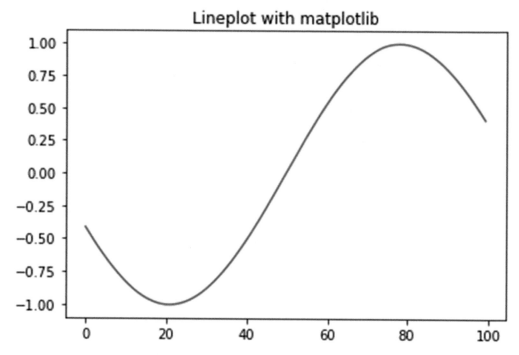

Figure 12-1. *Lineplot with Matplotlib*

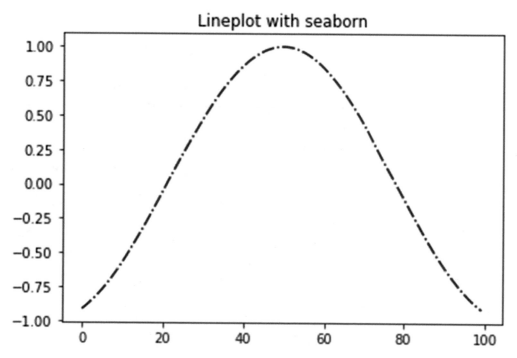

Figure 12-2. *Lineplot with seaborn*

Bar Plot

Let's create a simple bar plot using the bar method. The output with matplotlib is shown in Figure 12-3, and the output with seaborn is shown in Figure 12-4.

```
states = ["Cross River", "Lagos", "Rivers", "Kano"]
population = [3737517, 17552940, 5198716, 11058300]
# create barplot using matplotlib
plt.bar(states, population)
plt.show()
# create barplot using seaborn
sns.barplot(x=states, y=population)
plt.show()
```

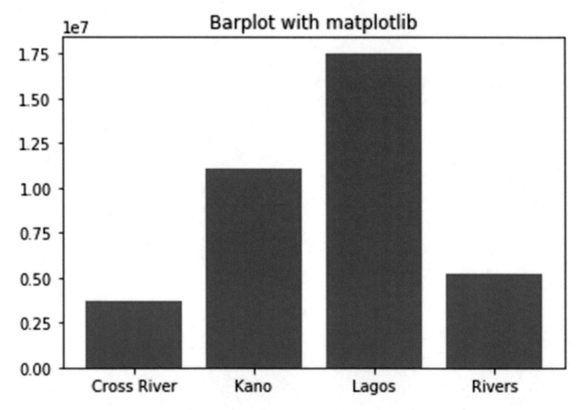

Figure 12-3. *Barplot with Matplotlib*

Figure 12-4. *Barplot with seaborn*

Histogram/Density Plots

Histogram and density plots are essential for examining the statistical distribution of a variable. For a simple histogram, we'll create a set of 100,000 points from the normal distribution. The outputs with matplotlib and seaborn are shown in Figure 12-5 and Figure 12-6, respectively.

```
# create 100000 data points from the normal distributions
data = np.random.randn(100000)
# create a histogram plot
plt.hist(data)
plt.show()
# crate a density plot using seaborn
my_fig = sns.distplot(data, hist=False)
plt.show()
```

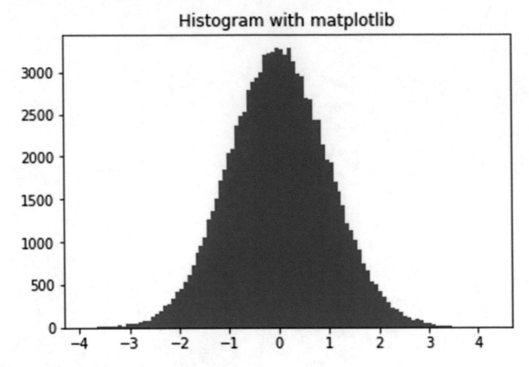

Figure 12-5. *Histogram with Matplotlib*

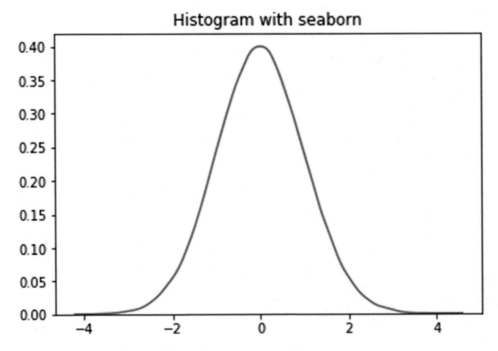

Figure 12-6. *Histogram with seaborn*

Box and Whisker Plots

Boxplot, also popularly called box and whisker plot, is another useful visualization technique for gaining insights into the underlying data distribution. The boxplot draws a box with the upper line representing the 75th percentile and the lower line the 25th percentile. A line is drawn at the center of the box indicating the 50th percentile or median value. The whiskers at both ends give an estimation of the spread or variance of the data values. The dots at the tail end of the whiskers represent possible outlier values. The outputs with matplotlib and seaborn are shown in Figure 12-7 and Figure 12-8, respectively.

```
# create data points
data = np.random.randn(1000)
## box plot with matplotlib
plt.boxplot(data)
plt.show()
## box plot with seaborn
sns.boxplot(data)
plt.show()
```

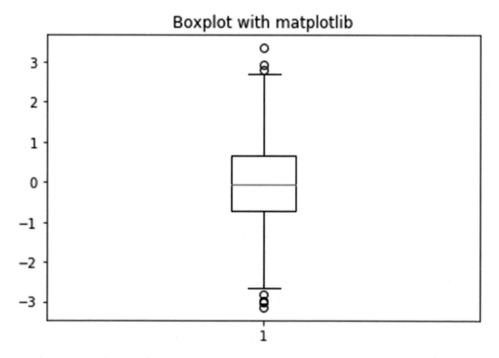

Figure 12-7. *Boxplot with Matplotlib*

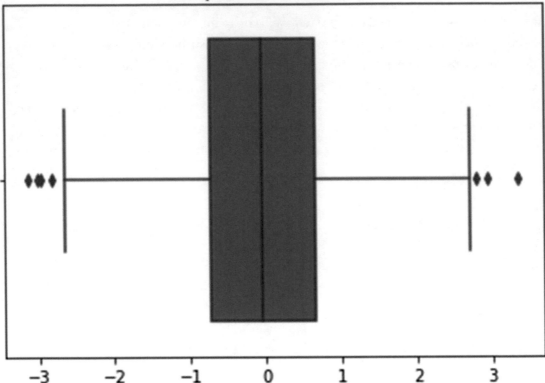

Figure 12-8. *Boxplot with seaborn*

Multivariate Plots

Common multivariate visualizations include the scatter plot and its extension the pairwise plot, parallel coordinate plots, and the covariance matrix plot.

Scatter Plot

Scatter plot exposes the relationships between two variables in a dataset. The outputs with matplotlib and seaborn are shown in Figure 12-9 and Figure 12-10, respectively.

```
# create the dataset
x = np.random.sample(100)
y = 0.9 * np.asarray(x) + 1 + np.random.uniform(0,0.8, size=(100,))
# scatter plot with matplotlib
```

```
plt.scatter(x,y)
plt.xlabel("x")
plt.ylabel("y")
plt.show()
# scatter plot with seaborn
sns.regplot(x=x, y=y, fit_reg=False)
plt.xlabel("x")
plt.ylabel("y")
plt.show()
```

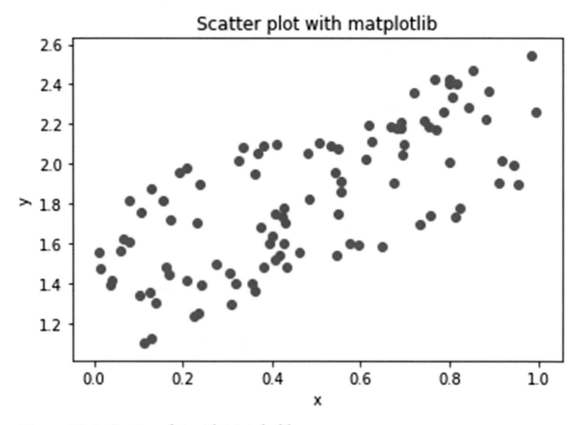

Figure 12-9. *Scatter plot with Matplotlib*

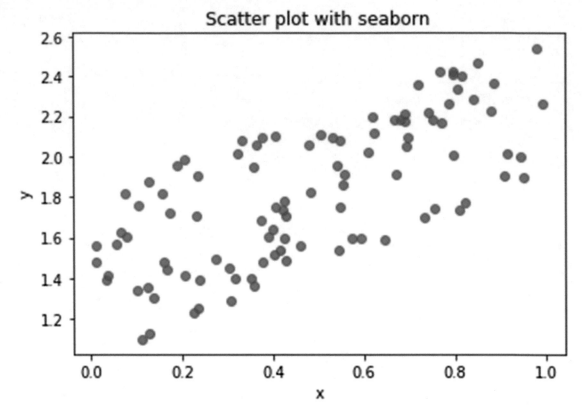

Figure 12-10. *Scatter plot with seaborn*

Pairwise Scatter Plot

Pairwise scatter plot is an effective window for visualizing the relationships among multiple variables within the same plot. However, with higher-dimension datasets, the plot may become clogged up, so use it with care. Let's see an example of this with Matplotlib and seaborn.

Here, we will use the method **scatter_matrix**, one of the plotting functions in Pandas to graph a pairwise scatter plot matrix. The outputs with matplotlib and seaborn are shown in Figure 12-11 and Figure 12-12, respectively.

```
# create the dataset
data = np.random.randn(1000,6)
# using Pandas scatter_matrix
pd.plotting.scatter_matrix(pd.DataFrame(data), alpha=0.5, figsize=(12, 12),
diagonal='kde')
plt.show()
```

Pair-wise scatter plot with Pandas 'scatter_matrix' method

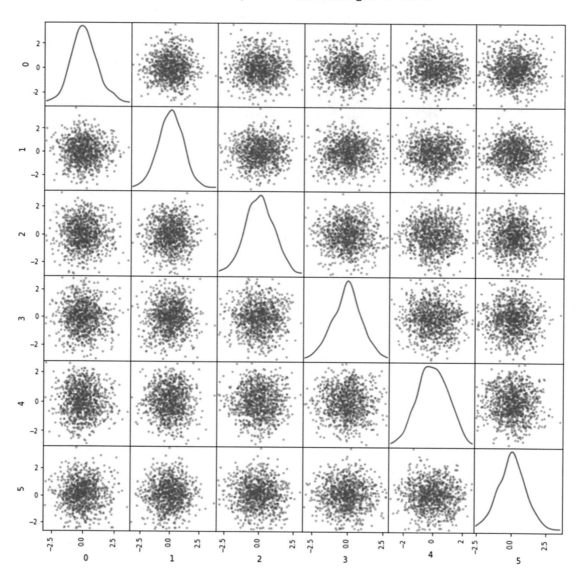

Figure 12-11. *Pairwise scatter plot with Pandas*

```
# pairwise scatter with seaborn
sns.pairplot(pd.DataFrame(data))
plt.show()
```

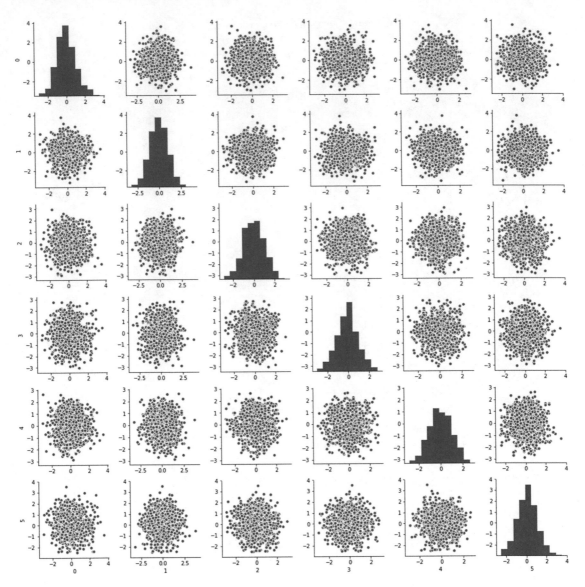

Figure 12-12. Pairwise scatter plot with seaborn

Correlation Matrix Plots

Again, correlation shows how much relationship exists between two variables. By plotting the correlation matrix, we get a visual representation of which variables in the dataset are highly correlated. Remember that parametric machine learning methods such as logistic and linear regression can take a performance hit when variables are highly correlated. Also, in practice, the correlation values that are greater than –0.7 or 0.7 are for the most part highly correlated. The outputs with matplotlib and seaborn are shown in Figure 12-13 and Figure 12-14, respectively.

```
# create the dataset
data = np.random.random([1000,6])
# plot covariance matrix using the Matplotlib matshow function
fig = plt.figure()
ax = fig.add_subplot(111)
my_plot = ax.matshow(pd.DataFrame(data).corr(), vmin=-1, vmax=1)
fig.colorbar(my_plot)
plt.show()
```

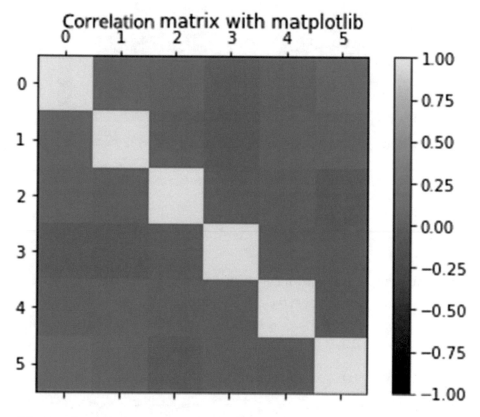

Figure 12-13. *Correlation matrix with Matplotlib*

```
# plot covariance matrix with seaborn heatmap function
sns.heatmap(pd.DataFrame(data).corr(), vmin=-1, vmax=1)
plt.show()
```

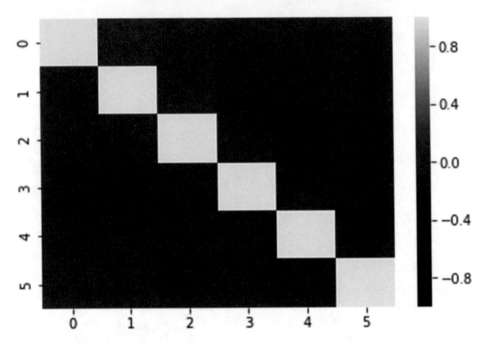

Figure 12-14. *Correlation matrix with seaborn*

Images

Matplotlib is also used to visualize images. This process is utilized when visualizing a dataset of image pixels. You will observe that image data is stored in the computer as an array of pixel intensity values ranging from 0 to 255 across three bands for colored images.

```
img = plt.imread('nigeria-coat-of-arms.png')
# check image dimension
img.shape
'Output': (232, 240, 3)
```

Note that the image contains 232 rows and 240 columns of pixel values across three channels (i.e., red, green, and blue).

Let's print the first row of the columns in the first channel of our image data. Remember that each pixel is an intensity value from 0 to 255. Values closer to 0 are black, while those closer to 255 are white. The output is shown in Figure 12-15.

```
img[0,:,0]
'Output':
array([0., 0., 0., ..., 0., 0., 0.], dtype=float32)
```

Now let's plot the image.

```
# plot image
plt.imshow(img)
plt.show()
```

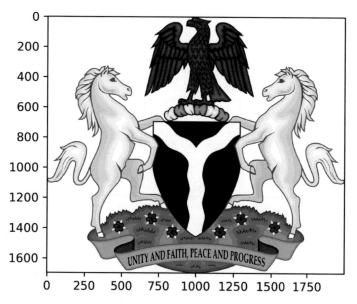

Figure 12-15. *Nigeria Coat of Arms*

This chapter completes Part 2 of this book, which provides the foundation to programming for data science using the Python data science stack. In the next segment, Part 3, containing Chapters 13–17, we will provide an introduction to the field of machine learning.

PART III

Introducing Machine Learning

CHAPTER 13

What Is Machine Learning?

Machine learning as a field grew out of the need to get computers to solve problems that are difficult to program as a sequence of instructions. Take, for example, that we want a computer to perform the task of recognizing faces in an image. One will realize that it is incredibly complicated, if not impossible to develop a precise instruction set that will satisfactorily perform this task. However, by drawing from the observation that humans improve on performing complex functions from past experiences, we can then attempt to develop algorithms and methods that enable the computer to establish a system for solving complex tasks based off prior experiences without being explicitly programmed. The set of methods and algorithms for discovering patterns in data is what is known as machine learning.

Two classical definitions of machine learning are that of Arthur Samuel in 1956 who described machine learning as "the ability for computers to learn without being explicitly programmed" and Tom Mitchell in 1997 who defined machine learning as "the process of teaching a computer to perform a particular task by improving its measure of performance with experience."

Machine learning is an interdisciplinary field of study that brings together techniques from the fields of computer science, statistics, mathematics, and the cognitive sciences which include biology, psychology, and linguistics, to mention just a few. While the idea of learning from data has been around the academic community for several decades, its entry into the mainstream technology industry began in the early 2000s. This growth coincided with the rise of humongous data as a result of the web explosion as people started sharing data over the Internet.

© Ekaba Bisong 2019
E. Bisong, *Building Machine Learning and Deep Learning Models on Google Cloud Platform*,
https://doi.org/10.1007/978-1-4842-4470-8_13

The Role of Data

Data is at the core of machine learning. It is central to the current evolution and further advancement of this field. Just as it is for humans, it is the same way for machines. Learning is not possible without data.

Humans learn how to perform tasks by collecting information from the Environment. This information is the data the brain uses to construct patterns and gain an understanding of the Environment. For a human being, data is captured through the sense organs. For example, the eyes capture visual data, the ears capture auditory data, the skin receives tactile data, while the nose and tongue detect olfactory and taste data, respectively.

As with humans, this same process of learning from data is replicated with machines. Let's take, for example, the task of identifying spam emails. In this example, the computer is provided email examples as data. It then uses an algorithm to learn to distinguish spam emails from regular emails.

The Cost of Data

Data is expensive to collect, and high-quality data is even more costly to capture due to the associated costs in storing and cleaning the data. Over the years, the paucity of data had limited the performance of machine learning methods. However, in the early 1990s, the Internet was born, and by the dawn of the century, it became a super highway for data distribution. As a result, large and diverse data became readily available for the research and development of machine learning products across various domains.

In this chapter, we covered the definition and history of machine learning and the importance of data. Next, we will take it further by discussing the principles of machine learning in Chapter 14.

CHAPTER 14

Principles of Learning

Machine learning is, for the most part, sub-divided into three components based on the approach to the learning problem. The three predominant categories of learning are the supervised, unsupervised, and reinforcement learning schemes. In this chapter, we will go over supervised learning schemes in detail and also touch upon unsupervised and reinforcement learning schemes to a lesser extent.

The focus on supervised learning is for a variety of reasons. Firstly, they are the predominant techniques used for building machine learning products in industry; secondly, as you will soon learn, they are easy to ground truth and assess their performances before being deployed as part of a large-scale production pipeline. Let's examine each of the three schemes.

Supervised Learning

To easily understand the concept of supervised learning, let's revisit the problem of identifying spam emails from a set of emails. We will use this example to understand key concepts that are central to the definition and the framing of a supervised learning problem, and they are

- Features
- Samples
- Targets

For this contrived example, let's assume that we have a dictionary of the top 4 words in the set of emails and we record the frequency of occurrence for each email sample. This information is represented in a tabular format, where each feature is a column and the rows are email samples. This tabular representation is called a dataset. Figure 14-1 illustrates this depiction.

© Ekaba Bisong 2019
E. Bisong, *Building Machine Learning and Deep Learning Models on Google Cloud Platform*,
https://doi.org/10.1007/978-1-4842-4470-8_14

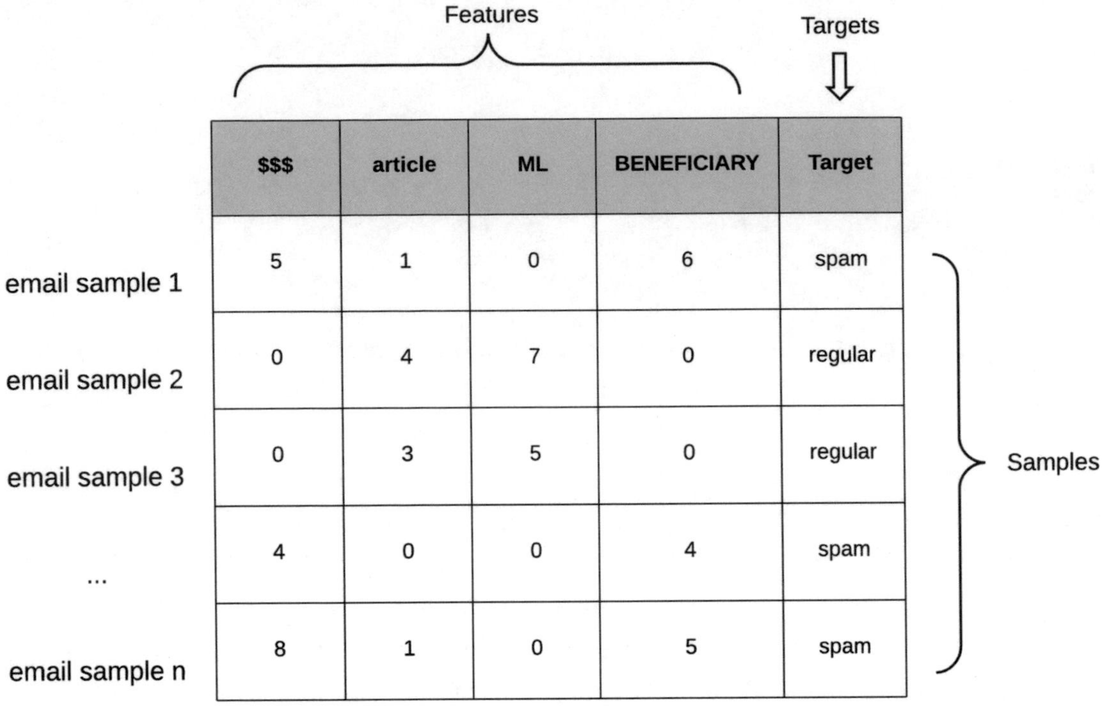

Figure 14-1. *Dataset representation*

The fundamental concept behind supervised machine learning is that each sample is associated with a target variable, and the goal is to teach the computer to learn the patterns from the dataset features that results in a target as a prediction outcome. The columns of a dataset in machine learning are referred to as features; other names you may find commonly used are **variables** or **attributes** of the dataset, but in this book, we will use the term features to describe the measurement units of a data sample. Moreover, the samples of a dataset are also referred to as rows, data points, or observations, but we will use the term samples throughout this book.

Hence, in supervised learning, a set of features are used to build a learning model that will predict the outcome of a target variable as shown in Figure 14-1.

Next, we will cover important modeling considerations for building supervised learning models.

Regression vs. Classification

In supervised learning, we typically have two types of modeling task, and they are **regression** and **classification**.

Regression

The supervised learning problem is a regression task when the values of the target variable are real-valued numbers.

Let's take, for example, that we are given a housing dataset and are asked to build a model that can predict the price of a house. The dataset, for example, has features such as the price of the house, the number of bedrooms, the number of bathrooms, and the total square feet. Let's illustrate how this dataset will look like with a contrived example in Figure 14-2.

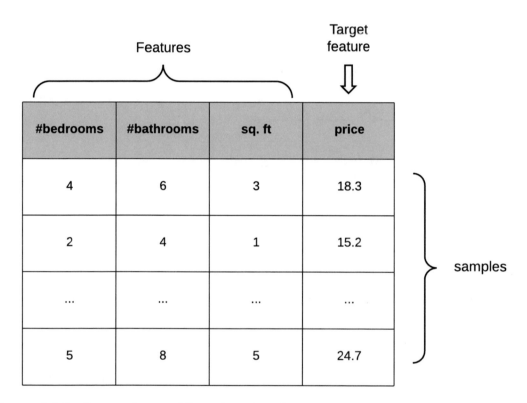

Figure 14-2. *Regression problem: housing dataset*

From the learning problem, the features of the dataset are the number of bedrooms, the number of bathrooms, and the square foot of the floor area, while the target feature is the price of the house. The use case presented in Figure 14-3 is framed as a **regression task** because the target feature is a **real-valued number**.

Classification

In a classification task, the target feature is a label denoting some sort of class membership. These labels are also called categorical variables, because they represent labels that belong to two or more categories. Also, no natural ordering exists between the categories or labels.

As an example, suppose we are given a dataset containing the heart disease diagnosis of patients, and we are asked to build a model to predict if a patient has a heart disease or not. Like the previous example, let's assume the dataset has features blood pressure, cholesterol level, heart rate, and heart disease diagnosis. A contrived illustration of this example is shown in Figure 14-3.

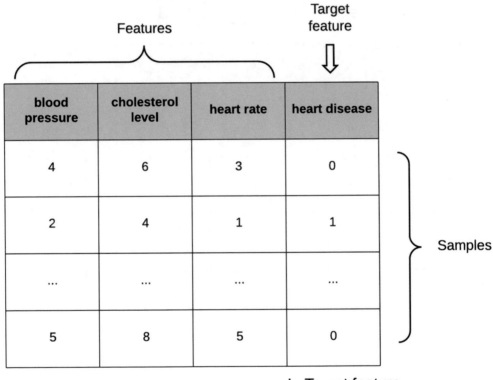

Figure 14-3. *Classification task: heart disease dataset*

From the table in Figure 14-3, the **target variable denotes a class membership of heart disease or no heart disease**; hence, the target is categorical and can be termed as a **classification problem**.

How Do We Know that Learning Has Occurred?

This question is vital to determine if the learning algorithm can learn a useful pattern between the input features and the targets. Let's create a scenario that will give us better insights into appraising the question of determining when learning has occurred.

Assume a teacher takes a physics class for 3 months, and at the end of each session, the teacher administers a test to ascertain if the student has learned anything.

Let's consider two different scenarios the teacher might use in evaluating the students:

1. The teacher evaluates the student with the exact word-for-word questions that were used as sample problems while teaching.

2. The teacher evaluates the student with an entirely different but similar set of sample problems that are based on the principles taught in class.

In which of these subplots can the teacher ascertain that the student has learned? To figure this out, we must consider the two norms of learning:

1. Memorization: In the first subplot, it will be incorrect for the teacher to form a basis for learning because the student has seen and most likely memorized the examples during the class sessions. Memorization is when the exact snapshot of a sample is stored for future recollection. Therefore, it is inaccurate to use samples used in training to carry out learning evaluation. In machine learning, this is known as *data snooping*.

2. Generalization: In the second subplot, the teacher can be confident that the assessment serves as an accurate test to evaluate if the student has learned from the session. The ability to use the principles learned to solve previously unseen samples is known as *generalization*.

Hence, we can conclude that learning is the ability to generalize to previously unseen samples.

Training, Test, and Validation Datasets

The goal of supervised machine learning is to be able to predict or classify the targets on unseen examples correctly. We can misjudge the performance of our learning models if we evaluate the model performance with the same samples used to train the model as explained previously.

To properly evaluate the performance of a learning algorithm, we need to set aside some data for testing purposes. This held-out data is called a **test set**.

Another situation arises when we have trained the model on a dataset, and we now need to improve the performance of the model by adjusting some of the learning algorithm's parameters.

We cannot use the test set for model tuning because if we do that, the model's parameters are trained with the test dataset rendering it unusable as an unseen held-out sample for model evaluation. Hence, it is typical to divide the entire dataset into

- The training set (to train the model)

- The validation set (to tune the model)

- The test set (to evaluate the effectiveness of the model)

A common and straightforward strategy is to split 60% of the dataset for training, 20% for validation, and the final 20% for testing. This strategy is popularly known as the 60/20/20 rule. We will discuss more sophisticated methods for resampling (i.e., using subsets of available data) for machine learning later in this chapter. See Figure 14-4.

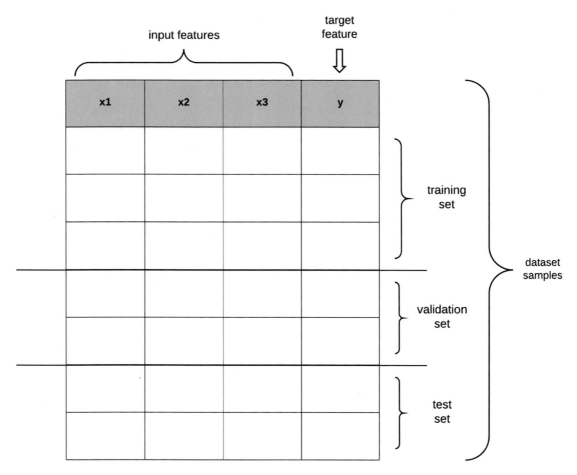

Figure 14-4. *Training, test, and validation set*

Bias vs. Variance Trade-Off

The concept of bias vs. variance is central to machine learning and is critical to understanding how the model is performing, as well as in suggesting the direction in which to improve the model.

A model is said to have high **bias** when it oversimplifies the learning problem or when the model fails to accurately capture the complex relationships that exist between the input features of the dataset. High bias makes the model unable to generalize to new examples.

High variance, on the other hand, is when the model learns too closely the intricate patterns of the dataset input features, and in the process, it *learns the irreducible noise of the dataset samples*. When the learning algorithm learns very closely the patterns of the training samples, including the noise, it will fail to generalize when exposed to previously unseen data.

Hence, we observe that there is a need to strike the right balance between bias and variance, and often it is down to the skill of the model builder to discover this middle ground. However, there exists practical rules of thumb for finding the right trade-off between bias and variance.

How Do We Recognize the Presence of Bias or Variance in the Results?

High bias is observed when the model performs poorly on the trained data. Of course, it will also perform poorly (or even worse) on the test data with high prediction errors. When high bias occurs, it can be said that the model has underfit the data. High variance is observed when the trained model learns the training data very well but performs poorly on unseen (test) data. In the event of high variance, we can say that the model has overfit the dataset.

The graph in Figure 14-5 illustrates the effect of bias and variance on the quality/performance of a machine learning model. In Figure 14-6, the reader will observe that there is a sweet spot somewhere in the middle where the model has good performances on both the training and the test datasets.

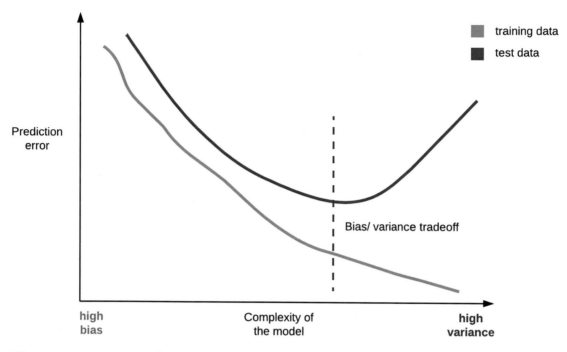

Figure 14-5. *Bias and variance*

To recap, our goal is to have a model that strikes a balance between high bias and high variance. Figure 14-6 provides further illustration on the effects of models with high bias and variance on a dataset. As seen in the image to the left of Figure 14-6, we want to have a model that can generalize to previously unseen example, such a model should have good prediction accuracy.

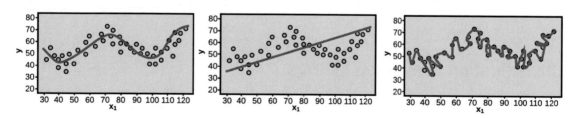

Figure 14-6. **Left:** *Good fit.* **Center:** *Underfit (high bias).* **Right:** *Overfit (high variance)*

Evaluating Model Quality

Evaluation metrics give us a way to quantitatively evaluate how well our model is performing. The model's performance on the training data is evaluated to get the training set accuracy, while its performance on the test data is evaluated to get the test data accuracy when the model predicts the targets of previously unseen examples. Evaluation on test data helps us to know the true performance measure of our model.

The learning problem determines the type of evaluation metric to use. As an example, for regression prediction problems, it is common to use the root mean squared error (RMSE) to evaluate the magnitude of the error made by the model. For classification problems, one of the common evaluation metrics is to use a confusion matrix to get a picture of how many samples are correctly classified or misclassified. From the confusion matrix, it is possible to derive other useful metrics for evaluating classification problems such as accuracy, precision, and recall.

The following are the evaluation metrics for machine learning that we will consider in this text:

Classification

- Confusion matrix

- Area under ROC curve (AUC-ROC)

Regression

- Root mean squared error (RMSE)

- R-squared (R^2)

Let's go through each.

Classification Evaluation Metrics

In this section, we'll briefly explain performance metrics for classification machine learning tasks.

Confusion Matrix

The confusion matrix is a popular evaluation metric for gleaning insights into the performance of a classification supervised machine learning model. It is represented as a table with grid-like cells. In the case of a two-class classification problem, the columns of the grid are the actual positive and negative class values of the target feature, while the rows are the predicted positive and negative class values of the targets. This is illustrated in Figure 14-7.

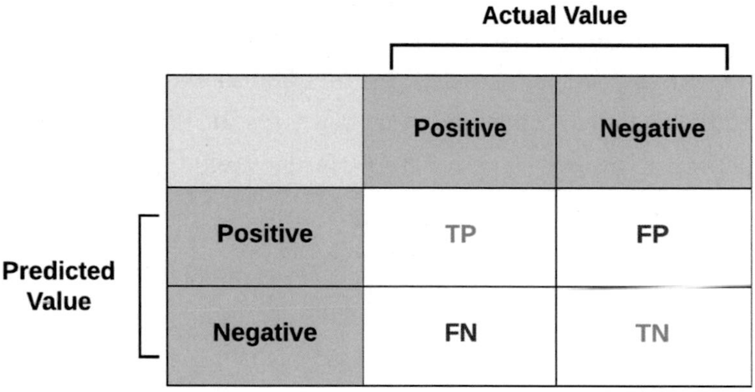

Figure 14-7. *Confusion matrix*

There are four primary values that can be gotten directly from examining the confusion matrix, and they are the ***true positive***, the ***false positive***, the ***true negative***, and the ***false negative*** *values*. Let's examine each of them briefly:

- True positive: True positive is the number of samples predicted to be positive (or true) when the actual class is positive.

- False positive: False positive is the number of samples predicted as positive (or true) when the actual class is negative.

- True negative: True negative is the number of samples predicted to be negative (or false) and the actual class is negative.

- False negative: False negative is the number of samples predicted to be negative (or false) when the actual class is positive.

From the four primary values, we have three other measures that provide more information on the performance of our model. These are accuracy, the positive predictive value (or precision), and sensitivity (or recall). Let's explain them briefly:

- Accuracy: Accuracy is the fraction of correct predictions made by the learning algorithm. It is represented as the ratio of the sum of true positive, *TP*, and true negative, *TN*, to the total population.

$$accuracy = \frac{TP + TN}{TP + FP + FN + TN}$$

- Precision or positive predictive value: Precision is the ratio of true positive, *TP*, to the sum of true positive, *TP*, and false positive, *FP*. In other words, precision measures the fraction of results that are correctly predicted as positive over all the results that the algorithm predicts as positive. The sum *TP* + *FP* is also called the predicted positive condition.

$$precision = \frac{TP}{TP + FP}$$

- Recall or sensitivity: Recall is the ratio of true positive, *TP*, to the sum of true positive, *TP*, and false negative, *FN*. In other words, recall retrieves the fraction of results that are correctly predicted as positive over all the results that are positive. The sum *TP* + *FN* is also known as condition positive.

$$recall = \frac{TP}{TP + FN}$$

To put this concept together, let's revisit the example heart disease dataset. Suppose we are to predict if a patient will be diagnosed with a heart disease or not, assume we have 50 samples in the dataset, of which 20 are diagnosed with heart disease and the remaining 30 are not. Of the 30 samples that do not have a disease diagnosis, the learning algorithm rightly identifies 25, while of the 20 samples that have a disease diagnosis, the learning algorithm correctly identifies 15.

Let's represent this information in a confusion matrix (see Figure 14-8) and calculate the necessary statistical measures to evaluate the algorithm performance.

Actual Value

	Positive [have disease]	Negative [no disease]
Positive [have disease]	15	5
Negative [no disease]	5	25

Predicted Value

Figure 14-8. *Confusion matrix example*

From the data in Figure 14-8, we can calculate the ***accuracy***, ***precision***, and ***recall***.

- Accuracy =

$$\frac{15+25}{15+5+5+25} = \frac{40}{50} = \frac{4}{5}$$

- Precision =

$$\frac{15}{15+5} = \frac{3}{4}$$

- Recall =

$$\frac{15}{15+5} = \frac{3}{4}$$

Hence, our algorithm is 80% accurate, with a precision of 75% and a recall of 75%.

Area Under the Receiver Operating Curve (AUC-ROC)

The area under the receiver operating characteristic (ROC) curve, also known as the AUC-ROC for short, is another widely used metric for evaluating classification machine learning problems. A significant feature of the AUC-ROC metric is that it can be a good metric for evaluating datasets with imbalanced classes.

An imbalanced class is when one of the outcome targets has far more samples than another target. A typical example of this can be seen in a fraud identification dataset where the samples of no fraud will be vastly more than the samples with fraud.

To better understand AUC-ROC, let us derive two (2) relevant formulas from our confusion matrix, and they are the **True negative rate (TNR)** (also known as **specificity**) and the **False positive rate** (also known as **fall-out**).

Specificity is the fraction of results that are **correctly predicted as negative** over all the results that are negative, whereas fall-out is the fraction of results that are **wrongly predicted as positive** over all the results that are negative. Fall-out is also represented as (1 – specificity).

This is further illustrated in Figure 14-9.

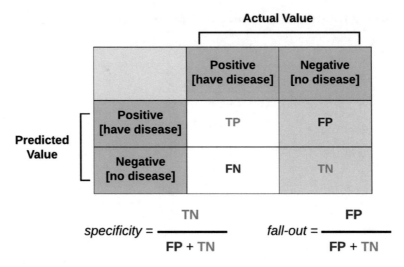

Figure 14-9. *Specificity and fall-out*

The ROC Space

The ROC or receiver operating characteristic space is a 2-D graph that **plots** the cumulative probability distribution of the sensitivity (i.e., the probability distribution of making the correct prediction) on the y axis and the cumulative probability distribution of the fall-out (i.e., the probability distribution of a false alarm) on the x axis.

A few notable details about the ROC space is that

- The area of the square space is 1 because the x and y axes range from 0 to 1, respectively.

- The diagonal line drawn from point $(x = 0, y = 0)$ to $(x = 1, y = 1)$ represented pure chance or a random guess. It is also known as the line of no discrimination.

These expressions are further illustrated in Figure 14-10.

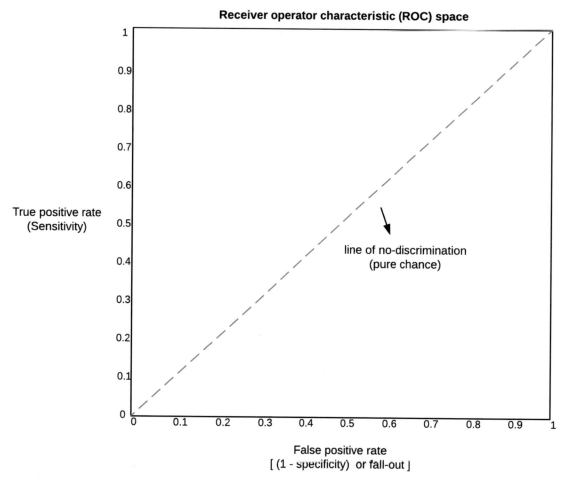

Figure 14-10. *Receiver operating characteristic (ROC) space*

The AUC-ROC Space

The ROC plot as shown in Figure 14-10 looks like a curve. So, the area under the curve, also known as AUC, is the area underneath the ROC curve. AUC provides a single floating-point number that describes the model's performance, and it is interpreted as follows:

- An AUC value below 0.5 indicates that the model's prediction is worse than a random guess of the targets.

- An AUC value closer to 1 signifies a model that is performing very well by generalizing to new examples on the test dataset.

A ROC curve that is closer to the top-left part of the ROC space (i.e., closer to the value 1) indicates that the model has a good classification accuracy.

The AUC-ROC curve is illustrated in Figure 14-11.

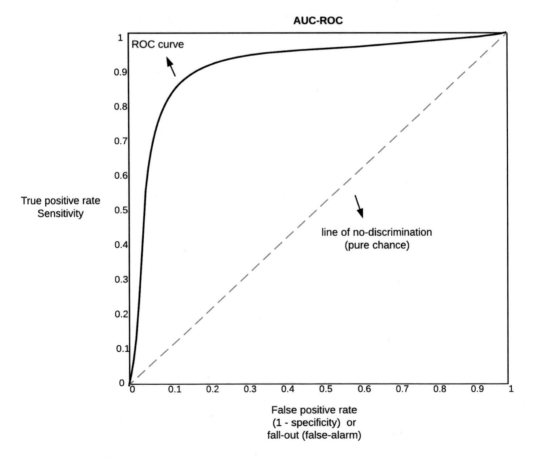

Figure 14-11. *AUC-ROC curve*

Regression Evaluation Metrics

In this section, we'll go through some of the metrics for evaluating regression machine learning tasks.

Root Mean Squared Error (RMSE)

Root mean squared error also known as RMSE for short is an important evaluation metric in supervised machine learning for regression problems. RMSE computes the error difference between the original value of the target feature and the value predicted by the learning algorithm. RMSE is formally represented by the following formula:

$$RMSE = \sqrt{\frac{\sum_{i=1}^{n}\left(y_i - \hat{y}_i\right)^2}{n}}$$

where

- n is the number of samples in the dataset

- y_i is the actual value of the target feature

- $\hat{y}_i$ is the target value predicted by the learning algorithm

Further notes on RMSE:

- Squaring the difference between the actual value and predicted value of the labels $\left(y_i - \hat{y}_i\right)^2$ gives the positive deviation (i.e., the magnitude) between the two numbers.

- Dividing by n gives the average of the sum of magnitudes. The square root ***returns the results in the same unit of measurement*** as the target feature.

An Example of Evaluation with RMSE

Assume we want to predict the price of houses (in thousands of Naira[1]), and we have the following dataset (see Figure 14-12).

---

[1]Naira is the currency of Nigeria. It is also symbolized by the code NGN and the sign ₦.

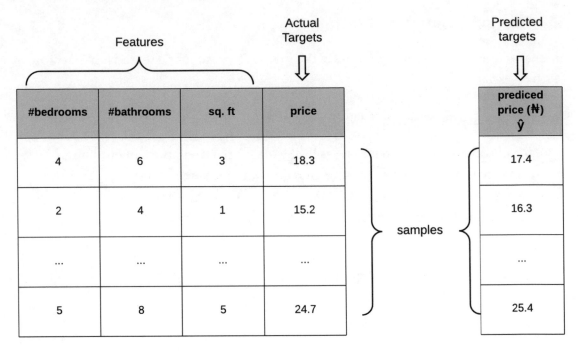

Figure 14-12. *RMSE illustration*

From the formula given, we calculate the RMSE as follows:

$$RMSE = \sqrt{\frac{(18.3-17.4)^2 + (15.2-16.3)^2 + \ldots + (24.7-25.4)^2}{3}}$$

$$= \sqrt{\frac{(0.9)^2 + (-1.1)^2 + \ldots + (-0.7)^2}{3}} = \sqrt{\frac{0.81+1.21+\ldots+0.49}{3}}$$

$$= \sqrt{\frac{2.51}{3}} = \sqrt{0.83666666666} = \sqrt{MSE} = 0.91469484893$$

The closer the RMSE is to 0, the better the performance of the model. Again, we are most interested in knowing the RMSE on the test data, as this gives us an accurate picture of the performance of our model. In this example, the error difference between the actual price and predicted price of houses made by our learning model is approximately NGN 910 (i.e., 0.91 * 1000).

Hence, we can calculate the percentage error as

$$\%error = \frac{error\ difference}{mean\ of\ the\ actual\ prices, y} = \frac{0.91}{19.4}$$

$$= 0.04690721649 = (0.04690721649 * 100) \approx 4\%$$

R-squared (R^2)

R-squared, written as R^2, is another regression error metric that gives us a different perspective into the performance of a learned model. R^2 is also known as the ***coefficient of determination***. The goal of R^2 is to tell us how much of the variance or the variability in the target feature, y, is explained or is captured by the model.

Recall that a model has high variance when it has learned closely the underlying structure of the targets in the dataset. ***Of course, we are mostly concerned with the*** R^2 ***metric on test data***. We typically want the R^2 value on test data to be high. It shows that our model generalizes well to new examples.

Interpretation of R^2

R^2 outputs a value between 0 and 1. Values close to 0 show that variability in the responses are not properly captured by the model, while values close to 1 indicate that the model explains the variability in the observed values. R^2 is calculated using the equation

$$R^2 = 1 - \frac{RSS}{TSS}$$

where

- *RSS* (i.e., the residual sum of squares) captures the error difference (or the variability) between the actual values and the values predicted by the learning algorithm. The formula is

$$RSS = \sum_{i=1}^{n}(y_i - \hat{y}_i)^2$$

- *TSS* (i.e., the total sum of squares), on the other hand, calculates the variability in the response variable, *y*. So, for each observation in the dataset, we find the squared difference from the mean of all observation, $\underline{y}$. The formula is

$$TSS = \sum_{i=1}^{n} \left(y_i - \underline{y}_i \right)^2$$

- Hence, $\dfrac{RSS}{TSS}$ gives us a ratio of how much of the variability in the response variable *y* **is not explained** by the model.

So, when we say $1 - \dfrac{RSS}{TSS}$, we reverse the definition to tell us the ratio of the variability in the response variable explained by the model.

An Example of Evaluating the Model Performance with R²

Using the dataset illustrated in Figure 14-12 and from the formula given earlier, we will calculate R^2 as follows:

$$RSS = \sum_{i=1}^{n} \left(y_i - \hat{y}_i \right)^2 = \left[(18.3 - 17.4)^2 + (15.2 - 16.3)^2 + \ldots + (24.7 - 25.4)^2 \right]$$

$$= \left[(0.9)^2 + (-1.1)^2 + \ldots + (-0.7)^2 \right] = [0.81 + 1.21 + \ldots + 0.49] = 2.51$$

while for *TSS*, we have that the mean of the response variable price, *y*, is

$$\frac{18.3 + 15.2 + 24.7}{3} = 19.4$$

$$TSS = \sum_{i=1}^{n} \left(y_i - \underline{y}_i \right)^2 = \left[(18.3 - 19.4)^2 + (15.2 - 19.4)^2 + \ldots + (24.7 - 19.4)^2 \right]$$

$$= \left[(-1.1)^2 + (-4.2)^2 + \ldots + (5.3)^2 \right] = [1.21 + 17.64 + \ldots + 28.09] = 46.94$$

Finally,

$$R^2 = 1 - \frac{RSS}{TSS} = 1 - \frac{2.51}{46.94} = 1 - (0.05347251811) = 0.94652748189$$

The result shows that the model does a good job in capturing the variability in the target feature. ***Of course, we want to have such good performances on the test dataset***.

Resampling Techniques

This section describes another vital concept for evaluating the performance of supervised learning methods. Resampling methods are a set of techniques that involve selecting a subset of the available dataset, training on that data subset, and then using the remainder of the data to evaluate the trained model.

This process involves creating subsets of the available data into a training set and a validation set. The training set is used to train the model, while the validation set will evaluate the performance of the learned model on unseen data. Typically, this process will be carried out repeatedly to get an approximate measure of the training and test errors.

We will examine three techniques for data resampling and also give some examples of when to use a particular technique. The techniques we'll examine are

- The validation set technique (or train-test split)

- The leave-one-out cross-validation (LOOCV) technique

- The k-fold cross-validation technique

The Validation Set

The validation set is the simplest approach for data resampling, and it involves randomly dividing the dataset into two parts; these are the training set and the validation set. The division can be into two equal sets if you have a big enough dataset, or it could be a 60/40 or 70/30 split.

After splitting, the model is trained on the training set, and its performance is evaluated on the validation set. This process is summarized in the list as follows:

1. Randomly split the dataset into

    - Training set

    - Validation set

2. Train the model on the training set.

3. Evaluate the model performance on the validation set using the appropriate error metric for a regression or classification problem.

4. No. 1 to No. 3 can be repeated.

5. Report the error metric to get the ensemble training and validation set error distribution.

A sample validation set is shown in Figure 14-13.

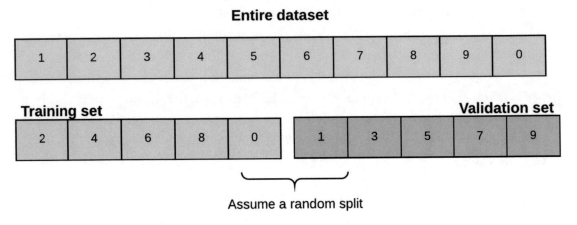

Figure 14-13. *Validation set*

Leave-One-Out Cross-Validation (LOOCV)

The leave-one-out cross-validation approach (commonly referred to as LOOCV) involves dividing the dataset into a training set and a test set. But unlike the validation approach, LOOCV assigns just one example to the test set, and trains the model on the remainder of the dataset. This process is repeated until all the examples in the dataset have been used for evaluating the model.

Assuming we have ten examples in a dataset (let n be used to denote the size of the dataset) to build a learning model. We will train the model using $n - 1$ examples and evaluate the model on just the single remaining example, hence the name leave-one-out. This process is repeated n times for all the examples in the dataset. At the end of the n iterations, we will report the average error estimate.

A sample LOOCV is shown in Figure 14-14.

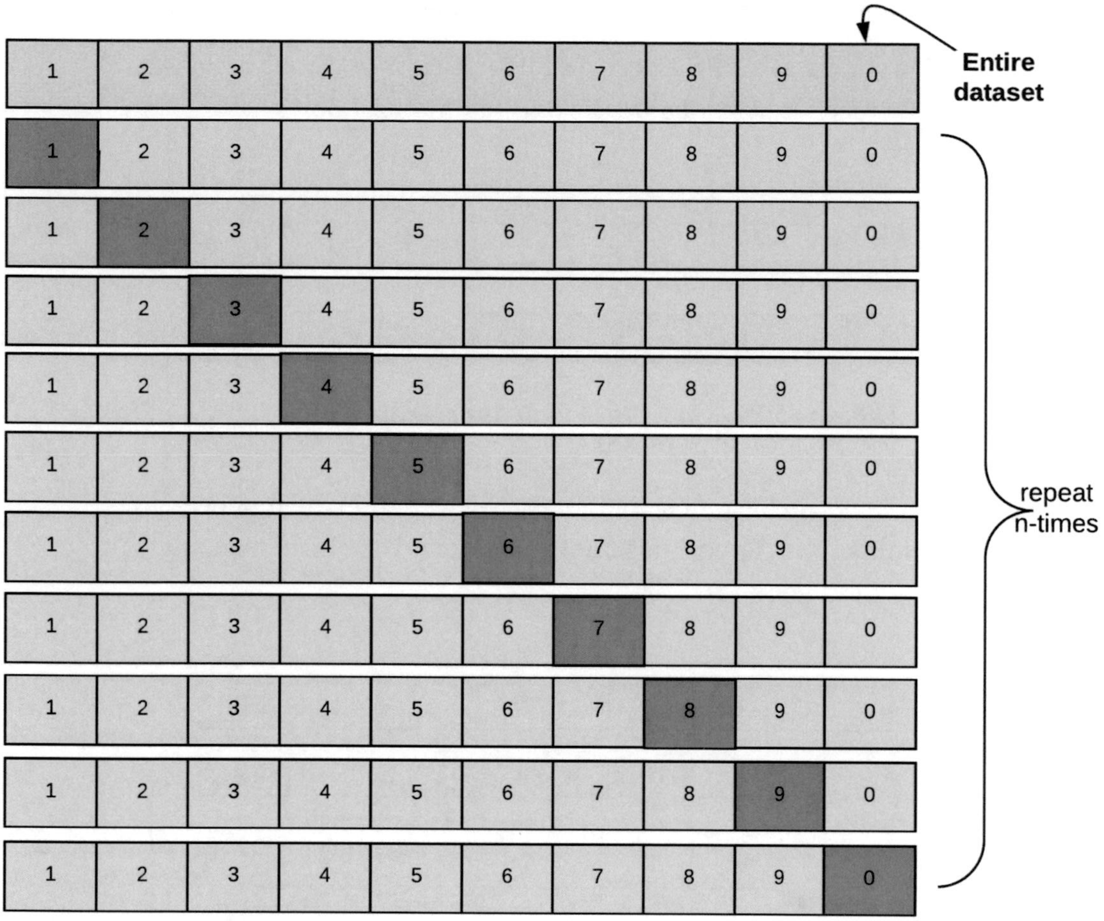

Figure 14-14. LOOCV

The main drawback to using LOOCV is that it is *computationally expensive*. The word *computationally expensive* is when a process takes a lot of computing time and memory to complete its execution.

k-Fold Cross-Validation

k-Fold cross-validation mitigates the computational cost of LOOCV while maintaining its benefits in terms of giving an unbiased estimate of the performance of the learned model when evaluated on validation data.

Let's use the following recipe to explain the idea behind k-fold CV:

- Divide the dataset into k parts or folds. Assume we have a dataset with 20 records; we'll divide the dataset into four parts. See Figure 14-15.

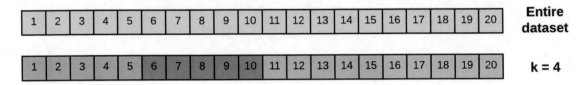

Figure 14-15. *Divide your dataset into k parts or folds*

- Hold out one of the four splits as a test set, and train the model on the remaining splits. Repeat this until all the splits have been held out for testing. See Figure 14-16.

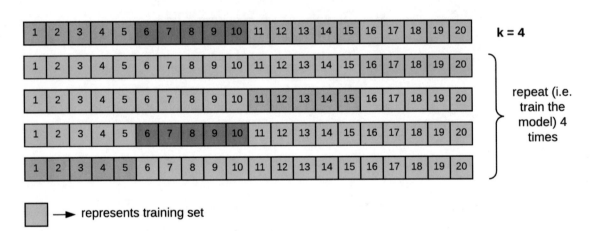

☐ —▶ represents training set

Figure 14-16. *Train the model using k − 1 example sets or splits*

- Report the ensemble error metric.

Note   From this explanation, we can see that LOOCV is a special case where $k = n$.

Improving Model Performance

To improve the performance of the model, a few of the techniques to consider are

1. Systematic feature engineering

2. Using ensemble learning methods (we'll discuss more on this in a later chapter)

3. Hyper-parameter tuning of the algorithm

Feature Engineering

In model building, a significant portion of time is spent on feature engineering. Feature engineering is the practice of systematically going through each feature in the dataset and investigating its relevance to the targets.

Through feature engineering, we can cleverly introduce new features by combining one or more existing features, and this can impact the prediction accuracy of the model. Feature engineering can sometimes be the difference between a decent learning model and a competition-winning model.

Ensemble Methods

Ensemble methods combine the output of weaker models to produce a better performing model. Two major classes of ensemble learning algorithms are

- Boosting

- Bagging

In practice, ensemble methods such as Random forests are known to do very well in various machine learning problems and are the algorithms of choice for machine learning competitions.

Hyper-parameter Tuning

When modeling with a learning algorithm, we can adjust certain configurations of the algorithm. These configurations are called hyper-parameters. Hyper-parameters are tuned to get the best settings of the algorithms that will optimize the performance of the model. One strategy is to use a grid search to adjust the hyper-parameters when fine-tuning the model.

Unsupervised Learning

In unsupervised learning, the goal is to build a model that captures the underlying distribution of the dataset. The dataset has no given targets for the input features (see Figure 14-17). So, it is not possible to learn a function that maps a relationship between the input features and the targets as we do in supervised learning.

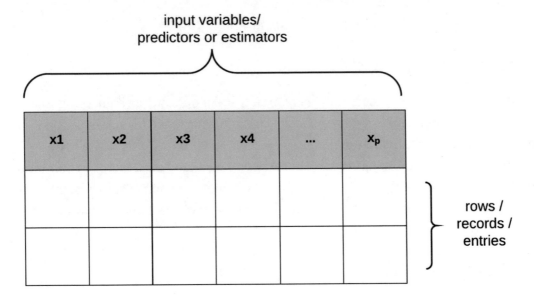

Figure 14-17. *Unsupervised dataset*

Rather, unsupervised learning algorithms attempt to determine the unknown structure of the dataset by grouping similar samples together.

Assume we have a dataset of patients with heart diseases; using unsupervised machine learning algorithms, we can find some hidden sub-groups of patients to help understand more about the disease patterns. This is known as *clustering*.

Also, we can use algorithms like *principal component analysis* (*PCA*) to compress a large number of features into principal components (that summarizes all the other features) for easy visualization. We will talk more about clustering and principal component analysis in later chapters.

Reinforcement Learning

Reinforcement learning presents an approach to learning that is quite different from what we have seen so far in supervised and unsupervised machine learning techniques. In reinforcement learning, an agent interacts with an environment in a feedback configuration and updates its strategy for choosing an action based on the responses it gets from the environment. An illustration of this scenario is shown in Figure 14-18.

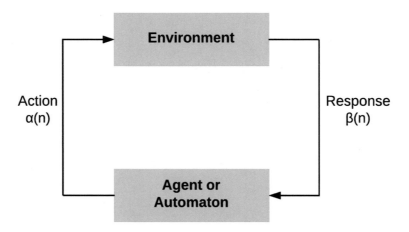

Figure 14-18. *Reinforcement learning model*

This book will not cover reinforcement learning techniques as it presents a different approach to the problem of learning from random environments that is distinct from the approach used in supervised and unsupervised learning problems.

In this chapter, we covered the three main components of machine learning, which are supervised, unsupervised, and reinforcement learning. The chapter largely focused on the principles for performing supervised machine learning such as framing a problem as a regression or classification task; splitting the dataset into training, test, and validation sets; understanding the bias/variance trade-off and consequently issues of overfitting and underfitting; and the evaluation metrics for assessing the performance of a learning model.

In the next chapter, we will briefly look at the differences between batch and online learning.

CHAPTER 15

Batch vs. Online Learning

Data is a vital component for building learning models. There are two design choices for how data is used in the modeling pipeline. The first is to build your learning model with data at rest (batch learning), and the other is when the data is flowing in streams into the learning algorithm (online learning). This flow can be as individual sample points in your dataset, or it can be in small batch sizes. Let's briefly discuss these concepts.

Batch Learning

In batch learning the machine learning model is trained using the entire dataset that is available at a certain point in time. Once we have a model that performs well on the test set, the model is shipped for production and thus learning ends. This process is also called *offline learning*. If in the process of time, new data becomes available, and there is need to update the model based on the new data, the model is trained from scratch all over again using both the previous data samples and the new data samples.

This pipeline is further illustrated in Figure 15-1.

© Ekaba Bisong 2019
E. Bisong, *Building Machine Learning and Deep Learning Models on Google Cloud Platform*,
https://doi.org/10.1007/978-1-4842-4470-8_15

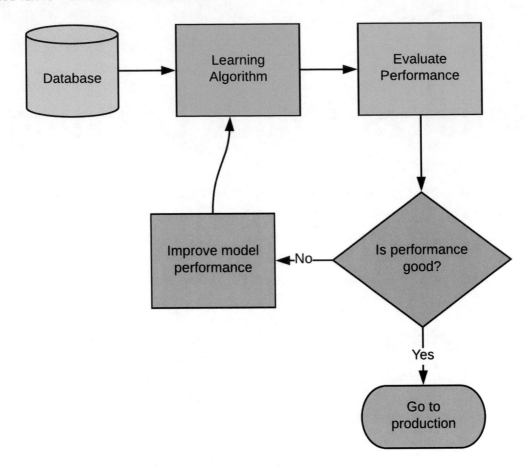

Figure 15-1. *Batch learning*

In a situation where there is a need to train the model with data that is generated continuously from the source, batch learning becomes inappropriate to deal with that situation. In such a circumstance, we want to be able to update our learning model on the go, based on the new data samples that are available.

Online Learning

In online learning, data *streams* (either individually or in mini-batches) into the learning algorithm and updates the model. Online learning is ideal in situations where data is generated continuously in time, and we need to use real-time data samples to build a prediction model. A typical example of this case is in stock market prediction.

Online learning is illustrated in Figure 15-2.

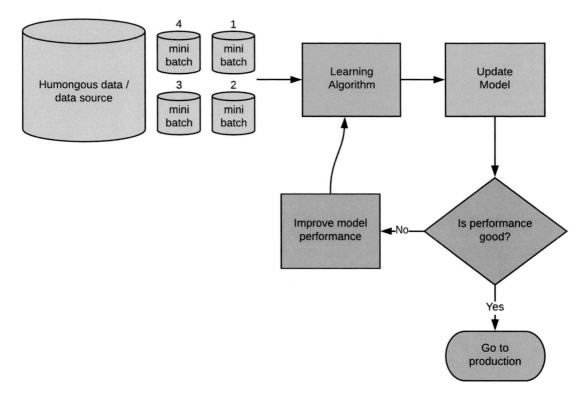

Figure 15-2. *Online learning*

This brief chapter explained the contrast between batch learning and online learning. In the next chapter, we will focus our attention on a vital optimization algorithm for machine learning, gradient descent.

CHAPTER 16

Optimization for Machine Learning: Gradient Descent

Gradient descent is an optimization algorithm that is used to minimize the cost function of a machine learning algorithm. Gradient descent is called an iterative optimization algorithm because, in a stepwise looping fashion, it tries to find an approximate solution by basing the next step off its present step until a terminating condition is reached that ends the loop.

Take the following convex function in Figure 16-1 as a visual of gradient descent finding the minimum point of a function space.

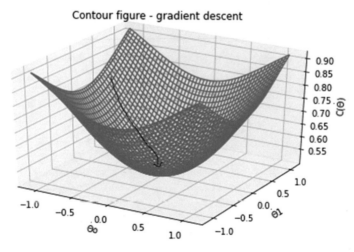

Figure 16-1. *Contour figure – gradient descent*

© Ekaba Bisong 2019
E. Bisong, *Building Machine Learning and Deep Learning Models on Google Cloud Platform*,
https://doi.org/10.1007/978-1-4842-4470-8_16

The image in Figure 16-1 is an example of a function space. This type of function is known as a ***convex or a bowl-shaped function***. The role of gradient descent in the function space is to find the set of values for the parameters of the function that minimizes the cost of the function and brings it to the global minimum. The global minimum is the lowest point of the function space.

For example, the mean squared error cost function for linear regression is nicely convex, so gradient descent is almost guaranteed to find the global minimum. However, this is not always the case for other types of non-convex function spaces. Remember, gradient descent is a global optimizer for minimizing any function space.

Some functions may have more than one minimum region; these regions are called local minima. The lowest region of the function space is called the global minimum.

The Learning Rate of Gradient Descent Algorithm

Learning rate is a hyper-parameter that controls how big a step the gradient descent algorithm takes when tracing its path in the direction of steepest descent in the function space.

If the learning rate is too large, the algorithm takes a large step as it goes downhill. In doing so, gradient descent runs faster, but it has a high propensity of missing the global minimum. An overly small learning rate makes the algorithm slow to converge (i.e., to reach the global minimum), but it is more likely to converge to the global minimum steadily. Empirically, examples of good learning rates are values in the range of 0.001, 0.01, and 0.1. In Figure 16-2, with a good learning rate, the cost function $C(\theta)$ should decrease after every iteration.

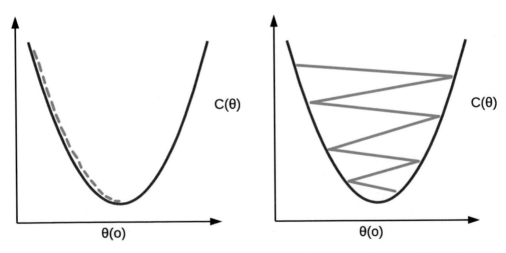

Figure 16-2. *Learning rates.* **Left:** *Good learning rate.* **Right:** *Bad learning rate.*

Classes of Gradient Descent Algorithm

The three types of gradient descent algorithms are

- Batch gradient descent

- Mini-batch gradient descent

- Stochastic gradient descent

The **batch gradient descent** algorithm uses the entire training data in computing each step of the gradient in the direction of steepest descent. Batch gradient descent is most likely to converge to the global minimum. However, the disadvantage of this method is that, for massive datasets, the optimization process can be prolonged.

In **stochastic gradient descent (SGD)**, the algorithm quickly learns the direction of steepest descent using a single example of the training set at each time step. While this method has the distinct advantage of being fast, it may never converge to the global minimum. However, it approximates the global minimum closely enough. In practice, SGD is enhanced by gradually reducing the learning rate over time as the algorithm converges. In doing this, we can take advantage of large step sizes to go downhill more quickly and then slow down so as not to miss the global minimum. Due to its speed when dealing with humongous datasets, SGD is often preferred to batch gradient descent.

Mini-batch gradient descent on the other hand randomly splits the dataset into manageable chunks called mini-batches. It operates on a mini-batch in each time step to learn the direction of steepest descent of the function. This method is a compromise between stochastic and batch gradient descent. Just like SGD, mini-batch gradient descent does not converge to the global minimum. However, it is more robust in avoiding local minimum. The advantage of mini-batch gradient descent over stochastic gradient descent is that it is more computational efficient by taking advantage of matrix vectorization under the hood to efficiently compute the algorithm updates.

Optimizing Gradient Descent with Feature Scaling

This process involves making sure that the features in the dataset are all on the same scale. Typically all real-valued features in the dataset should lie between $-1 \leq x_i \leq 1$ or a range around that region. Any range too large or arbitrarily too small can generate a contour plot that is too narrow and hence will take a longer time for gradient descent to converge to the optimal solution. The plot in Figure 16-3 is called a contour plot. Contour plots are used to represent 3-D surfaces on a 2-D plane. The smaller circles represent the lowest point (or the global optimum) of the convex function.

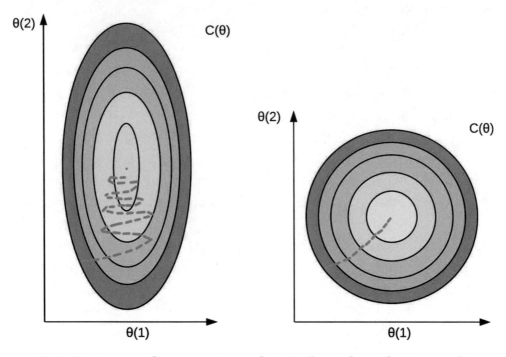

Figure 16-3. *Feature scaling – contour plots. **Left:** without feature scaling. **Right:** with feature scaling*

A popular technique for feature scaling is called mean normalization. In mean normalization, for each feature, the mean of the feature is subtracted from each record and divided by the feature's range (i.e., the difference between the maximum and minimum elements in the feature). Alternatively, it can be divided by the standard deviation of the features. Feature scaling is formally written as

$$x_i = \frac{x_i - \mu_i}{\max - \min} \text{ divided by range} \quad x_i = \frac{x_i - \mu_i}{\sigma} \text{ divided by standard deviation}$$

Figure 16-4 is an example of a dataset with feature scaling.

Normal feature	Feature scaled by range		Feature scaled by standard deviation	
x1	x1	x1	x1	x1
40	(40 - 49.83)/ 58	-0.17	(40 - 49.83)/22.23	-0.44
31	(31 - 49.83)/ 58	-0.32	(31 - 49.83)/ 22.23	-0.85
81	(81 - 49.83)/ 58	0.54	(81 - 49.83)/ 22.23	1.40
58	(58 - 49.83)/ 58	0.14	(58 - 49.83)/ 22.23	0.37
23	(23 - 49.83)/ 58	-0.46	(23 - 49.83)/ 22.23	-1.21
66	(66 - 49.83)/ 58	0.28	(66 - 49.83)/ 22.23	0.73

Figure 16-4. *Feature scaling example*

In this chapter, we discussed gradient descent, an important algorithm for optimizing machine learning models. In the next chapter, we will introduce a suite of supervised and unsupervised machine learning algorithms.

CHAPTER 17

Learning Algorithms

In this section, we introduce a variety of supervised and unsupervised machine learning algorithms. The algorithms presented here provide a foundation for understanding other machine learning methods (e.g., linear and logistic regression), and others like Random forests and Extreme Stochastic Gradient Boosting (XGBoost) are widely used in applied machine learning.

We will survey the various learning algorithms from a conceptual level. In general, the discussion will cut across

- What a particular algorithm is all about and how it works.

- How we interpret the results of the learning algorithm.

- What various ways it can be optimized to improve performance in certain circumstances.

Classes of Supervised Algorithms

Supervised machine learning algorithms are broadly classified into

- Linear

- Non-linear

- Ensemble methods

© Ekaba Bisong 2019
E. Bisong, *Building Machine Learning and Deep Learning Models on Google Cloud Platform*,
https://doi.org/10.1007/978-1-4842-4470-8_17

Let's briefly go through them:

- *Linear methods* are also known as *parametric methods* or algorithms. Linear methods assume that the underlying structure of the data is linear, put in another form, that there exists a linear interaction between the features of the dataset. Examples of linear algorithms are

    - Linear regression

    - Logistic regression

    - Support vector machines

- *Non-linear methods* (also known as *non-parametric methods*) do not assume any parametric or structural form of the dataset. Instead, they attempt to learn the internal relationships or representation between the features of the dataset. Examples of non-linear algorithms are

    - K-nearest neighbors

    - Classification and regression trees (they form the foundation for ensemble methods such as boosting and bagging)

    - Support vector machines

    - Neural networks

- *Ensemble methods* combine the output of multiple algorithms to build a better model estimator that generalizes to unseen examples. Two major classes of ensemble methods are

    - Boosting (stochastic gradient boosting)

    - Bagging (Random forests)

As we can see from the preceding list, some algorithms can function as both a linear and non-linear model. An example is support vector machine (SVM) which applies the so-called kernel trick to use it as a non-linear classification algorithm (more on this later).

Supervised machine learning algorithms can also be grouped as regression or classification algorithms. As we saw in Chapter 14 on regression vs. classification, regression is when the target variable is real-valued and classification is when the target variable is class labels.

Unsupervised Algorithms

Examples of unsupervised learning include

- Clustering

- Principal component analysis

In the later chapters, we will survey the preceding unsupervised learning algorithms for learning from non-labeled datasets. Clustering is an algorithm for grouping homogeneous samples into partitions called clusters. Principal component analysis is a method for finding low-dimensional feature sub-spaces that capture as much information as possible from the original higher-dimensional features of the dataset.

This chapter provides an overview of the machine learning algorithms that we'll discuss together with code examples in Part 4 of this book.

PART IV

Machine Learning in Practice

## CHAPTER 18

Introduction to Scikit-learn

Scikit-learn is a Python library that provides a standard interface for implementing machine learning algorithms. It includes other ancillary functions that are integral to the machine learning pipeline such as data preprocessing steps, data resampling techniques, evaluation parameters, and search interfaces for tuning/optimizing an algorithm's performance.

This section will go through the functions for implementing a typical machine learning pipeline with Scikit-learn. Since, Scikit-learn has a variety of packages and modules that are called depending on the use case, we'll import a module directly from a package if and when needed using the **from** keyword. Again the goal of this material is to provide the foundation to be able to comb through the exhaustive Scikit-learn library and be able to use the right tool or function to get the job done.

Loading Sample Datasets from Scikit-learn

Scikit-learn comes with a set of small standard datasets for quickly testing and prototyping machine learning models. These datasets are ideal for learning purposes when starting off working with machine learning or even trying out the performance of some new model. They save a bit of the time required to identify, download, and clean up a dataset obtained from the wild. However, these datasets are small and well curated, they do not represent real-world scenarios.

Five popular sample datasets are

- Boston house-prices dataset

- Diabetes dataset

© Ekaba Bisong 2019
E. Bisong, *Building Machine Learning and Deep Learning Models on Google Cloud Platform*,
https://doi.org/10.1007/978-1-4842-4470-8_18

- Iris dataset

- Wisconsin breast cancer dataset

- Wine dataset

Table 18-1 summarizes the properties of these datasets.

Table 18-1. *Scikit-learn Sample Dataset Properties*

Dataset name	Observations	Dimensions	Features	Targets
Boston house-prices dataset (regression)	506	13	real, positive	real 5.–50.
Diabetes dataset (regression)	442	10	real, −.2 < x < .2	integer 25–346
Iris dataset (classification)	150	4	real, positive	3 classes
Wisconsin breast cancer dataset (classification)	569	30	real, positive	2 classes
Wine dataset (classification)	178	13	real, positive	3 classes

To load the sample dataset, we'll run

```
# load library
from sklearn import datasets
import numpy as np
```

Load the Iris dataset

```
# load iris
iris = datasets.load_iris()
iris.data.shape
'Output': (150, 4)
iris.feature_names
'Output':
['sepal length (cm)',
 'sepal width (cm)',
 'petal length (cm)',
 'petal width (cm)']
```

Methods for loading other datasets:

- Boston house-prices dataset – **datasets.load_boston()**

- Diabetes dataset – **datasets.load_diabetes()**

- Wisconsin breast cancer dataset – **datasets.load_breast_cancer()**

- Wine dataset – **datasets.load_wine()**

Splitting the Dataset into Training and Test Sets

A core practice in machine learning is to split the dataset into different partitions for training and testing. Scikit-learn has a convenient method to assist in that process called **train_test_split(X, y, test_size=0.25)**, where **X** is the design matrix or dataset of predictors and **y** is the target variable. The split size is controlled using the attribute **test_size**. By default, test_size is set to 25% of the dataset size. It is standard practice to shuffle the dataset before splitting by setting the attribute **shuffle=True**.

```
# import module
from sklearn.model_selection import train_test_split
# split in train and test sets
X_train, X_test, y_train, y_test = train_test_split(iris.data, iris.target,
shuffle=True)
```

```
X_train.shape
'Output': (112, 4)
X_test.shape
'Output': (38, 4)
y_train.shape
'Output': (112,)
y_test.shape
'Output': (38,)
```

Preprocessing the Data for Model Fitting

Before a dataset is trained or fitted with a machine learning model, it necessarily undergoes some vital transformations. These transformations have a huge effect on the performance of the learning model. Transformations in Scikit-learn have a **fit()** and **transform()** method, or a **fit_transform()** method.

Depending on the use case, the **fit()** method can be used to learn the parameters of the dataset, while the **transform()** method applies the data transform based on the learned parameters to the same dataset and also to the test or validation datasets before modeling. Also, the **fit_transform()** method can be used to learn and apply the transformation to the same dataset in a one-off fashion. Data transformation packages are found in the **sklearn.preprocessing** package.

This section will cover some critical transformation for numeric and categorical variables. They include

- Data rescaling

- Standardization

- Normalization

- Binarization

- Encoding categorical variables

- Input missing data

- Generating higher-order polynomial features

Data Rescaling

It is often the case that the features of the dataset contain data with different scales. In other words, the data in column A can be in the range of 1–5, while the data in column B is in the range of 1000–9000. This different scale for units of observations in the same dataset can have an adverse effect for certain machine learning models, especially when minimizing the cost function of the algorithm because it shrinks the function space and makes it difficult for an optimization algorithm like gradient descent to find the global minimum.

When performing data rescaling, usually the attributes are rescaled with the range of 0 and 1. Data rescaling is implemented in Scikit-learn using the **MinMaxScaler** module. Let's see an example.

```
# import packages
from sklearn import datasets
from sklearn.preprocessing import MinMaxScaler
```

```
# load dataset
data = datasets.load_iris()
# separate features and target
X = data.data
y = data.target

# print first 5 rows of X before rescaling
X[0:5,:]
'Output':
array([[5.1, 3.5, 1.4, 0.2],
       [4.9, 3. , 1.4, 0.2],
       [4.7, 3.2, 1.3, 0.2],
       [4.6, 3.1, 1.5, 0.2],
       [5. , 3.6, 1.4, 0.2]])

# rescale X
scaler = MinMaxScaler(feature_range=(0, 1))
rescaled_X = scaler.fit_transform(X)

# print first 5 rows of X after rescaling
rescaled_X[0:5,:]
'Output':
array([[0.22222222, 0.625     , 0.06779661, 0.04166667],
       [0.16666667, 0.41666667, 0.06779661, 0.04166667],
       [0.11111111, 0.5       , 0.05084746, 0.04166667],
       [0.08333333, 0.45833333, 0.08474576, 0.04166667],
       [0.19444444, 0.66666667, 0.06779661, 0.04166667]])
```

Standardization

Linear machine learning algorithms such as linear regression and logistic regression make an assumption that the observations of the dataset are normally distributed with a mean of 0 and standard deviation of 1. However, this is often not the case with real-world datasets as features are often skewed with differing means and standard deviations.

Applying the technique of standardization to the datasets transforms the features into a standard Gaussian (or normal) distribution with a mean of 0 and standard deviation of 1. Scikit-learn implements data standardization in the **StandardScaler** module. Let's look at an example.

```
# import packages
from sklearn import datasets
from sklearn.preprocessing import StandardScaler

# load dataset
data = datasets.load_iris()
# separate features and target
X = data.data
y = data.target

# print first 5 rows of X before standardization
X[0:5,:]
'Output':
array([[5.1, 3.5, 1.4, 0.2],
       [4.9, 3. , 1.4, 0.2],
       [4.7, 3.2, 1.3, 0.2],
       [4.6, 3.1, 1.5, 0.2],
       [5. , 3.6, 1.4, 0.2]])

# standardize X
scaler = StandardScaler().fit(X)
standardize_X = scaler.transform(X)

# print first 5 rows of X after standardization
standardize_X[0:5,:]
'Output':
array([[-0.90068117,  1.03205722, -1.3412724 , -1.31297673],
       [-1.14301691, -0.1249576 , -1.3412724 , -1.31297673],
       [-1.38535265,  0.33784833, -1.39813811, -1.31297673],
       [-1.50652052,  0.10644536, -1.2844067 , -1.31297673],
       [-1.02184904,  1.26346019, -1.3412724 , -1.31297673]])
```

Normalization

Data normalization involves transforming the observations in the dataset so that it has a unit norm or has magnitude or length of 1. The length of a vector is the square root of the sum of squares of the vector elements. A unit vector (or unit norm) is obtained by dividing the vector by its length. Normalizing the dataset is particularly useful in scenarios where the dataset is sparse (i.e., a large number of observations are zeros) and also has differing scales. Normalization in Scikit-learn is implemented in the **Normalizer** module.

```
# import packages
from sklearn import datasets
from sklearn.preprocessing import Normalizer

# load dataset
data = datasets.load_iris()
# separate features and target
X = data.data
y = data.target

# print first 5 rows of X before normalization
X[0:5,:]
'Output':
array([[5.1, 3.5, 1.4, 0.2],
       [4.9, 3. , 1.4, 0.2],
       [4.7, 3.2, 1.3, 0.2],
       [4.6, 3.1, 1.5, 0.2],
       [5. , 3.6, 1.4, 0.2]])

# normalize X
scaler = Normalizer().fit(X)
normalize_X = scaler.transform(X)

# print first 5 rows of X after normalization
normalize_X[0:5,:]
'Output':
```

```
array([[0.80377277, 0.55160877, 0.22064351, 0.0315205 ],
       [0.82813287, 0.50702013, 0.23660939, 0.03380134],
       [0.80533308, 0.54831188, 0.2227517 , 0.03426949],
       [0.80003025, 0.53915082, 0.26087943, 0.03478392],
       [0.790965  , 0.5694948 , 0.2214702 , 0.0316386 ]])
```

Binarization

Binarization is a transformation technique for converting a dataset into binary values by setting a cutoff or threshold. All values above the threshold are set to 1, while those below are set to 0. This technique is useful for converting a dataset of probabilities into integer values or in transforming a feature to reflect some categorization. Scikit-learn implements binarization with the **Binarizer** module.

```
# import packages
from sklearn import datasets
from sklearn.preprocessing import Binarizer

# load dataset
data = datasets.load_iris()
# separate features and target
X = data.data
y = data.target

# print first 5 rows of X before binarization
X[0:5,:]
'Output':
array([[5.1, 3.5, 1.4, 0.2],
       [4.9, 3. , 1.4, 0.2],
       [4.7, 3.2, 1.3, 0.2],
       [4.6, 3.1, 1.5, 0.2],
       [5. , 3.6, 1.4, 0.2]])

# binarize X
scaler = Binarizer(threshold = 1.5).fit(X)
binarize_X = scaler.transform(X)
```

```
# print first 5 rows of X after binarization
binarize_X[0:5,:]
'Output':
array([[1., 1., 0., 0.],
       [1., 1., 0., 0.],
       [1., 1., 0., 0.],
       [1., 1., 0., 0.],
       [1., 1., 0., 0.]])
```

Encoding Categorical Variables

Most machine learning algorithms do not compute with non-numerical or categorical variables. Hence, encoding categorical variables is the technique for converting non-numerical features with labels into a numerical representation for use in machine learning modeling. Scikit-learn provides modules for encoding categorical variables including the **LabelEncoder** for encoding labels as integers, **OneHotEncoder** for converting categorical features into a matrix of integers, and **LabelBinarizer** for creating a one-hot encoding of target labels.

LabelEncoder is typically used on the target variable to transform a vector of hashable categories (or labels) into an integer representation by encoding label with values between 0 and the number of categories minus 1. This is further illustrated in Figure 18-1.

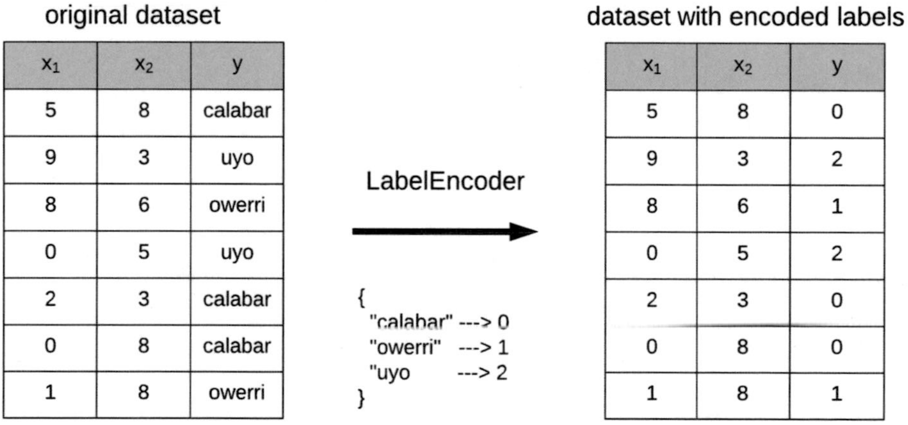

Figure 18-1. *LabelEncoder*

Let's see an example of **LabelEncoder**.

```
# import packages
from sklearn.preprocessing import LabelEncoder

# create dataset
data = np.array([[5,8,"calabar"],[9,3,"uyo"],[8,6,"owerri"],
                [0,5,"uyo"],[2,3,"calabar"],[0,8,"calabar"],
                [1,8,"owerri"]])
data
'Output':
array([['5', '8', 'calabar'],
       ['9', '3', 'uyo'],
       ['8', '6', 'owerri'],
       ['0', '5', 'uyo'],
       ['2', '3', 'calabar'],
       ['0', '8', 'calabar'],
       ['1', '8', 'owerri']], dtype='<U21')

# separate features and target
X = data[:,:2]
y = data[:,-1]

# encode y
encoder = LabelEncoder()
encode_y = encoder.fit_transform(y)

# adjust dataset with encoded targets
data[:,-1] = encode_y
data
'Output':
array([['5', '8', '0'],
       ['9', '3', '2'],
       ['8', '6', '1'],
       ['0', '5', '2'],
       ['2', '3', '0'],
       ['0', '8', '0'],
       ['1', '8', '1']], dtype='<U21')
```

OneHotEncoder is used to transform a categorical feature variable in a matrix of integers. This matrix is a sparse matrix with each column corresponding to one possible value of a category. This is further illustrated in Figure 18-2.

original dataset

x_1	x_2	x_3	y
5	efik	8	calabar
9	ibibio	3	uyo
8	igbo	6	owerri
0	ibibio	5	uyo
2	efik	3	calabar
0	efik	8	calabar
1	igbo	8	owerri

LabelEncoder

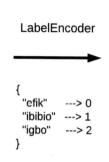

{
 "efik"   ---> 0
 "ibibio" ---> 1
 "igbo"   ---> 2
}

encode categorical feature

x_1	x_2	x_3	y
5	0	8	calabar
9	1	3	uyo
8	2	6	owerri
0	1	5	uyo
2	0	3	calabar
0	0	8	calabar
1	2	8	owerri

OneHotEncoder

one-hot encoding of feature x_2

x_1	x_3	$x_{2,0}$	$x_{2,1}$	$x_{2,2}$	y
5	8	1	0	0	calabar
9	3	0	1	0	uyo
8	6	0	0	1	owerri
0	5	0	1	0	uyo
2	3	1	0	0	calabar
0	8	1	0	0	calabar
1	8	0	0	1	owerri

Figure 18-2. *OneHotEncoder*

Let's see an example of **OneHotEncoder**.

```
# import packages
from sklearn.preprocessing import OneHotEncoder
```

```
# create dataset
data = np.array([[5,"efik", 8,"calabar"],[9,"ibibio",3,"uyo"],[8,"igbo",
6,"owerri"],[0,"ibibio",5,"uyo"],[2,"efik",3,"calabar"],[0,"efik",
8,"calabar"],[1,"igbo",8,"owerri"]])

# separate features and target
X = data[:,:3]
y = data[:,-1]

# print the feature or design matrix X
X
'Output':
array([['5', 'efik', '8'],
       ['9', 'ibibio', '3'],
       ['8', 'igbo', '6'],
       ['0', 'ibibio', '5'],
       ['2', 'efik', '3'],
       ['0', 'efik', '8'],
       ['1', 'igbo', '8']], dtype='<U21')

# one_hot_encode X
one_hot_encoder = OneHotEncoder(handle_unknown='ignore')
encode_categorical = X[:,1].reshape(len(X[:,1]), 1)
one_hot_encode_X = one_hot_encoder.fit_transform(encode_categorical)

# print one_hot encoded matrix - use todense() to print sparse matrix
# or convert to array with toarray()
one_hot_encode_X.todense()
'Output':
matrix([[1., 0., 0.],
        [0., 1., 0.],
        [0., 0., 1.],
        [0., 1., 0.],
        [1., 0., 0.],
        [1., 0., 0.],
        [0., 0., 1.]])
```

```
# remove categorical label
X = np.delete(X, 1, axis=1)
# append encoded matrix
X = np.append(X, one_hot_encode_X.toarray(), axis=1)
X
'Output':
array([['5', '8', '1.0', '0.0', '0.0'],
       ['9', '3', '0.0', '1.0', '0.0'],
       ['8', '6', '0.0', '0.0', '1.0'],
       ['0', '5', '0.0', '1.0', '0.0'],
       ['2', '3', '1.0', '0.0', '0.0'],
       ['0', '8', '1.0', '0.0', '0.0'],
       ['1', '8', '0.0', '0.0', '1.0']], dtype='<U32')
```

Input Missing Data

It is often the case that a dataset contains several missing observations. Scikit-learn implements the **Imputer** module for completing missing values.

```
# import packages
from sklearn. impute import SimpleImputer

# create dataset
data = np.array([[5,np.nan,8],[9,3,5],[8,6,4],
                [np.nan,5,2],[2,3,9],[np.nan,8,7],
                [1,np.nan,5]])
data
'Output':
array([[ 5., nan,  8.],
       [ 9.,  3.,  5.],
       [ 8.,  6.,  4.],
       [nan,  5.,  2.],
       [ 2.,  3.,  9.],
       [nan,  8.,  7.],
       [ 1., nan,  5.]])
```

```
# impute missing values - axis=0: impute along columns
imputer = SimpleImputer(missing_values=np.nan, strategy='mean')
imputer.fit_transform(data)
'Output':
array([[5., 5., 8.],
       [9., 3., 5.],
       [8., 6., 4.],
       [5., 5., 2.],
       [2., 3., 9.],
       [5., 8., 7.],
       [1., 5., 5.]])
```

Generating Higher-Order Polynomial Features

Scikit-learn has a module called PolynomialFeatures for generating a new dataset containing high-order polynomial and interaction features based off the features in the original dataset. For example, if the original dataset has two dimensions [a, b], the second-degree polynomial transformation of the features will result in $[1, a, b, a^2, ab, b^2]$.

```
# import packages
from sklearn.preprocessing import PolynomialFeatures

# create dataset
data = np.array([[5,8],[9,3],[8,6],
                [5,2],[3,9],[8,7],
                [1,5]])
data
'Output':
array([[5, 8],
       [9, 3],
       [8, 6],
       [5, 2],
       [3, 9],
       [8, 7],
       [1, 5]])
```

```
# create polynomial features
polynomial_features = PolynomialFeatures(2)
data = polynomial_features.fit_transform(data)
data
'Output':
array([[ 1.,   5.,   8.,  25.,  40.,  64.],
       [ 1.,   9.,   3.,  81.,  27.,   9.],
       [ 1.,   8.,   6.,  64.,  48.,  36.],
       [ 1.,   5.,   2.,  25.,  10.,   4.],
       [ 1.,   3.,   9.,   9.,  27.,  81.],
       [ 1.,   8.,   7.,  64.,  56.,  49.],
       [ 1.,   1.,   5.,   1.,   5.,  25.]])
```

Machine Learning Algorithms

This chapter provides an introduction to working with the Scikit-learn library for implementing machine learning algorithms.

In the next chapters, we'll implement supervised and unsupervised machine learning models using Scikit-learn. Scikit-learn provides a consistent set of methods, which are the **fit()** method for fitting models to the training dataset and the **predict()** method for using the fitted parameters to make a prediction on the test dataset. The examples are geared at explaining working with Scikit-learn; hence, we are not so keen on the performance of the model.

## CHAPTER 19

Linear Regression

The fundamental idea behind the linear regression algorithm is that it assumes a linear relationship between the features of the dataset. As a result of the pre-defined structure that is imposed on the parameters of the model, it is also called a parametric learning algorithm. Linear regression is used to predict targets that contain real values. As we will see later in Chapter 20 on logistic regression, the linear regression model is not adequate to deal with learning problems whose targets are categorical.

The Regression Model

In linear regression, the prevailing assumption is that the target variable (i.e., the unit that we want to predict) can be modeled as a linear combination of the features.

A linear combination is simply the addition of a certain number of vectors that are scaled (or adjusted) by some arbitrary constant. A vector is a mathematical construct for representing a set of numbers.

For example, let us assume a randomly generated dataset consisting of two features and a target variable. The dataset has 50 observations (see Figure 19-1).

© Ekaba Bisong 2019
E. Bisong, *Building Machine Learning and Deep Learning Models on Google Cloud Platform*,
https://doi.org/10.1007/978-1-4842-4470-8_19

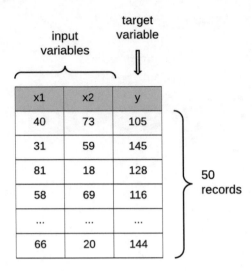

Figure 19-1. *Sample dataset*

The vectors of this dataset are

$$x1 = \begin{bmatrix} 40\,31\,81\,57...66 \end{bmatrix}, \quad x2 = \begin{bmatrix} 73\,59\,18\,69...20 \end{bmatrix}, \quad y = \begin{bmatrix} 105\,145\,128\,116...144 \end{bmatrix}$$

In a linear regression model, every feature has an assigned "weight." We can say that the weight parameterizes each feature in the dataset. The weights in the dataset are adjusted to take on values that capture the underlying relationship between the features that optimally approximate the target variable. The linear regression model is formally written as

$$\hat{y} = \theta_0 + \theta_1 x_1 + \theta_2 x_2 + ... + \theta_n x_n$$

where

- $\hat{y}$ (pronounced y-hat) is the approximate value of the output y that we want to predict.

- θ_i, where $i = \{1, 2, ...n\}$, is the weight assigned to each feature in the dataset. The notation n is the size of features of the dataset.

- θ_0 represents the "bias" term.

A Visual Representation of Linear Regression

To provide more intuition, let us draw a 2-D plot of the first feature x_1 and the target variable y of the dataset with all 50 records. We are using just one feature in this illustration because it is easier to visualize with a 2-D scatter plot (see Figure 19-2).

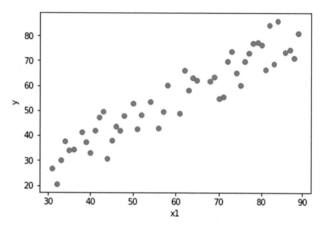

Figure 19-2. *Scatter plot of x_1 (on the x axis) and y (on the y axis)*

The goal of the linear model is to find a line that gives the best approximation or best fit to the data points. When found, this line will look like something in Figure 19-3. The line of best fit is known as the regression line.

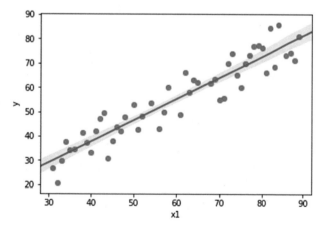

Figure 19-3. *Scatter plot of x_1 (on the x axis) and y (on the y axis) with regression line*

Finding the Regression Line – How Do We Optimize the Parameters of the Linear Model?

To find the regression line, we need to define the cost function, which is also called the loss function. Remember that the cost in machine learning is the error measure that the learning algorithm minimizes. We can also define the cost as the penalty when the model outputs an incorrect prediction.

In the case of the linear regression model, the cost function is defined as half the sum of the squared difference between the predicted value and the actual value. The linear regression cost function is called the **_squared error cost function_** and is written as

$$C(\theta) = \frac{1}{2} \Sigma (\hat{y} - y)^2$$

To put it more simply, the closer the approximate value of the target variable $\hat{y}$ is to the actual variable y, the lower our cost and the better our model.

Having defined the cost function, an optimization algorithm such as gradient descent is used to minimize the cost $C(\theta)$ by updating the weights of the linear regression model.

How Do We Interpret the Linear Regression Model?

In machine learning, the focus of linear regression differs slightly from traditional statistics. In statistics, the goal of a regression model is to understand the relationships between the features and targets by interpreting p-values, whereas in machine learning, the goal of the linear regression model is to predict the targets given new samples.

Figure 19-4 shows a regression model with a line of best fit that optimizes the squared difference between the data features and the targets. This difference is also called the residuals (shown as the purple vertical lines in Figure 19-4). What we care about in a linear regression model is to minimize the error between the predicted labels and the actual labels in the dataset.

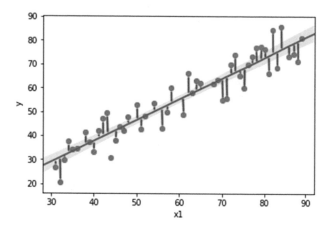

Figure 19-4. *Linear regression model showing residuals*

If all the points in Figure 19-4 entirely fall on the predicted regression line, then the error will be 0. In interpreting the regression model, we want the error measure to be as low as possible.

However, our emphasis is to obtain a low error measure when we evaluate our model on the test dataset. Recall that the test of learning is when a model can generalize to examples that it was not exposed to during training.

Linear Regression with Scikit-learn

In this example, we will implement a linear regression model with Scikit-learn. The model will predict house prices from the Boston house-prices dataset. The dataset contains 506 observations and 13 features.

We begin by importing the following packages:

`sklearn.linear_model.LinearRegression`: function that implements the LinearRegression model.
`sklearn.datasets`: function to load sample datasets integrated with scikit-learn for experimental and learning purposes.
`sklearn.model_selection.train_test_split`: function that partitions the dataset into train and test splits.
`sklearn.metrics.mean_squared_error`: function to load the evaluation metric for checking the performance of the model.

math.sqrt: imports the square-root math function. It is used later to calculate the RMSE when evaluating the model.

```
# import packages
from sklearn.linear_model import LinearRegression
from sklearn import datasets
from sklearn.model_selection import train_test_split
from sklearn.metrics import mean_squared_error
from math import sqrt

# load dataset
data = datasets.load_boston()
# separate features and target
X = data.data
y = data.target

# split in train and test sets
X_train, X_test, y_train, y_test = train_test_split(X, y, shuffle=True)

# create the model
# setting normalize to true normalizes the dataset before fitting the model
linear_reg = LinearRegression(normalize = True)

# fit the model on the training set
linear_reg.fit(X_train, y_train)
'Output': LinearRegression(copy_X=True, fit_intercept=True, n_jobs=1,
normalize=True)

# make predictions on the test set
predictions = linear_reg.predict(X_test)

# evaluate the model performance using the root mean square error metric
print("Root mean squared error (RMSE): %.2f" % sqrt(mean_squared_error(y_
test, predictions)))
'Output':
Root mean squared error (RMSE): 4.33
```

In the preceding code, using the *train_test_split()* function, the dataset is split into training and testing sets. The linear regression algorithm is applied to the training dataset to find the optimal values that parameterize the weights of the model. The model is evaluated by calling the *.predict()* function on the test set.

The error of the model is evaluated using the RMSE error metric (discussed in Chapter 14).

Adapting to Non-linearity

Although linear regression has the premise that the underlying structure of the dataset features is linear, this is, however, not the case for most datasets. It is nevertheless possible to adapt linear regression to fit or build a model for non-linear datasets. This process of adding non-linearity to linear models is called ***polynomial regression***.

Polynomial regression fits a non-linear relationship to the data by adding higher-order polynomial terms of existing data features as new features in the dataset. More of this is visualized in Figure 19-5.

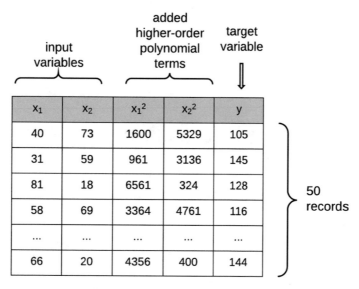

Figure 19-5. *Adding polynomial features to the dataset*

It is important to note that from a statistical point of view, when approximating the optimal values of the weights to minimize the model, the underlying assumption of the interactions of the parameters is linear. Non-linear regression models may tend to overfit the data, but this can be mitigated by adding regularization to the model. Here is a formal example of the polynomial regression model.

$$\hat{y} = \theta_0 + \theta_1 x_1 + \theta_2 x_1^2 + \theta_3 x_2 + \theta_4 x_2^2 + \ldots + \theta_n x_n + \theta_n x_n^2$$

An illustration of polynomial regression is shown in Figure 19-6.

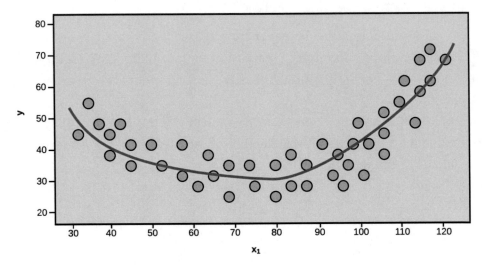

Figure 19-6. *Fitting a non-linear model with polynomial regression*

Higher-Order Linear Regression with Scikit-learn

In this example, we will create higher-order polynomials from the dataset features in hope of fitting a more flexible model that may better capture the variance in the dataset. As seen in Chapter 18, we will use the PolynomialFeatures method to create these higher-order polynomial and interaction features. The following code example is similar to the previous code example except where it extends the feature matrix with higher-order features.

```
# import packages
from sklearn.linear_model import LinearRegression
from sklearn import datasets
```

```python
from sklearn.model_selection import train_test_split
from sklearn.metrics import mean_squared_error
from math import sqrt
from sklearn.preprocessing import PolynomialFeatures

# load dataset
data = datasets.load_boston()

# separate features and target
X = data.data
y = data.target

# create polynomial features
polynomial_features = PolynomialFeatures(2)
X_higher_order = polynomial_features.fit_transform(X)

# split in train and test sets
X_train, X_test, y_train, y_test = train_test_split(X_higher_order, y,
shuffle=True)

# create the model
# setting normalize to true normalizes the dataset before fitting the model
linear_reg = LinearRegression(normalize = True)

# fit the model on the training set
linear_reg.fit(X_train, y_train)
'Output': LinearRegression(copy_X=True, fit_intercept=True, n_jobs=None,
normalize=True)

# make predictions on the test set
predictions = linear_reg.predict(X_test)

# evaluate the model performance using the root mean square error metric
print("Root mean squared error (RMSE): %.2f" % sqrt(mean_squared_error(y_
test, predictions)))

'Output':
Root mean squared error (RMSE): 3.01
```

From the example, we can observe a slight improvement in the error score of the model with added higher-order features. This result is similar to what may most likely be observed in practice. It is rare to find datasets from real-world events where the features have a perfectly underlying linear structure. So adding higher-order terms is most likely to improve the model performance. But we must watch out to avoid overfitting the model.

Improving the Performance of a Linear Regression Model

The following techniques are options that can be explored to improve the performance of a linear regression model.

In the case of Bias (i.e., poor MSE on training data)

- Perform feature selection to reduce the parameter space. Feature selection is the process of eliminating variables that do not contribute to learning the prediction model. There are various automatic methods for feature selection with linear regression. A couple of them are backward selection, forward propagation, and stepwise regression. Features can also be pruned manually by systematically going through each feature in the dataset and determining its relevance to the learning problem.

- Remove features with high correlation. Correlation occurs when two predictor features are strongly dependent on one another. Empirically, highly correlated features in the datasets may hurt the model accuracy.

- Use higher-order features. A more flexible fit may better capture the variance in the dataset.

- Rescale your data before training. Unscaled features negatively affect the prediction quality of a regression model. Because of the different feature scales in multi-dimensional space, it becomes difficult for the model to find the optimal weights that capture the learning problem. As mentioned in Chapter 16, gradient descent performs better with feature scaling.

- In a rare case, we may need to collect more data. However, this is potentially costly.

In the case of variance (i.e., the MSE is good when evaluated on training data, but poor on the test data)

- A standard practice, in this case, is to apply regularization (more on this in Chapter 21) to the regression model. This can do a good job at preventing overfitting.

This chapter provides an overview on the linear regression machine learning algorithm for learning real-valued targets. Also, the chapter provided practical steps for implementing linear regression models with Scikit-learn. In the next chapter, we will examine logistic regression for learning classification problems.

## CHAPTER 20

Logistic Regression

Logistic regression is a supervised machine learning algorithm developed for learning classification problems. A classification learning problem is when the target variable is categorical. The goal of logistic regression is to map a function from the features of the dataset to the targets to predict the probability that a new example belongs to one of the target classes. Figure 20-1 is an example of a dataset with categorical targets.

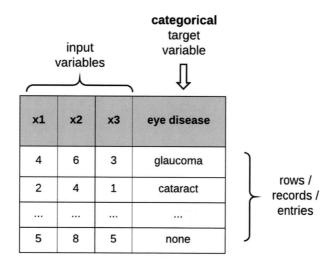

Figure 20-1. *Dataset with qualitative variables as output*

Why Logistic Regression?

To develop our understanding of classification with logistic regression and why linear regression is unsuitable for learning categorical outputs, let us consider a binary or two-class classification problem. The dataset illustrated in Figure 20-2 has the output *y* (i.e., eye disease) = {disease, no-disease} is an example of dataset with binary targets.

© Ekaba Bisong 2019

E. Bisong, *Building Machine Learning and Deep Learning Models on Google Cloud Platform*, https://doi.org/10.1007/978-1-4842-4470-8_20

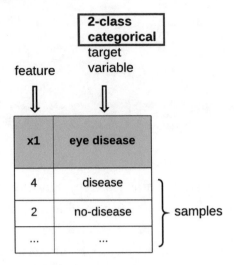

Figure 20-2. *Two-class classification problem*

From the illustration in Figure 20-3, the linear regression algorithm is susceptible to plot inaccurate decision boundaries especially in the presence of outliers (as seen toward the far right of the graph in Figure 20-3). Moreover, the linear regression model will be looking to learn a real-valued output, whereas a classification learning problem predicts the class membership of an observation using probability estimates.

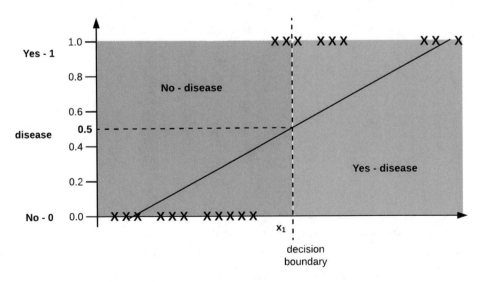

Figure 20-3. *Linear regression on a classification dataset*

Introducing the Logit or Sigmoid Model

The logistic function, also known as the logit or the sigmoid function, is responsible for constraining the output of the cost function so that it becomes a probability output between 0 and 1. The sigmoid function is formally written as

$$h(t) = \frac{1}{1 + e^{-t}}$$

The logistic regression model is formally similar to the linear regression model except that it is acted upon by the sigmoid model. The following is the formal representation:

$$\hat{y} = \theta_0 + \theta_1 x_1 + \theta_2 x_2 + \ldots + \theta_n x_n$$

$$h(\hat{y}) = \frac{1}{1 + e^{-\hat{y}}}$$

where $0 \leq h(t) \leq 1$. The sigmoid function is graphically shown in Figure 20-4.

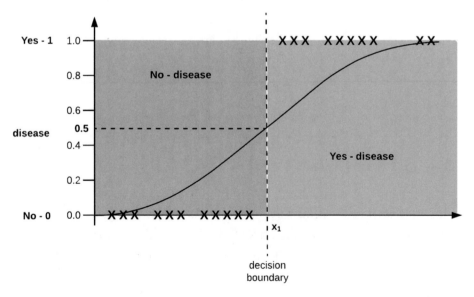

Figure 20-4. *Logistic function*

The sigmoid function, which looks like an S curve, rises from 0 and plateaus at 1. From the sigmoid function shown in Figure 20-4, as $\hat{y}$ increases to positive infinity, the sigmoid output gets closer to 1, and as t decreases toward negative infinity, the sigmoid function outputs 0.

Training the Logistic Regression Model

The logistic regression cost function is formally written as

$$Cost\big(h(t),y\big)=\{-\log\big(h(t)\big)\,if\;\;y=1-\log\big(1-h(t)\big)\,if\;\;y=0$$

The cost function also known as **log-loss** is set up in this form to output the penalty of the algorithm if the model predicts a wrong class. To give more intuition, take, for example, a plot of $-\,log\,(h(t))$ when $y=1$ in Figure 20-5.

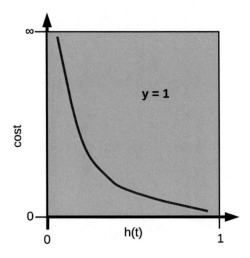

Figure 20-5. *Plot of h(t) when y = 1*

In Figure 20-5, if the algorithm correctly predicts that the target is 1, then the cost tends toward 0. However, if the algorithm $h(t)$ predicts incorrectly the target as 0, then the cost on the model grows exponentially large. The converse is the case with the plot of $-\,log\,(1-h(t))$ when y = 0.

The logistic model is optimized using gradient descent to find the optimal values of the parameter θ that minimizes the cost function to predict the class with the highest probability estimate.

Multi-class Classification/Multinomial Logistic Regression

In multi-class or multinomial logistic regression, the labels of the dataset contain more than 2 classes. The multinomial logistic regression setup (i.e., the cost function and optimization procedure) is structurally similar to logistic regression; the only difference is that the output of logistic regression is 2 classes, while multinomial has greater than 2 classes (see Figure 20-6).

In Figure 20-6, the multi-class logistic regression builds a one-vs.-rest classifier to construct decision boundaries for the different class memberships.

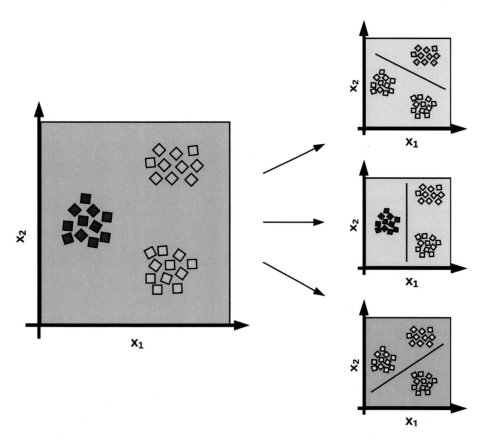

Figure 20-6. *An illustration of multinomial regression*

At this point, we introduce a critical function in machine learning called the softmax function. The softmax function is used to compute the probability that an instance belongs to one of the K classes when $K > 2$. We will see the softmax function show up again when we discuss (artificial) neural networks.

In order to build a classification model with k classes, the multinomial logistic model is formally defined as

$$\hat{y}(k) = \theta_0^k + \theta_1^k x_1 + \theta_2^k x_2 + \ldots + \theta_n^k x_n$$

The preceding model takes into consideration the parameters for the k different classes. The softmax function is formally written as

$$p(k) = \sigma\left(\hat{y}(k)\right)_i = \frac{e^{\hat{y}(k)_i}}{\sum_{j=1}^{K} e^{\hat{y}(k)_j \hat{}(k_j)}}$$

where

- $i = \{1, \ldots, K\}$ classes.

- $\sigma\left(\hat{y}(k)\right)_i$ outputs the probability estimates that an example in the training dataset belongs to one of the K classes.

The cost function for learning the class labels in a multinomial logistic regression model is called the ***cross-entropy*** cost function. Gradient descent is used to find the optimal values of the parameter θ that will minimize the cost function to ***predict the class with the highest probability estimate accurately***.

Logistic Regression with Scikit-learn

In this example, we will implement a multi-class logistic regression model with Scikit-learn. The model will predict the three species of flowers from the Iris dataset. The dataset contains 150 observations and 4 features. For this example, we use the accuracy metric and confusion matrix to access the model's performance.

```
# import packages
from sklearn.linear_model import LogisticRegression
from sklearn import datasets
```

```
from sklearn.model_selection import train_test_split
from sklearn.metrics import accuracy_score
from sklearn.metrics import multilabel_confusion_matrix

# load dataset
data = datasets.load_iris()
# separate features and target
X = data.data
y = data.target

# split in train and test sets
X_train, X_test, y_train, y_test = train_test_split(X, y, shuffle=True)

# create the model
logistic_reg = LogisticRegression(solver='lbfgs', multi_class='ovr')

# fit the model on the training set
logistic_reg.fit(X_train, y_train)

# make predictions on the test set
predictions = logistic_reg.predict(X_test)

# evaluate the model performance using accuracy metric
print("Accuracy: %.2f" % accuracy_score(y_test, predictions))

'Output':
Accuracy: 0.97

# print the confusion matrix
multilabel_confusion_matrix(y_test, predictions)

'Output':
array([[[26,  0],
        [ 0, 12]],

       [[25,  0],
        [ 1, 12]],

       [[24,  1],
        [ 0, 13]]])
```

Take note of the following in the preceding code block:

- The logistic regression model is initialized by calling the method Logi sticRegression(solver='lbfgs', multi_class='ovr'). The attribute 'multi_class' is set to 'ovr' to create a one-vs.-rest classifier.

- The confusion matrix for a multi-class learning problem uses the `multilabel_confusion_matrix' to calculate classwise confusion matrices where the labels are binned in a one-vs.-rest manner. As an example, the first matrix is interpreted as the difference between the actual and predicted targets for class 1 against other classes.

Optimizing the Logistic Regression Model

This section surveys a few techniques to consider in optimizing/improving the performance of logistic regression models.

In the case of Bias (i.e., when the accuracy is poor with training data)

- Remove highly correlated features. Logistic regression is susceptible to degraded performance when highly correlated features are present in the dataset.

- Logistic regression will benefit from standardizing the predictors by applying feature scaling.

- Good feature engineering to remove redundant features or recombine features based on intuition into the learning problem can improve the classification model.

- Applying log transforms to normalize the dataset can boost logistic regression classification accuracy.

In the case of variance (i.e., when the accuracy is good with training data, but poor on test data)

Applying regularization (more on this in Chapter 21) is a good technique to prevent overfitting.

This chapter provides a brief overview of logistic regression for building classification models. The chapter includes practical steps for implementing a logistic regression classifier with Scikit-learn. In the next chapter, we will examine the concept of applying regularization to linear models to mitigate the problem of overfitting.

Regularization for Linear Models

Regularization is the technique of adding a parameter, λ, to the loss function of a learning algorithm to improve its ability to generalize to new examples by reducing overfitting. The role of the extra regularization parameter is to shrink or to minimize the measure of the weights (or parameters) of the other features in the model.

Regularization is applied to linear models such as polynomial linear regression and logistic regression which are susceptible to overfitting when high-order polynomial features are added to the set of features.

How Does Regularization Work

During model building, the regularization parameter λ is calibrated to determine how much the magnitude of other features in the model is adjusted when training the model. The higher the value of the regularization, the more the magnitude of the feature weights is reduced.

If the regularization parameter is set too close to zero, it reduces the regularization effect on the feature weights of the model. At zero, the penalty the regularization term imposes is virtually non-existent, and the model is as if the regularization term was never present.

Effects of Regularization on Bias vs. Variance

The higher the value of λ (i.e., the regularization parameter), the more restricted the coefficients (or weights) of the cost function. Hence, if the value of λ is high, the model can result in a learning bias (i.e., it underfits the dataset).

© Ekaba Bisong 2019
E. Bisong, *Building Machine Learning and Deep Learning Models on Google Cloud Platform*,
https://doi.org/10.1007/978-1-4842-4470-8_21

However, if the value of λ approaches zero, the regularization parameter has negligible effects on the model, hence resulting in overfitting the model. Regularization is an important technique and should be used when injecting polynomial features into linear or logistic regression classifiers to learn non-linear relationships.

Applying Regularization to Models with Scikit-learn

The technique of adding a penalty to restrain the values of the parameters of the model is also known as Ridge regression or Tikhonov regularization. In this section we will build a linear and logistic regression model with regularization.

Linear Regression with Regularization

This code block is similar to the polynomial linear regression example in Chapter 19. The model will predict house prices from the Boston house-prices dataset. However, this model includes regularization.

```
# import packages
from sklearn.linear_model import Ridge
from sklearn import datasets
from sklearn.model_selection import train_test_split
from sklearn.metrics import mean_squared_error
from math import sqrt
from sklearn.preprocessing import PolynomialFeatures

# load dataset
data = datasets.load_boston()

# separate features and target
X = data.data
y = data.target

# create polynomial features
polynomial_features = PolynomialFeatures(2)
X_higher_order = polynomial_features.fit_transform(X)
```

```
# split in train and test sets
X_train, X_test, y_train, y_test = train_test_split(X_higher_order, y,
shuffle=True)

# create the model. The parameter alpha represents the regularization
magnitude
linear_reg = Ridge(alpha=1.0)

# fit the model on the training set
linear_reg.fit(X_train, y_train)

# make predictions on the test set
predictions = linear_reg.predict(X_test)

# evaluate the model performance using the root mean square error metric
print("Root mean squared error (RMSE): %.2f" % sqrt(mean_squared_error(y_
test, predictions)))

'Output':
Root mean squared error (RMSE): 3.74
```

Take note of the following:

- The method Ridge(alpha=1.0) initializes a linear regression
  model with regularization, where the attribute 'alpha' controls the
  magnitude of the regularization parameter.

Logistic Regression with Regularization

This code block here is also similar to the example in Chapter 20 on logistic regression. The model will predict the three species of flowers from the Iris dataset. The addition to this code segment is the inclusion of a regularization term to the logistic model using the 'RidgeClassifier' package.

```
# import packages
from sklearn.linear_model import RidgeClassifier
from sklearn import datasets
from sklearn.model_selection import train_test_split
from sklearn.metrics import accuracy_score
```

```
# load dataset
data = datasets.load_iris()

# separate features and target
X = data.data
y = data.target

# split in train and test sets
X_train, X_test, y_train, y_test = train_test_split(X, y, shuffle=True)

# create the logistic regression model
logistic_reg = RidgeClassifier()

# fit the model on the training set
logistic_reg.fit(X_train, y_train)

# make predictions on the test set
predictions = logistic_reg.predict(X_test)

# evaluate the model performance using accuracy metric
print("Accuracy: %.2f" % accuracy_score(y_test, predictions))

'Output':
Accuracy: 0.76
```

In the preceding code block, logistic regression with regularization is implemented by the method 'RidgeClassifier()'. The reduced accuracy observed in this example when regularization is applied to logistic regression is because the algorithm is restricting the values of the model parameters to prevent high variance on a dataset that is fairly simplistic and already has high accuracy on test samples without regularization.

This chapter discusses the role of regularization in linear models like linear and logistic regression. Other forms of regularization exist for other model types such as early stopping for neural networks (to be discussed later in Chapter 34). Regularization is an important technique when designing machine learning models. The next chapter will discuss and implement another important machine learning algorithm known as support vector machines.

CHAPTER 22

Support Vector Machines

Support vector machine (SVM) is a machine learning algorithm for learning classification and regression models. To build intuition, we will consider the case of learning a classification model with SVM. Given a dataset with two target classes that are linearly separable, it turns out that there exists an infinite number of lines that can discriminate between the two classes (see Figure 22-1). The goal of the SVM is to find the best line that separates the two classes. In higher dimensions, this line is called a hyperplane.

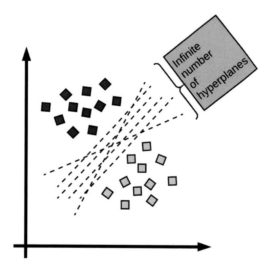

Figure 22-1. *Infinite set of discriminants*

What Is a Hyperplane?

A hyperplane is a line or more technically called a discriminant that separates two classes in n-dimensional space. When a hyperplane is drawn in 2-D space, it is called a line. In 3-D space, it is called a plane, and in dimensions greater than 3, the discriminant is called a hyperplane (see Figure 22-2). For any n-dimensional world, we have n-1 hyperplanes.

© Ekaba Bisong 2019

E. Bisong, *Building Machine Learning and Deep Learning Models on Google Cloud Platform*, https://doi.org/10.1007/978-1-4842-4470-8_22

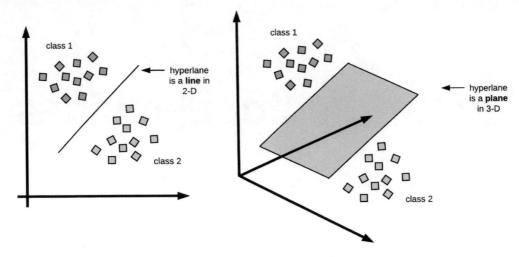

Figure 22-2. *Left: A hyperplane in 2-D is a line. Right: A hyperplane in 3-D is a plane. For dimension greater than 3, visualization becomes difficult.*

Finding the Optimal Hyperplane

The best hyperplane that linearly separates two classes is identified as the line lying at the largest margin from the nearest vectors at the boundary of the two classes.

In Figure 22-3, we observe that the best hyperplane is the line at the exact center of the two classes and constitutes the largest margin between both classes. Hence, this optimal hyperplane is also known as the largest margin classifier.

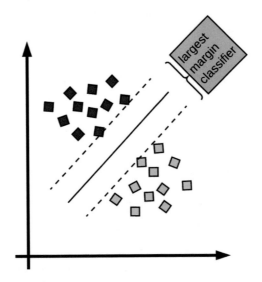

Figure 22-3. *The largest margin classifier*

The boundary points of the respective classes which are known as the support vectors are essential in finding the optimal hyperplane. The support vectors are illustrated in Figure 22-4. The boundary points are called support vectors because they are used to determine the maximum distance between the class they belong to and the discriminant function separating the classes.

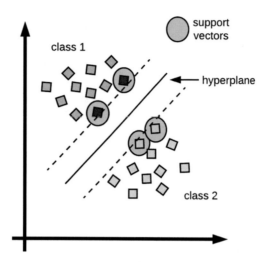

Figure 22-4. *Support vectors*

The mathematical formulation for finding the margin and consequently the hyperplane that maximizes the margin is beyond the scope of this book, but suffice to say this technique involves the Lagrange multiplier.

The Support Vector Classifier

In the real world, it is difficult to find data points that are precisely linearly separable and for which exists a large margin hyperplane. In Figure 22-5, the left image represents the data points for two classes in a dataset. Observe that there readily exists a linear separator between those two classes. Now, suppose we have an additional point from class 1 adjusted in such a way that it is much closer to class 2, we see that this point upsets the location of the hyperplane as seen in the right image of Figure 22-5. This reveals the sensitivity of the hyperplane to an additional data point that may result in a very narrow margin.

This sensitivity to data samples has significant drawbacks, the first being that the distance between the support vectors and the hyperplane reflects the confidence in the classification accuracy. Also, the drastic change in the position of the hyperplane due to a single additional point shows that the classifier is susceptible to high variability and can overfit the training data.

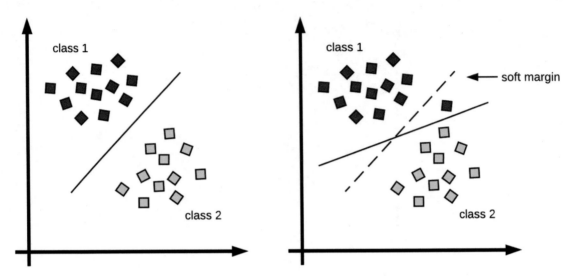

Figure 22-5. *Left: A linearly separable data distribution with a large margin. Right: The data point distribution makes it more difficult to find a large margin classifier that linearly separates the two classes*

The goal of the support vector classifier is to find a hyperplane that nearly discriminates between the two classes. This technique is also called a soft margin. A soft margin is tuned to ignore a degree of error when finding the separating hyperplane. This concept of a soft margin is how we generalize the support vector classifier to find a hyperplane in datasets that are not readily linearly separable. The margin is called soft because some examples are purposefully misclassified.

In such cases, as outlined in Figure 22-5, a soft margin classifier is preferred as it is more insensitive to individual data points and overall will have a better chance of generalizing to new examples. Howbeit, this might misclassify a couple of examples while training, but this is overall beneficial to the quality of the classifier as it generalizes to new samples.

Again, the margin is called soft because some examples are allowed to violate the margin or even be misclassified by the hyperplane to preserve overall generalizability. This is illustrated in Figure 22-6.

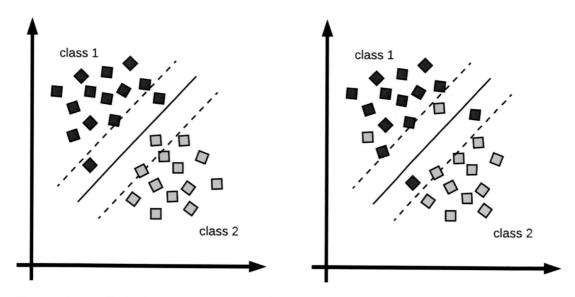

Figure 22-6. *Left: An example of a soft margin with points allowed to violate the margin. Right: An example with some points intentionally misclassified.*

The C Parameter

The C parameter is the hyper-parameter that is responsible for controlling the degree of violations to the margins or the number of intentionally misclassified points allowed by the support vector classifier. The C hyper-parameter is a non-negative real number. When this C parameter is set to 0, the classifier becomes the large margin classifier.

In a soft margin classifier, the C parameter is tuned by adjusting its values to control the tolerance of the margin. With larger values of C, the classifier margins become wider and more tolerant to violations and misclassifications. However, with smaller values of C, the margins become narrower and are less tolerant of violations and misclassified points.

Observe that the C hyper-parameter is vital for regulating the bias/variance trade-off of the support vector classifier. The higher the value of C, our classifier is more prone to variability in the data points and can under-simplify the learning problem. Also, if C is set closer to zero, it results in a much narrower margin, and this can overfit the classifier, leading to high variance – and this will likely fail to generalize to new examples (see Figure 22-7).

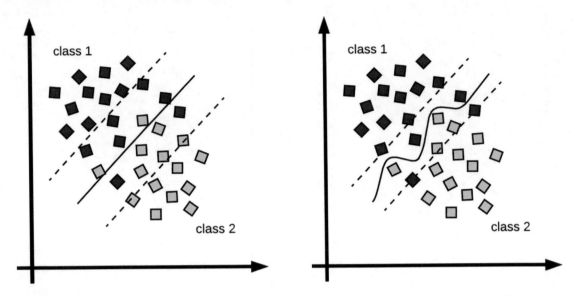

Figure 22-7. *Left: Higher values of C result in wider margins with more tolerance. Right: Lower values of C result in narrower margins with less tolerance*

Multi-class Classification

Previously, we have used the SVC to build a discriminant classifier for binary classes. What happens when we have more than two classes of outputs in the dataset, which is often the case in practice? The SVM can be extended for classifying k classes within a dataset, where k > 2. This extension is, however, not trivial with the SVM. There exist two standard approaches for addressing this problem. The first is the one-vs.-one (OVO) multi-class classification, while the other is the one-vs.-all (OVA) or one-vs.rest (OVR) multi-class classification technique.

One-vs.-One (OVO)

In the one-vs.-one approach, when the number of classes, k, is greater than 2, the algorithm constructs "k combination 2", $\binom{k}{2}$ classifiers, where each classifier is for a pair of classes. So if we have 10 classes in our dataset, a total of 45 classifiers is constructed or trained for every pair of classes. This is illustrated with four classes in Figure 22-8.

After training, the classifiers are evaluated by comparing examples from the test set against each of the $\binom{k}{2}$ classifiers. The predicted class is then determined by choosing the highest number of times an example is assigned to a particular class.

The one-vs.-one multi-class technique can potentially lead to a large number of constructed classifiers and hence can result in slower processing time. Conversely, the classifiers are more robust to class imbalances when training each classifier.

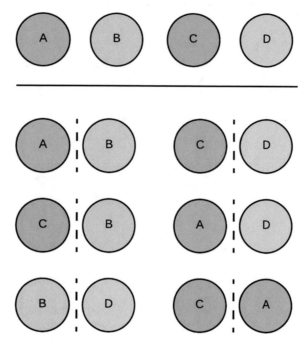

Figure 22-8. *Suppose we have four classes in the dataset labeled A to D, this will result in six different classifiers*

One-vs.-All (OVA)

The one-vs.-all method for fitting an SVM to a multi-classification problem where the number of classes k is greater than 2 consists of fitting each k class against the remaining k – 1 classes. Suppose we have ten classes, each of the classes will be classified against the remaining nine classes. This example is illustrated with four classes in Figure 22-9.

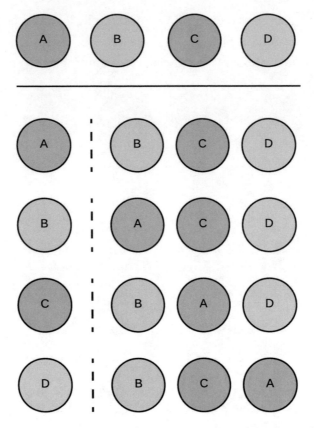

Figure 22-9. *Given four classes in a dataset, we construct four classifiers, with each class fitted against the rest*

The classifiers are evaluated by comparing a test example to each fitted classifier. The classifier for which the margin of the hyperplane is the largest is chosen as the predicted classification target because the classifier margin size is indicative of high confidence of class membership.

The Kernel Trick: Fitting Non-linear Decision Boundaries

Non-linear datasets occur more often than not in real world scenarios.

Technically speaking, the name support vector machine is when a support vector classifier is used with a non-linear kernel to learn non-linear decision boundaries.

SVM uses an essential technique for extending the feature space of a dataset to construct a non-linear classifier. This technique is called kernel and is popularly known as the kernel trick. Figure 22-10 illustrates the kernel trick as an extra dimension is added to the feature space.

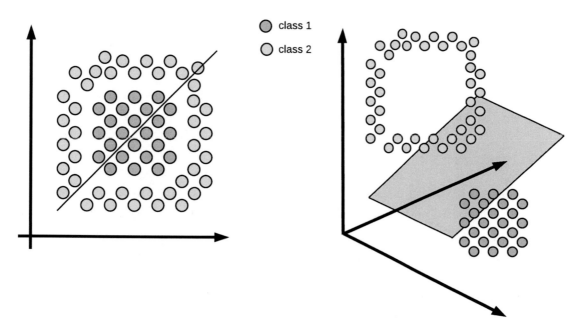

Figure 22-10. *Left: Linear discriminant to non-linear data. Right: By using the kernel trick, we can linearly separate a non-linear dataset by adding an extra dimension to the feature space.*

Adding Polynomial Features

The feature space of the dataset can be extended by adding higher-order polynomial terms or interaction terms. For example, instead of training the classifier with linear features, we can add polynomial features or add interaction terms to our model.

Depending on the dimensions of the dataset, the combinations for extending the feature space can quickly become unmanageable, and this can easily lead to a model that overfits the test set and also become expensive to compute with a larger feature space.

Kernels

Kernel is a mathematical procedure for extending the feature space of a dataset to learn non-linear decision boundaries between different classes. The mathematical details of kernels are beyond the scope of this text. Suffice to say that a kernel can be seen as a mathematical function that captures similarity between data samples.

Linear Kernel

The support vector classifier is the same as a linear kernel. It is also known as a linear kernel because the feature space of the support vector classifier is linear.

Polynomial Kernel

The kernel can also be expressed as a polynomial. With this, a support vector classifier is trained on higher-dimensional polynomial features without manually adding an exponential number of polynomial features to the dataset. Adding a polynomial kernel to the support vector classifier enables the classifier to learn a non-linear decision boundary.

Radial Basis Function or the Radial Kernel

The radial basis function or radial kernel is another non-linear kernel that enables the support vector classifier to learn a non-linear decision boundary. The radial kernel is similar to adding multiple similarity features to the space. For the radial basis function, a hyper-parameter called gamma, γ, is used to control the flexibility of the non-linear decision boundary. The smaller the gamma value, the less complex (or flexible) the non-linear discriminant becomes, but a larger value for gamma leads to a more flexible and sophisticated decision boundary that tightly fits the non-linearity in the data, which can inadvertently lead to overfitting. This is illustrated in Figure 22-11. RBF is a popular kernel option used in practice.

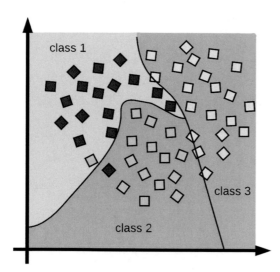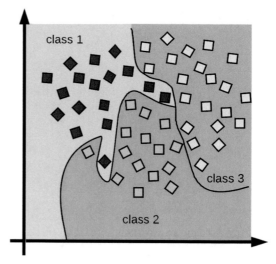

Figure 22-11. *An illustration of adjusting the radial basis function γ parameter, together with the C parameter of the support vector classifier to fit a non-linear decision boundary. Left: RBF kernel with C = 1 and γ = 10⁻³. Right: RBF kernel with C = 1 and γ = 10⁻⁵.*

When using the radial kernel with the support vector classifier, the values of C and gamma are hyper-parameters that are tuned to find an appropriate level of model flexibility that generalizes to new examples when deployed.

In practice, a linear kernel or support vector classifier sometimes surprisingly performs well when used to map a function to non-linear data. This observation follows Occam's razor which suggests that it is advantageous to select the simplest hypothesis to solve a problem in the presence of more complex options.

Also, with regard to choosing the best set of C and gamma, γ, to avoid overfitting, a grid search is used to explore a range of values for the hyper-parameters and come up with the combination that performs best on test data. The grid search is used in conjunction with cross-validation approaches. However, the grid search procedure can be potentially computationally expensive.

Support vector machines perform well with high-dimensional data. However, they are preferred for small or medium-sized datasets. For humongous datasets, SVMs become computationally infeasible. Another limitation is that the performance of SVMs is known to plateau at some point, even when there exist large training samples. This is one of the motivations and advantages of deep neural networks.

Support Vector Machines with Scikit-learn

In Scikit-learn, **SVC** is the SVM package for classification, while **SVR** is the SVM package for regression. The attribute 'gamma' in both the SVC and SVR methods controls the flexibility of the decision boundary, and the default kernel is the radial basis function (rbf).

SVM for Classification

In this code example, we will build an SVM classification model to predict the three species of flowers from the Iris dataset.

```
# import packages
from sklearn.svm import SVC
from sklearn import datasets
from sklearn.model_selection import train_test_split
from sklearn.metrics import accuracy_score
from math import sqrt

# load dataset
data = datasets.load_iris()

# separate features and target
X = data.data
y = data.target

# split in train and test sets
X_train, X_test, y_train, y_test = train_test_split(X, y, shuffle=True)

# create the model
svc_model = SVC(gamma='scale')

# fit the model on the training set
svc_model.fit(X_train, y_train)

# make predictions on the test set
predictions = svc_model.predict(X_test)
```

```
# evaluate the model performance using accuracy metric
print("Accuracy: %.2f" % accuracy_score(y_test, predictions))

'Output':
Accuracy: 0.95
```

SVM for Regression

In this code example, we will build an SVM regression model to predict house prices from the Boston house-prices dataset.

```
# import packages
from sklearn.svm import SVR
from sklearn import datasets
from sklearn.model_selection import train_test_split
from sklearn.metrics import mean_squared_error

# load dataset
data = datasets.load_boston()

# separate features and target
X = data.data
y = data.target

# split in train and test sets
X_train, X_test, y_train, y_test = train_test_split(X, y, shuffle=True)

# create the model
svr_model = SVR(gamma='scale')

# fit the model on the training set
svr_model.fit(X_train, y_train)

# make predictions on the test set
predictions = svr_model.predict(X_test)

# evaluate the model performance using the root mean squared error metric
print("Mean squared error: %.2f" % sqrt(mean_squared_error(y_test,
predictions)))
```

```
'Output':
Root mean squared error: 7.58
```

In this chapter, we surveyed the support vector machine algorithm and its implementation with Scikit-learn. In the next chapter, we will discuss on ensemble methods that combine outputs of multiple classifiers or weak learners to build better prediction models.

CHAPTER 23

Ensemble Methods

Ensemble learning is a technique that combines the output of multiple classifiers also called weak learners to build a more robust prediction model. Ensemble methods work by combining a group of classifiers (or models) to get an enhanced prediction accuracy. The idea behind an "ensemble" is that the performance from the average of a group of classifiers will be better than each classifier on its own. So each classifier is called a "weak" learner.

Ensemble learners are usually high-performing algorithms for both classification and regression tasks and are mostly competition-winning algorithms. Examples of ensemble learning algorithms are Random Forest (RF) and Stochastic Gradient Boosting (SGB). We will motivate our discussion of ensemble methods by first discussing decision trees because ensemble classifiers such as RF and SGB are built by combining several decision tree classifiers.

Decision Trees

Decision trees, more popularly known as classification and regression trees (CART), can be visualized as a graph or flowchart of decisions. A branch connects the nodes in the graph, the last node of the graph is called a terminal node, and the topmost node is called the root. As seen in Figure 23-1, when constructing a decision tree, the root is at the top, while the branches connect nodes at lower layers until the terminal node.

© Ekaba Bisong 2019

E. Bisong, *Building Machine Learning and Deep Learning Models on Google Cloud Platform*, https://doi.org/10.1007/978-1-4842-4470-8_23

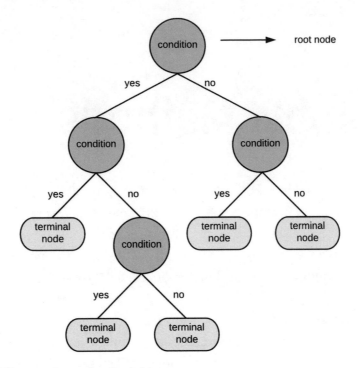

Figure 23-1. *Illustration of a decision tree*

On Regression and Classification with CART

A classification or regression tree is built by randomly splitting the set of attributes of the given dataset into distinct regions. The data points that fall within a particular region are used to form the predictor from the means of the targets in the regression case and the highest occurring class in the classification setting.

Thus, if an unseen observation or test data falls within a region, the mean or modal class is used to predict the output for regression and classification problems, respectively. In regression trees, the output variable is continuous, whereas in classification trees, the output variable is categorical. The terminal node of a regression tree takes the average of the samples in that region, while the terminal node of a classification tree is the highest occurring class in that area.

The process of splitting the features of the dataset into regions is by a greedy algorithm called recursive binary splitting. This strategy works by continuously dividing the feature space into two new branches or regions until a stopping criterion is reached.

Growing a Regression Tree

In regression trees, the recursive binary splitting technique is used to divide a particular feature in the dataset into two regions. The splitting is carried out by choosing a value of the feature that minimizes the regression error measure. This step is done for all the predictors in the dataset by finding a value that reduces the squared error of the final tree. This process is repeated continuously for every sub-tree or sub-region until a stopping criterion is reached. For example, we can stop the algorithm when no region contains less than ten observations. An example of a tree resulting from the splitting of a feature space into six regions is shown in Figure 23-2.

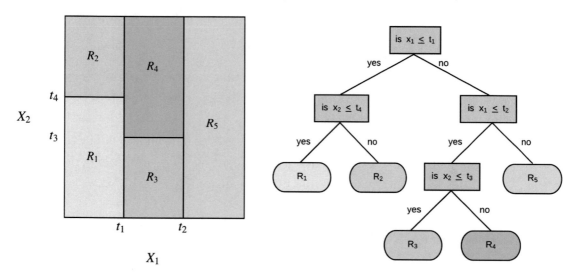

Figure 23-2. *Left: An example of splitting a 2-D dataset into sub-trees/regions using the recursive binary splitting technique. Right: The resulting tree from the partitioning on the left.*

Growing a Classification Tree

Growing a classification tree is very similar to the regression tree setting described in Figure 23-2. The difference here is that the error measure to minimize is no longer the squared error, but the misclassification error. This is because a classification tree is for predicting a qualitative response, where a data point is assigned to a particular region based on the modal value or the highest occurring class in that region.

Two algorithms for selecting which value to use for splitting the feature space in a classification setting are the Gini index and entropy; further discussions on these are beyond the scope of this chapter.

Tree Pruning

Tree pruning is a technique for dealing with model overfitting when growing trees. Fully grown trees have a high tendency to overfit with high variances when applied to unseen samples.

Pruning involves growing a large tree and then pruning or clipping it to create a sub-tree. By doing so, we can have a full picture of the tree performance and then select a sub-tree that results in a minimized error measure on the test dataset. The technique for selecting the best sub-tree is called the cost complexity pruning or the weakest link pruning.

Strengths and Weaknesses of CART

One of the significant advantages of CART models is that they perform well on linear and non-linear datasets. Moreover, CART models implicitly take care of feature selection and work well with high-dimensional datasets.

On the flip side, CART models can very easily overfit the dataset and fail to generalize to new examples. This downside is mitigated by aggregating a large number of decision trees in techniques like Random forests and boosting ensemble algorithms.

CART with Scikit-learn

In this section, we will implement a classification and regression decision tree classifier with Scikit-learn.

Classification Tree with Scikit-learn

In this code example, we will build a classification decision tree classifier to predict the species of flowers from the Iris dataset.

```
# import packages
from sklearn.tree import DecisionTreeClassifier
from sklearn import datasets
from sklearn.model_selection import train_test_split
from sklearn.metrics import accuracy_score
```

```
# load dataset
data = datasets.load_iris()

# separate features and target
X = data.data
y = data.target

# split in train and test sets
X_train, X_test, y_train, y_test = train_test_split(X, y, shuffle=True)

# create the model
tree_classifier = DecisionTreeClassifier()

# fit the model on the training set
tree_classifier.fit(X_train, y_train)

# make predictions on the test set
predictions = tree_classifier.predict(X_test)

# evaluate the model performance using accuracy metric
print("Accuracy: %.2f" % accuracy_score(y_test, predictions))

'Output':
Accuracy: 0.97
```

Regression Tree with Scikit-learn

In this code example, we will build a regression decision tree classifier to predict house prices from the Boston house-prices dataset.

```
# import packages
from sklearn.tree import DecisionTreeRegressor
from sklearn import datasets
from sklearn.model_selection import train_test_split
from sklearn.metrics import mean_squared_error
from math import sqrt
```

```python
# load dataset
data = datasets.load_boston()

# separate features and target
X = data.data
y = data.target

# split in train and test sets
X_train, X_test, y_train, y_test = train_test_split(X, y, shuffle=True)

# create the model
tree_reg = DecisionTreeRegressor()

# fit the model on the training set
tree_reg.fit(X_train, y_train)

# make predictions on the test set
predictions = tree_reg.predict(X_test)

# evaluate the model performance using the root mean square error metric
print("Root mean squared error: %.2f" % sqrt(mean_squared_error(y_test,
predictions)))

'Output':
Root mean squared error: 4.93
```

Random Forests

Random forest is a robust machine learning algorithm and is often the algorithm of choice for many classification and regression problems. It is a popular algorithm in machine learning competitions.

Random forest builds an ensemble classifier from a combination of several decision tree classifiers. This does an excellent job of reducing the variance that may be found in a single decision tree classifier.

Random forest is an improvement on the bagging ensemble algorithm (also known as bootstrap aggregation) which involves creating a large number of fully grown decision trees by repeatedly selecting random samples from the training dataset (also called bootstrapping). The result of these trees is then averaged to smoothen out the variance.

Random forest improves this bagging procedure by using only a subset of the features or attributes in the training dataset on each tree split. In doing this, Random forest creates trees whose average is more robust and less prone to high variances.

Observe that the principal distinction between bagging and Random forests is the choice of features when splitting the feature space or when building the tree. Bagging makes use of the entire features in the dataset, whereas Random forest imposes a constraint on the number of features and uses only a subset of features on each tree split to reduce the correlation of each sub-tree. Empirically, the size of features for each tree split using Random forests is the square root of the original number of predictors.

Making Predictions with Random Forests

In order to make a prediction using Random forest, the test example is passed through each trained decision tree. For the regression case, a prediction is made for a new example by taking the average of the outputs of the different trees. In the case of classification problems, the prediction is the class with the most votes from all other trees in the forest. This is best illustrated in Figure 23-3.

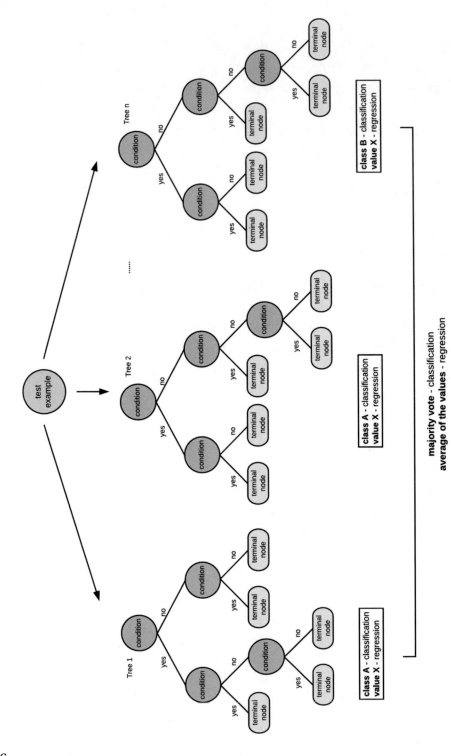

Figure 23-3. *Take a majority vote to determine the final class in the classification case and the average of the values in each tree to determine the predicted value in the regression case*

Random Forests with Scikit-learn

This section will implement Random forests with Scikit-learn for both regression and classification use cases.

Random Forests for Classification

In this code example, we will build a Random forest classification model to predict the species of flowers from the Iris dataset.

```
# import packages
from sklearn.ensemble import RandomForestClassifier
from sklearn import datasets
from sklearn.model_selection import train_test_split
from sklearn.metrics import accuracy_score

# load dataset
data = datasets.load_iris()

# separate features and target
X = data.data
y = data.target

# split in train and test sets
X_train, X_test, y_train, y_test = train_test_split(X, y, shuffle=True)

# create the model
rf_classifier = RandomForestClassifier()

# fit the model on the training set
rf_classifier.fit(X_train, y_train)
```

```
# make predictions on the test set
predictions = rf_classifier.predict(X_test)

# evaluate the model performance using accuracy metric
print("Accuracy: %.2f" % accuracy_score(y_test, predictions))

'Output":
Accuracy: 1.00
```

Random Forests for Regression

In this code example, we will build a Random forest regression model to predict house prices from the Boston house-prices dataset.

```
# import packages
from sklearn.ensemble import RandomForestRegressor
from sklearn import datasets
from sklearn.model_selection import train_test_split
from sklearn.metrics import mean_squared_error
from math import sqrt

# load dataset
data = datasets.load_boston()

# separate features and target
X = data.data
y = data.target

# split in train and test sets
X_train, X_test, y_train, y_test = train_test_split(X, y, shuffle=True)
```

```
# create the model
rf_reg = RandomForestRegressor()

# fit the model on the training set
rf_reg.fit(X_train, y_train)

# make predictions on the test set
predictions = rf_reg.predict(X_test)

# evaluate the model performance using the root mean square error metric
print("Root mean squared error: %.2f" % sqrt(mean_squared_error(y_test,
predictions)))

'Output':
Root mean squared error: 2.96
```

Stochastic Gradient Boosting (SGB)

Boosting involves growing trees in succession using knowledge from the residuals of the previously grown tree. In this case, each successive tree works to improve the model of the previous tree by boosting the areas in which the previous tree did not perform so well without affecting the areas of high performance. By doing this, we iteratively create a model that reduces the residual variance when generalizing to test examples. Boosting is illustrated in Figure 23-4.

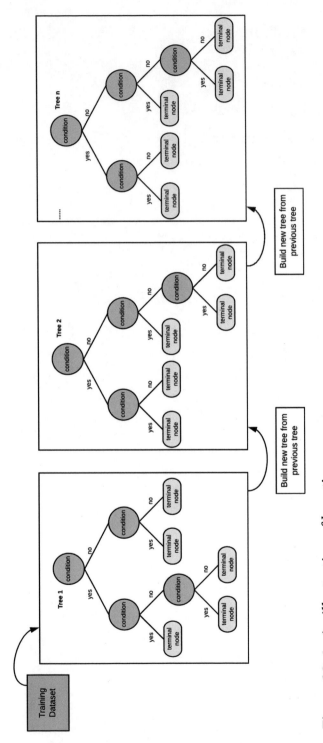

Figure 23-4. An illustration of boosting

Gradient boosting evaluates the difference of the residuals for each tree and then uses that information to determine how to split the feature space in the successive tree.

Gradient boosting employs a pseudo-gradient in computing the residuals. This gradient is the direction of quickest improvement to the loss function. The residual variance is minimized as the gradient moves in the direction of steepest descent. This movement is the same as the stochastic gradient descent algorithm discussed in Chapter 16.

Tree Depth/Number of Trees

Gradient boosting can be controlled by choosing the tree depth as a hyper-parameter to the model. In practice, a tree depth of 1 performs well, as each tree consists of just a single split. Also, the number of trees can affect the model accuracy, because gradient boosting can overfit if the number of successive trees is vast.

Shrinkage

The shrinkage hyper-parameter λ controls the learning rate of the gradient boosting model. An arbitrarily small value of λ may necessitate a larger number of trees to obtain a good model performance. However, with a small shrinkage size and tree depth $d = 1$, the residuals slowly improve by creating more varied trees to improve the worst performing areas of the model. Rule of thumb: shrinkage size is 0.01 or 0.001.

Stochastic Gradient Boosting with Scikit-learn

This section will implement SGB with Scikit-learn for both regression and classification use cases.

SGB for Classification

In this code example, we will build a SGB classification model to predict the species of flowers from the Iris dataset.

```
# import packages
from sklearn.ensemble import GradientBoostingClassifier
from sklearn import datasets
from sklearn.model_selection import train_test_split
from sklearn.metrics import accuracy_score

# load dataset
data = datasets.load_iris()

# separate features and target
X = data.data
y = data.target

# split in train and test sets
X_train, X_test, y_train, y_test = train_test_split(X, y, shuffle=True)

# create the model
sgb_classifier = GradientBoostingClassifier()

# fit the model on the training set
sgb_classifier.fit(X_train, y_train)

# make predictions on the test set
predictions = sgb_classifier.predict(X_test)

# evaluate the model performance using accuracy metric
print("Accuracy: %.2f" % accuracy_score(y_test, predictions))

'Output':
Accuracy: 0.92
```

SGB for Regression

In this code example, we will build a SGB regression model to predict house prices from the Boston house-prices dataset.

```
# import packages
from sklearn.ensemble import GradientBoostingRegressor
from sklearn import datasets
from sklearn.model_selection import train_test_split
from sklearn.metrics import mean_squared_error
from math import sqrt

# load dataset
data = datasets.load_boston()

# separate features and target
X = data.data
y = data.target

# split in train and test sets
X_train, X_test, y_train, y_test = train_test_split(X, y, shuffle=True)

# create the model
sgb_reg = GradientBoostingRegressor ()

# fit the model on the training set
sgb_reg.fit(X_train, y_train)

# make predictions on the test set
predictions = sgb_reg.predict(X_test)

# evaluate the model performance using the root mean square error metric
print("Root mean squared error: %.2f" % sqrt(mean_squared_error(y_test,
predictions)))

'Output':
Root mean squared error: 2.86
```

XGBoost (Extreme Gradient Boosting)

XGBoost which is short for Extreme Gradient Boosting makes a couple of computational and algorithmic modifications to the stochastic gradient boosting algorithm. This enhanced algorithm is a favorite in machine learning practice due to its speed and has been the winning algorithm in many machine learning competitions. Let's go through some of the modifications made by the XGBoost algorithm.

1. Parallel training: XGBoost supports parallel training over multiple cores. This has made XGBoost extremely fast compared to other machine learning algorithms.

2. Out of core computation: XGBoost facilitates training from data not loaded into memory. This feature is a huge advantage when you're dealing with large datasets that may not necessarily fit into the RAM of the computer.

3. Sparse data optimization: XGBoost is optimized to handle and speed up computation with sparse matrices. Sparse matrices contain lots of zeros in its cells.

XGBoost with Scikit-learn

This section will implement XGBoost with Scikit-learn for both regression and classification use cases.

XGBoost for Classification

In this code example, we will build a XGBoost classification model to predict the species of flowers from the Iris dataset.

```
# import packages
from xgboost import XGBClassifier
from sklearn import datasets
from sklearn.model_selection import train_test_split
from sklearn.metrics import accuracy_score

# load dataset
data = datasets.load_iris()
```

```
# separate features and target
X = data.data
y = data.target

# split in train and test sets
X_train, X_test, y_train, y_test = train_test_split(X, y, shuffle=True)

# create the model
xgboost_classifier = XGBClassifier()

# fit the model on the training set
xgboost_classifier.fit(X_train, y_train)

# make predictions on the test set
predictions = xgboost_classifier.predict(X_test)

# evaluate the model performance using accuracy metric
print("Accuracy: %.2f" % accuracy_score(y_test, predictions))

'Output":
Accuracy: 0.95
```

XGBoost for Regression

In this code example, we will build a XGBoost regression model to predict house prices from the Boston house-prices dataset.

```
# import packages
from xgboost import XGBRegressor
from sklearn import datasets
from sklearn.model_selection import train_test_split
from sklearn.metrics import mean_squared_error
from math import sqrt

# load dataset
data = datasets.load_boston()

# separate features and target
X = data.data
y = data.target
```

```
# split in train and test sets
X_train, X_test, y_train, y_test = train_test_split(X, y, shuffle=True)

# create the model
xgboost_reg = XGBRegressor()

# fit the model on the training set
xgboost_reg.fit(X_train, y_train)

# make predictions on the test set
predictions = xgboost_reg.predict(X_test)

# evaluate the model performance using the root mean square error metric
print("Root mean squared error: %.2f" % sqrt(mean_squared_error(y_test,
predictions)))

'Output':
Root mean squared error: 3.69
```

In this chapter, we surveyed and implemented ensemble machine learning algorithms that combine weak decision tree learners to create a strong classifier for learning regression and classification problems. In the next chapter, we will discuss more techniques for implementing supervised machine learning models with Scikit-learn.

## CHAPTER 24

More Supervised Machine Learning Techniques with Scikit-learn

This chapter will cover using Scikit-learn to implement machine learning models using techniques such as

- Feature engineering

- Resampling methods

- Model evaluation methods

- Pipelines for streamlining machine learning workflows

- Techniques for model tuning

Feature Engineering

Feature engineering is the process of systematically choosing the set of features in the dataset that are useful and relevant to the learning problem. It is often the case that irrelevant features negatively affect the performance of the model. This section will review some techniques implemented in Scikit-learn for selecting relevant features from a dataset. The techniques surveyed include

- Statistical tests to select the best k features using the **SelectKBest** module

- Recursive feature elimination (RFE) to recursively remove irrelevant features from the dataset

© Ekaba Bisong 2019

E. Bisong, *Building Machine Learning and Deep Learning Models on Google Cloud Platform*, https://doi.org/10.1007/978-1-4842-4470-8_24

- Principal component analysis to select the components that account for the variation in the dataset

- Feature importances using ensembled or tree classifiers

Statistical Tests to Select the Best *k* Features Using the SelectKBest Module

The following list is a selection of statistical tests to use with **SelectKBest**. The choice depends if the dataset target variable is numerical or categorical:

- ANOVA F-value, **f_classif** (classification)

- Chi-squared stats of non-negative features, **chi2** (classification)

- F-value, **f_regression** (regression)

- Mutual information for a continuous target, **mutual_info_regression** (regression)

Let's see an example using chi-squared test to select the best variables.

```
# import packages
from sklearn import datasets
from sklearn.feature_selection import SelectKBest
from sklearn.feature_selection import chi2

# load dataset
data = datasets.load_iris()

# separate features and target
X = data.data
y = data.target

# display first 5 rows
X[0:5,:]

# feature engineering. Let's see the best 3 features by setting k = 3
kBest_chi = SelectKBest(score_func=chi2, k=3)
fit_test = kBest_chi.fit(X, y)
```

```
# print test scores
fit_test.scores_
'Output': array([ 10.81782088,    3.59449902, 116.16984746,  67.24482759])
```

From the test scores, the top 3 important features in the dataset are ranked from feature 3 to 4 to 1 and to 2 in order. The data scientist can choose to drop the second column and observe the effect on the model performance.

We can transform the dataset to subset only the important features.

```
adjusted_features = fit_test.transform(X)
adjusted_features[0:5,:]
'Output':
array([[5.1, 1.4, 0.2],
       [4.9, 1.4, 0.2],
       [4.7, 1.3, 0.2],
       [4.6, 1.5, 0.2],
       [5. , 1.4, 0.2]])
```

The result drops the second column of the dataset.

Recursive Feature Elimination (RFE)

RFE is used together with a learning model to recursively select the desired number of top performing features.

Let's use RFE with **LinearRegression**.

```
# import packages
from sklearn.feature_selection import RFE
from sklearn.linear_model import LinearRegression
from sklearn import datasets

# load dataset
data = datasets.load_boston()

# separate features and target
X = data.data
y = data.target
```

```
# feature engineering
linear_reg = LinearRegression()
rfe = RFE(estimator=linear_reg, n_features_to_select=6)
rfe_fit = rfe.fit(X, y)

# print the feature ranking
rfe_fit.ranking_
'Output': array([3, 5, 4, 1, 1, 1, 8, 1, 2, 6, 1, 7, 1])
```

From the result, the 4th, 5th, 6th, 8th, 11th, and 13th features are the top 6 features in the Boston dataset.

Feature Importances

Tree-based or ensemble methods in Scikit-learn have a **feature_importances_** attribute which can be used to drop irrelevant features in the dataset using the **SelectFromModel** module contained in the **sklearn.feature_selection** package.

Let's used the ensemble method **AdaBoostClassifier** in this example.

```
# import packages
from sklearn.ensemble import AdaBoostClassifier
from sklearn.feature_selection import SelectFromModel
from sklearn import datasets

# load dataset
data = datasets.load_iris()

# separate features and target
X = data.data
y = data.target

# original data shape
X.shape

# feature engineering
ada_boost_classifier = AdaBoostClassifier()
ada_boost_classifier.fit(X, y)
```

```
'Output':
AdaBoostClassifier(algorithm='SAMME.R', base_estimator=None,
          learning_rate=1.0, n_estimators=50, random_state=None)

# print the feature importances
ada_boost_classifier.feature_importances_
'Output': array([0.  , 0.  , 0.58, 0.42])

# create a subset of data based on the relevant features
model = SelectFromModel(ada_boost_classifier, prefit=True)
new_data = model.transform(X)

# the irrelevant features have been removed
new_data.shape
'Output': (150, 2)
```

Resampling Methods

Resampling methods are a set of techniques that involve selecting a subset of the available dataset, training on that data subset, and using the remainder of the data to evaluate the trained model. Let's review the techniques for resampling using Scikit-learn. This section covers

- k-Fold cross-validation

- Leave-one-out cross-validation

k-Fold Cross-Validation

In k-fold cross validation, the dataset is divided into k-parts or folds. The model is trained using $k - 1$ folds and evaluated on the remaining kth fold. This process is repeated k-times so that each fold can serve as a test set. At the end of the process, k-fold averages the result and reports a mean score with a standard deviation. Scikit-learn implements K-fold CV in the module **KFold**. The module **cross_val_score** is used to evaluate the cross-validation score using the splitting strategy, which is **KFold** in this case.

Let's see an example of this using the k-nearest neighbors (kNN) classification algorithm. When initializing **KFold**, it is standard practice to shuffle the data before splitting.

```
from sklearn.model_selection import KFold
from sklearn.model_selection import cross_val_score
from sklearn.neighbors import KNeighborsClassifier

# load dataset
data = datasets.load_iris()

# separate features and target
X = data.data
y = data.target

# initialize KFold - with shuffle = True, shuffle the data before splitting
kfold = KFold(n_splits=3, shuffle=True)

# create the model
knn_clf = KNeighborsClassifier(n_neighbors=3)

# fit the model using cross validation
cv_result = cross_val_score(knn_clf, X, y, cv=kfold)

# evaluate the model performance using accuracy metric
print("Accuracy: %.3f%% (%.3f%%)" % (cv_result.mean()*100.0, cv_result.
std()*100.0))
'Output':
Accuracy: 93.333% (2.494%)
```

Leave-One-Out Cross-Validation (LOOCV)

In LOOCV just one example is assigned to the test set, and the model is trained on the remainder of the dataset. This process is repeated for all the examples in the dataset. This process is repeated until all the examples in the dataset have been used for evaluating the model.

```
from sklearn.model_selection import LeaveOneOut
from sklearn.model_selection import cross_val_score
from sklearn.neighbors import KNeighborsClassifier
```

```
# load dataset
data = datasets.load_iris()

# separate features and target
X = data.data
y = data.target

# initialize LOOCV
loocv = LeaveOneOut()

# create the model
knn_clf = KNeighborsClassifier(n_neighbors=3)

# fit the model using cross validation
cv_result = cross_val_score(knn_clf, X, y, cv=loocv)

# evaluate the model performance using accuracy metric
print("Accuracy: %.3f%% (%.3f%%)" % (cv_result.mean()*100.0, cv_result.
std()*100.0))
'Output':
Accuracy: 96.000% (19.596%)
```

Model Evaluation

This chapter has already used a couple of evaluation metrics for assessing the quality of the fitted models. In this section, we survey a couple of other metrics for regression and classification use cases and how to implement them using Scikit-learn. For each metric, we show how to use them as stand-alone implementations, as well as together with cross-validation using the **cross_val_score** method.

What we'll cover here includes

Regression evaluation metrics

- Mean squared error (MSE): The average sum of squared difference between the predicted label, ŷ, and the true label, y. A score of 0 indicates a perfect prediction without errors.

- Mean absolute error (MAE): The average absolute difference between the predicted label, ŷ, and the true label, y. A score of 0 indicates a perfect prediction without errors.

- R^2: The amount of variance or variability in the dataset explained by the model. The score of 1 means that the model perfectly captures the variability in the dataset.

Classification evaluation metrics

- Accuracy: Is the ratio of correct predictions to the total number of predictions. The bigger the accuracy, the better the model.

- Logarithmic loss (a.k.a logistic loss or cross-entropy loss): Is the probability that an observation is correctly assigned to a class label. By minimizing the log-loss, conversely, the accuracy is maximized. So with this metric, values closer to zero are good.

- Area under the ROC curve (AUC-ROC): Used in the binary classification case. Implementation is not provided, but very similar in style to the others.

- Confusion matrix: More intuitive in the binary classification case. Implementation is not provided, but very similar in style to the others.

- Classification report: It returns a text report of the main classification metrics.

Regression Evaluation Metrics

The following code is an example of regression evaluation metrics implemented stand-alone.

```
# import packages
from sklearn.linear_model import LinearRegression
from sklearn import datasets
from sklearn.model_selection import train_test_split
from sklearn.metrics import mean_squared_error
from sklearn.metrics import mean_absolute_error
from sklearn.metrics import r2_score

# load dataset
data = datasets.load_boston()
```

```
# separate features and target
X = data.data
y = data.target

# split in train and test sets
X_train, X_test, y_train, y_test = train_test_split(X, y, shuffle=True)

# create the model
# setting normalize to true normalizes the dataset before fitting the model
linear_reg = LinearRegression(normalize = True)

# fit the model on the training set
linear_reg.fit(X_train, y_train)
'Output': LinearRegression(copy_X=True, fit_intercept=True, n_jobs=1,
normalize=True)

# make predictions on the test set
predictions = linear_reg.predict(X_test)
# evaluate the model performance using mean square error metric
print("Mean squared error: %.2f" % mean_squared_error(y_test, predictions))
'Output':
Mean squared error: 14.46

# evaluate the model performance using mean absolute error metric
print("Mean absolute error: %.2f" % mean_absolute_error(y_test,
predictions))
'Output':
Mean absolute error: 3.63

# evaluate the model performance using r-squared error metric
print("R-squared score: %.2f" % r2_score(y_test, predictions))
'Output':
R-squared score: 0.69
```

The following code is an example of regression evaluation metrics implemented with cross-validation. The MSE and MAE metrics for cross-validation are implemented with the sign inverted. The simple way to interpret this is to have it in mind that the closer the values are to zero, the better the model.

```python
from sklearn.linear_model import LinearRegression
from sklearn.model_selection import KFold
from sklearn.model_selection import cross_val_score

# load dataset
data = datasets.load_boston()

# separate features and target
X = data.data
y = data.target

# initialize KFold - with shuffle = True, shuffle the data before splitting
kfold = KFold(n_splits=3, shuffle=True)

# create the model
linear_reg = LinearRegression(normalize = True)

# fit the model using cross validation - score with Mean square error (MSE)
mse_cv_result = cross_val_score(linear_reg, X, y, cv=kfold, scoring="neg_
mean_squared_error")
# print mse cross validation output
print("Negative Mean squared error: %.3f%% (%.3f%%)" % (mse_cv_result.
mean(), mse_cv_result.std()))
'Output':
Negtive Mean squared error: -24.275% (4.093%)

# fit the model using cross validation - score with Mean absolute error (MAE)
mae_cv_result = cross_val_score(linear_reg, X, y, cv=kfold, scoring="neg_
mean_absolute_error")
# print mse cross validation output
print("Negtive Mean absolute error: %.3f%% (%.3f%%)" % (mae_cv_result.
mean(), mae_cv_result.std()))
'Output':
Negtive Mean absolute error: -3.442% (4.093%)

# fit the model using cross validation - score with R-squared
r2_cv_result = cross_val_score(linear_reg, X, y, cv=kfold, scoring="r2")
# print mse cross validation output
```

```
print("R-squared score: %.3f%% (%.3f%%)" % (r2_cv_result.mean(), r2_cv_
result.std()))
'Output':
R-squared score: 0.707% (0.030%)
```

Classification Evaluation Metrics

The following code is an example of classification evaluation metrics implemented stand-alone.

```
# import packages
from sklearn.linear_model import LogisticRegression
from sklearn import datasets
from sklearn.model_selection import train_test_split
from sklearn.metrics import accuracy_score
from sklearn.metrics import log_loss
from sklearn.metrics import classification_report

# load dataset
data = datasets.load_iris()

# separate features and target
X = data.data
y = data.target

# split in train and test sets
X_train, X_test, y_train, y_test = train_test_split(X, y, shuffle=True)

# create the model
logistic_reg = LogisticRegression()

# fit the model on the training set
logistic_reg.fit(X_train, y_train)
'Output':
LogisticRegression(C=1.0, class_weight=None, dual=False, fit_intercept=True,
          intercept_scaling=1, max_iter=100, multi_class='ovr', n_jobs=1,
          penalty='l2', random_state=None, solver='liblinear', tol=0.0001,
          verbose=0, warm_start=False)
```

```python
# make predictions on the test set
predictions = logistic_reg.predict(X_test)

# evaluate the model performance using accuracy
print("Accuracy score: %.2f" % accuracy_score(y_test, predictions))
'Output':
Accuracy score: 0.89

# evaluate the model performance using log loss

### output the probabilities of assigning an observation to a class
predictions_probabilities = logistic_reg.predict_proba(X_test)
print("Log-Loss likelihood: %.2f" % log_loss(y_test, predictions_
probabilities))
'Output':
Log-Loss likelihood: 0.39

# evaluate the model performance using classification report
print("Classification report: \n", classification_report(y_test,
predictions, target_names=data.target_names))
'Output':
Classification report:
```

	precision	recall	f1-score	support
setosa	1.00	1.00	1.00	12
versicolor	0.85	0.85	0.85	13
virginica	0.85	0.85	0.85	13
avg / total	0.89	0.89	0.89	38

Let's see an example of classification evaluation metrics implemented with cross-validation. Evaluation metrics for log-loss using cross-validation is implemented with the sign inverted. The simple way to interpret this is to have it in mind that the closer the values are to zero, the better the model.

```python
from sklearn.linear_model import LogisticRegression
from sklearn.model_selection import KFold
from sklearn.model_selection import cross_val_score
```

```
# load dataset
data = datasets.load_iris()

# separate features and target
X = data.data
y = data.target

# initialize KFold - with shuffle = True, shuffle the data before splitting
kfold = KFold(n_splits=3, shuffle=True)

# create the model
logistic_reg = LogisticRegression()

# fit the model using cross validation - score with accuracy
accuracy_cv_result = cross_val_score(logistic_reg, X, y, cv=kfold,
scoring="accuracy")
# print accuracy cross validation output
print("Accuracy: %.3f%% (%.3f%%)" % (accuracy_cv_result.mean(), accuracy_
cv_result.std()))
'Output':
Accuracy: 0.953% (0.025%)

# fit the model using cross validation - score with Log-Loss
logloss_cv_result = cross_val_score(logistic_reg, X, y, cv=kfold,
scoring="neg_log_loss")
# print mse cross validation output
print("Log-Loss likelihood: %.3f%% (%.3f%%)" % (logloss_cv_result.mean(),
logloss_cv_result.std()))
'Output':
Log-Loss likelihood: -0.348% (0.027%)
```

Pipelines: Streamlining Machine Learning Workflows

The concept of pipelines in Scikit-learn is a compelling tool for chaining a bunch of operations together to form a tidy process flow of data transforms from one state to another. The operations that constitute a pipeline can be any of Scikit-learn's transformers (i.e., modules with a **fit** and **transform** method, or a **fit_transform** method) or classifiers (i.e., modules with a **fit** and **predict** method, or a **fit_predict** method). Classifiers are also called predictors.

For a typical machine learning workflow, the steps taken may involve cleaning the data, feature engineering, scaling the dataset, and then fitting a model. Pipelines can be used in this case to chain these operations together into a coherent workflow. They have the advantage of providing a convenient and consistent interface for calling at once a sequence of operations.

These transformers or predictors are collectively called estimators in Scikit-learn terminology. In the last two paragraphs, we called them operations.

Another advantage of pipelines is that it safeguards against accidentally fitting a transform on the entire dataset and thereby leaking statistics influenced by the test data to the machine learning model while training. For example, if a standardizer is fitted on the whole dataset, the test set will be compromised because the test observations have contributed in estimating the mean and standard deviation for scaling the training set before fitting the model.

Finally, only the last step of the pipeline can be a classifier or predictor. All the stages of the pipeline must contain a **transform** method except the final stage, which can be a transformer or a classifier.

To begin using Scikit-learn pipelines, first import

```
from sklearn.pipeline import Pipeline
```

Let's see some examples of working with Pipelines in Scikit-learn. In the following example, we'll apply a scaling transform to standardize our dataset and then use a support vector classifier to train the model.

```
# import packages
from sklearn.svm import SVC
from sklearn import datasets
from sklearn.model_selection import KFold
from sklearn.model_selection import cross_val_score
from sklearn.preprocessing import StandardScaler

from sklearn.pipeline import Pipeline

# load dataset
data = datasets.load_iris()
```

```
# separate features and target
X = data.data
y = data.target

# create the pipeline
estimators = [
    ('standardize' , StandardScaler()),
    ('svc', SVC())
]

# build the pipeline model
pipe = Pipeline(estimators)

# run the pipeline
kfold = KFold(n_splits=3, shuffle=True)
cv_result = cross_val_score(pipe, X, y, cv=kfold)

# evaluate the model performance
print("Accuracy: %.3f%% (%.3f%%)" % (cv_result.mean()*100.0, cv_result.
std()*100.0))
'Output':
Accuracy: 94.667% (0.943%)
```

Pipelines Using make_pipeline

Another method for building machine learning pipelines is by using the **make_pipeline**
method. For the next example, we use PCA to select the best six features and reduce the
dimensionality of the dataset, and then we'll fit the model using Random forests for
regression.

```
from sklearn.pipeline import make_pipeline
from sklearn.svm import SVR
from sklearn import datasets
from sklearn.model_selection import KFold
from sklearn.model_selection import cross_val_score
from sklearn.decomposition import PCA
from sklearn.pipeline import Pipeline
from sklearn.ensemble import RandomForestRegressor
```

```
# load dataset
data = datasets.load_boston()

# separate features and target
X = data.data
y = data.target

# build the pipeline model
pipe = make_pipeline(
    PCA(n_components=9),
    RandomForestRegressor()
)

# run the pipeline
kfold = KFold(n_splits=4, shuffle=True)
cv_result = cross_val_score(pipe, X, y, cv=kfold)

# evaluate the model performance
print("Accuracy: %.3f%% (%.3f%%)" % (cv_result.mean()*100.0, cv_result.
std()*100.0))
'Output':
Accuracy: 73.750% (2.489%)
```

Pipelines Using FeatureUnion

Scikit-learn provides a module for merging the output of several transformers called
feature_union. It does this by fitting each transformer independently to the dataset, and
then their respective outputs are combined to form a transformed dataset for training
the model.

FeatureUnion works in the same way as a Pipeline, and in many ways can be thought
of as a means of building complex pipelines within a Pipeline.

Let's see an example using FeatureUnion. Here, we will combine the output of
recursive feature elimination (RFE) and PCA for feature engineering, and then we'll apply
the Stochastic Gradient Boosting (SGB) ensemble model for regression to train the model.

```
from sklearn.ensemble import GradientBoostingRegressor
from sklearn.ensemble import RandomForestRegressor
from sklearn import datasets
```

```python
from sklearn.model_selection import KFold
from sklearn.model_selection import cross_val_score
from sklearn.feature_selection import RFE
from sklearn.decomposition import PCA
from sklearn.pipeline import Pipeline
from sklearn.pipeline import make_pipeline
from sklearn.pipeline import make_union

# load dataset
data = datasets.load_boston()

# separate features and target
X = data.data
y = data.target

# construct pipeline for feature engineering - make_union similar to make_
pipeline
feature_engr = make_union(
    RFE(estimator=RandomForestRegressor(n_estimators=100), n_features_to_
                select=6),
    PCA(n_components=9)
)

# build the pipeline model
pipe = make_pipeline(
    feature_engr,
    GradientBoostingRegressor(n_estimators=100)
)

# run the pipeline
kfold = KFold(n_splits=4, shuffle=True)
cv_result = cross_val_score(pipe, X, y, cv=kfold)

# evaluate the model performance
print("Accuracy: %.3f%% (%.3f%%)" % (cv_result.mean()*100.0, cv_result.
std()*100.0))
'Output':
Accuracy: 88.956% (1.493%)
```

Model Tuning

Each machine learning model has a set of options or configurations that can be tuned to optimize the model when fitting to data. These configurations are called **hyper-parameters**. Hence, for each hyper-parameter, there exist a range of values that can be chosen. Taking into consideration the number of hyper-parameters that an algorithm has, the entire space can become exponentially large and infeasible to explore all of them. Scikit-learn provides two convenient modules for searching through the hyper-parameter space of an algorithm to find the best values for each hyper-parameter that optimizes the model.

These modules are the

- Grid search

- Randomized search

Grid Search

Grid search comprehensively explores all the specified hyper-parameter values for an estimator. It is implemented using the **GridSearchCV** module. Let's see an example using the Random forest for regression. The hyper-parameters we'll search over are

- The number of trees in the forest, **n_estimators**

- The maximum depth of the tree, **max_depth**

- The minimum number of samples required to split an internal node, **min_samples_leaf**

```
from sklearn.model_selection import GridSearchCV
from sklearn.ensemble import RandomForestRegressor
from sklearn import datasets

# load dataset
data = datasets.load_boston()

# separate features and target
X = data.data
y = data.target
```

```python
# construct grid search parameters in a dictionary
parameters = {
    'n_estimators': [2, 4, 6, 8, 10, 12, 14, 16],
    'max_depth': [2, 4, 6, 8],
    'min_samples_leaf': [1,2,3,4,5]
    }

# create the model
rf_model = RandomForestRegressor()

# run the grid search
grid_search = GridSearchCV(estimator=rf_model, param_grid=parameters)

# fit the model
grid_search.fit(X,y)
'Output':
GridSearchCV(cv=None, error_score='raise',
        estimator=RandomForestRegressor(bootstrap=True, criterion='mse',
        max_depth=None,
            max_features='auto', max_leaf_nodes=None,
            min_impurity_decrease=0.0, min_impurity_split=None,
            min_samples_leaf=1, min_samples_split=2,
            min_weight_fraction_leaf=0.0, n_estimators=10, n_jobs=1,
            oob_score=False, random_state=None, verbose=0, warm_
            start=False),
        fit_params=None, iid=True, n_jobs=1,
        param_grid={'n_estimators': [2, 4, 6, 8, 10, 12, 14, 16],
        'max_depth': [2, 4, 6, 8], 'min_samples_leaf': [1, 2, 3, 4, 5]},
        pre_dispatch='2*n_jobs', refit=True, return_train_score='warn',
        scoring=None, verbose=0)

# evaluate the model performance
print("Best Accuracy: %.3f%%" %  (grid_search.best_score_*100.0))
'Output':
Best Accuracy: 57.917%
```

```
# best set of hyper-parameter values
print("Best n_estimators: %d \nBest max_depth: %d \nBest min_samples_leaf:
%d " % \
            (grid_search.best_estimator_.n_estimators, \
            grid_search.best_estimator_.max_depth, \
            grid_search.best_estimator_.min_samples_leaf))
'Output':
Best n_estimators: 14
Best max_depth: 8
Best min_samples_leaf: 1
```

Randomized Search

As opposed to grid search, not all the provided hyper-parameter values are evaluated, but rather a determined number of hyper-parameter values are sampled from a random uniform distribution. The number of hyper-parameter values that can be evaluated is determined by the **n_iter** attribute of the **RandomizedSearchCV** module.

In this example, we will use the same scenario as in the grid search case.

```
from sklearn.model_selection import RandomizedSearchCV
from sklearn.ensemble import RandomForestRegressor
from sklearn import datasets

# load dataset
data = datasets.load_boston()

# separate features and target
X = data.data
y = data.target

# construct grid search parameters in a dictionary
parameters = {
    'n_estimators': [2, 4, 6, 8, 10, 12, 14, 16],
    'max_depth': [2, 4, 6, 8],
    'min_samples_leaf': [1,2,3,4,5]
    }
```

```
# create the model
rf_model = RandomForestRegressor()

# run the grid search
randomized_search = RandomizedSearchCV(estimator=rf_model, param_
distributions=parameters, n_iter=10)
# fit the model
randomized_search.fit(X,y)
'Output':
RandomizedSearchCV(cv=None, error_score='raise',
          estimator=RandomForestRegressor(bootstrap=True, criterion='mse',
          max_depth=None,
           max_features='auto', max_leaf_nodes=None,
           min_impurity_decrease=0.0, min_impurity_split=None,
           min_samples_leaf=1, min_samples_split=2,
           min_weight_fraction_leaf=0.0, n_estimators=10, n_jobs=1,
           oob_score=False, random_state=None, verbose=0, warm_
           start=False),
          fit_params=None, iid=True, n_iter=10, n_jobs=1,
          param_distributions={'n_estimators': [2, 4, 6, 8, 10, 12, 14, 16],
          'max_depth': [2, 4, 6, 8], 'min_samples_leaf': [1, 2, 3, 4, 5]},
          pre_dispatch='2*n_jobs', random_state=None, refit=True,
          return_train_score='warn', scoring=None, verbose=0)

# evaluate the model performance
print("Best Accuracy: %.3f%%" % (randomized_search.best_score_*100.0))
'Output':
Best Accuracy: 57.856%

# best set of hyper-parameter values
print("Best n_estimators: %d \nBest max_depth: %d \nBest min_samples_leaf:
%d " % \
            (randomized_search.best_estimator_.n_estimators, \
            randomized_search.best_estimator_.max_depth, \
            randomized_search.best_estimator_.min_samples_leaf))
```

```
'Output':
Best n_estimators: 12
Best max_depth: 6
Best min_samples_leaf: 5
```

This chapter further explored using Scikit-learn to incorporate other machine learning techniques such as feature selection and resampling methods to develop a more robust machine learning method. In the next chapter, we will examine our first unsupervised machine learning method, clustering, and its implementation with Scikit-learn.

CHAPTER 25

Clustering

Clustering is an unsupervised machine learning technique for grouping homogeneous data points into partitions called clusters. In the example dataset illustrated in Figure 25-1, suppose we have a set of n points and 2 features. A clustering algorithm can be applied to determine the number of distinct subclasses or groups among the data samples.

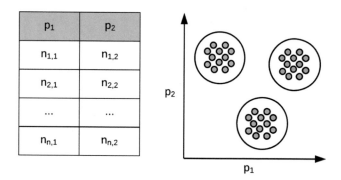

Figure 25-1. *An illustration of clustering in a 2-D space*

Clustering a 2-D dataset as seen in Figure 25-1 is relatively trivial. The real challenge arises when we have to perform clustering in higher-dimensional spaces. The question now is how do we ascertain or find out if a set of points are similar or if a set of points should be in the same group? In this section, we would cover two essential types of clustering algorithms known as k-means clustering and hierarchical clustering.

K-means clustering is used when the number of anticipated distinct classes or sub-groups is known in advance. In hierarchical clustering, the exact number of clusters is not known, and the algorithm is tasked to find the optimal number of heterogeneous sub-groups in the dataset.

© Ekaba Bisong 2019

E. Bisong, *Building Machine Learning and Deep Learning Models on Google Cloud Platform*, https://doi.org/10.1007/978-1-4842-4470-8_25

K-Means Clustering

k-Means clustering is one of the most famous and widely used clustering algorithms in practice. It works by using a distance measurement (most commonly the Euclidean distance) to iteratively assign data points in a hyperspace to a set of non-overlapping clusters.

In *K*-means, the anticipated number of clusters, *K*, is chosen at the onset. The clusters are initialized by arbitrarily selecting at random one of the data points as an initial cluster for each *K*. The algorithm now works by iteratively assigning each point in the space to the cluster centroid that it is nearest to using the distance measurement.

After all the points have been assigned to their closest cluster point, the cluster centroid is adjusted to find a new center among the points in the cluster. This process is repeated until the algorithm converges, that is, when the cluster centroids stabilize and points do not readily swap clusters after every reassignment. These steps are illustrated in Figure 25-2.

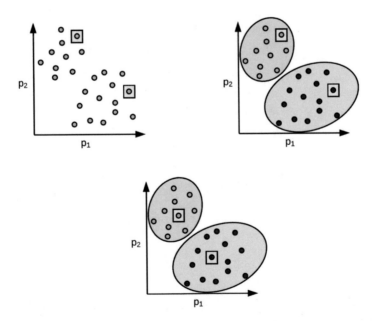

Figure 25-2. *An illustration of k-means clustering with k = 2. Top left: Randomly pick a point for each k. Top right: Iteratively assign each point to its closest cluster centroid. Bottom: Update the cluster centroids for each of the k clusters. Typically, we repeat the iterative assignment of all the points and update the cluster centroid until the algorithm resolves in a stable clustering.*

Considerations for Selecting *K*

There's really no way of telling the number of clusters in a dataset from the onset. The best way of selecting *k* is to try out different values of *K* to see what works best in creating distinct clusters.

Another strategy, which is widely employed in practice, is to compute the average distance of the points in the cluster to the cluster centroid for all clusters. This estimate is plotted on a graph as we progressively increase the value of *K*. We observe that as *K* increases, the distance of points from the centroid of its cluster gradually reduces, and the generated curve resembles the elbow of an arm. From practice, we choose the value of *K* just after the elbow as the best *K* value for that dataset. This method is called the elbow method for selecting *K* as is illustrated in Figure 25-3.

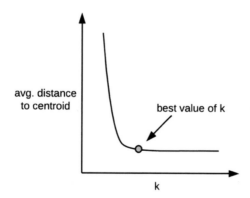

Figure 25-3. *The elbow method for choosing the best value of k*

Considerations for Assigning the Initial *K* Points

The points that determine the initial value of *K* are important in finding a good set of clusters. By selecting the point for *K* at random, two or more points may reside in the same cluster, and this will invariably lead to sub-par results. To mitigate this from occurring, we can employ more sophisticated approaches to selecting the value of *K*. A common strategy is to randomly select the first *K* point and then select the next point as the point that is farthest from the first chosen point. This strategy is repeated until all *K* points have been selected. Another approach is to run hierarchical clustering on a sub-sample of the dataset (this is because hierarchical clustering is a computationally expensive algorithm) and use the number of clusters after cutting off the dendrogram as the value of *K*.

311

K-Means Clustering with Scikit-learn

This example implements K-means clustering with Scikit-learn. Since this is an unsupervised learning use case, we use just the features of the Iris dataset to cluster the observations into labels.

```
# import packages
import matplotlib.pyplot as plt
from sklearn.cluster import KMeans
from sklearn import datasets
from sklearn.model_selection import train_test_split

# load dataset
data = datasets.load_iris()

# get the dataset features
X = data.data

# create the model. Since we know that the Iris dataset has 3 classes, we
set n_clusters = 3
kmeans = KMeans(n_clusters=3, random_state=0)

# fit the model on the training set
kmeans.fit(X)
'Output':
KMeans(algorithm='auto', copy_x=True, init='k-means++', max_iter=300,
    n_clusters=3, n_init=10, n_jobs=1, precompute_distances='auto',
    random_state=0, tol=0.0001, verbose=0)

# predict the closest cluster each sample in X belongs to.
y_kmeans = kmeans.predict(X)

# plot clustered labels
plt.scatter(X[:, 0], X[:, 1], c=y_kmeans, cmap='viridis')

# plot cluster centers
centers = kmeans.cluster_centers_
plt.scatter(centers[:, 0], centers[:, 1], c='black', s=200, alpha=0.7);
plt.show()
```

The code to plot the clustered labels and the cluster centers should be executed in the same notebook. The plot of clusters made by the K-means algorithm is shown in Figure 25-4.

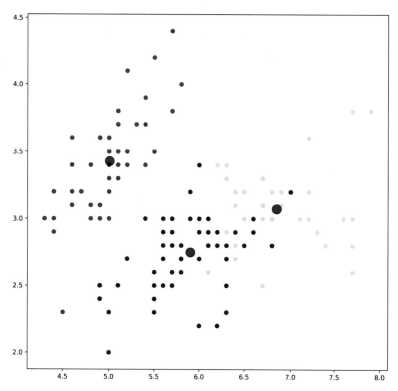

Figure 25-4. *Plot of K-means clusters and their cluster centers*

Hierarchical Clustering

Hierarchical clustering is another clustering algorithm for finding homogeneous sub-groups or classes within a dataset. However, as opposed to k-means, we do not need to make an a priori assumption of the number of clusters in the dataset before running the algorithm.

The two main techniques for performing hierarchical clustering are

- Bottom-up or agglomerative

- Top-down or divisive

In the bottom-up or agglomerative method, each data point is initially designated as a cluster. Clusters are iteratively combined based on homogeneity that is determined by some distance measure. On the other hand, the divisive or top-down approach starts with a cluster and subsequently splits into homogeneous sub-groups.

Hierarchical clustering creates a tree-like representation of the partitioning called a dendrogram. A dendrogram is drawn somewhat similar to a binary tree with the root at the top and the leaves at the bottom. The leaf on the dendrogram represents a data sample. The dendrogram is constructed by iteratively combining the leaves based on homogeneity to form clusters moving up the tree. An illustration of hierarchical clustering is shown in Figure 25-5.

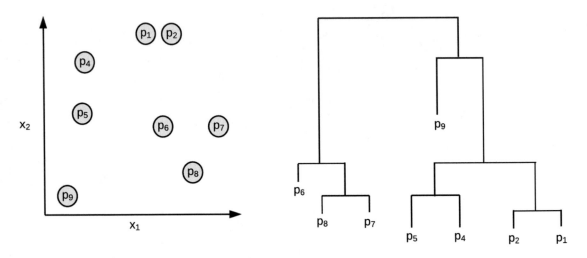

Figure 25-5. *An illustration of hierarchical clustering of data points in a 2-D feature space. Left: The spatial representation of points in 2-D space. Right: A hierarchical cluster of points represented by a dendrogram.*

How Are Clusters Formed

Clusters are formed by computing the nearness between each pair of data points. The notion of nearness is most popularly calculated using the Euclidean distance measure. Beginning at the leaves of the dendrogram, we iteratively combine those data points that are closer to one another in the multi-dimensional vector space until all the homogeneous points are placed into a single group or cluster.

The Euclidean distance is used to compute the nearness between *n* data points. After each pair of data points has combined to form a cluster, the new cluster pairs are then pulled into groups going up the tree, with the tree branch or dendrogram height reflecting the dissimilarity between the clusters.

Dissimilarity computes how different each cluster of data is from one another. The notion of dissimilarity between two clusters or groups is described in terms of *linkage*. Four types of linkage exist for grouping clusters in hierarchical clustering. They are centroid, complete, average, and single.

The centroid linkage computes the dissimilarity between two clusters using the geometric centroid of the clusters. The complete linkage uses the two farthest data points between the two clusters to compute the dissimilarity (see Figure 25-6).

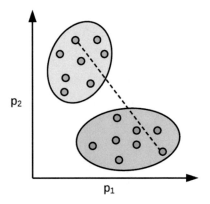

Figure 25-6. *Complete linkage*

The average linkage finds the means of points within the pair of clusters and uses that new artificial point to calculate the dissimilarity (see Figure 25-7).

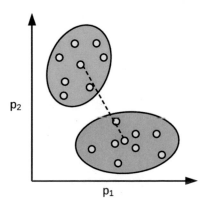

Figure 25-7. *Average linkage*

The single linkage uses the closest data point between the cluster pairs to compute the dissimilarity measure (see Figure 25-8).

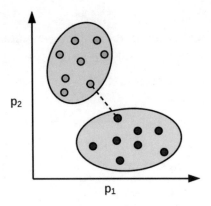

Figure 25-8. *Single linkage*

Empirically, the complete and average linkages are preferred in practice because they yield more balanced dendrograms. Other dissimilarity measures exist for evaluating the nearness or homogeneity of data points. One of such is the Manhattan distance, another distance-based measure, or the correlation-based distance which groups pairs of data samples with highly correlated features. A correlated-based dissimilarity measure may be more useful in datasets where proximity in multi-dimensional spaces is not as useful a metric for homogeneity as compared to the correlation of their features in the space. A choice of calculating dissimilarity has a significant impact on the ensuring dendrogram.

After running the algorithm, the dendrogram is cut at a particular height, and the number of distinct lines or branches after the cut is circumscribed as the number of clusters in the dataset. An illustration of cutting the dendrogram is shown in Figure 25-9.

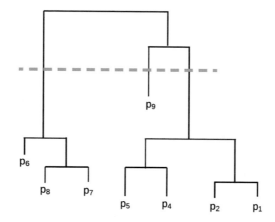

Figure 25-9. *Dendrogram cut*

Hierarchical Clustering with the SciPy Package

This example implements hierarchical or agglomerative clustering with SciPy. The
'scipy.cluster.hierarchy' package has simple methods for performing hierarchical
clustering and plotting dendrograms. This example uses the 'complete' linkage method.
The plot of the dendrogram is shown in Figure 25-10.

```
# import packages
import matplotlib.pyplot as plt
from scipy.cluster.hierarchy import dendrogram
from scipy.cluster import hierarchy

Z = hierarchy.linkage(X, method='complete')

plt.figure()
dn = hierarchy.dendrogram(Z, truncate_mode='lastp')
```

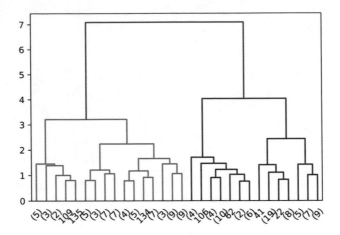

Figure 25-10. *Dendrogram produced by hierarchical clustering*

This chapter reviewed the pros and cons of K-means and hierarchical clustering. Both hierarchical and K-means are susceptible to perturbations in the dataset and can give very different results if a few data points are removed or added. Also, it is crucial to standardize the dataset features (i.e., to subtract each element in the feature from its mean and divide by its standard deviation or by the range) before performing clustering. This ensures that the features are within similar numeric bounds and have tempered or measured distances in the feature space.

The results of these clustering algorithms also depend on a wide range of considerations such as the choice of K for K-means, and for hierarchical clustering, the choice of dissimilarity measure, the type of linkage, and where to cut the dendrogram all affect the final result of the clusters. Hence, to get the best out of clustering, it is best to perform a grid search and try out all these different configurations in order to get a measured view on the robustness of the results before applying into your learning pipeline or using as a model to explain the dataset.

In the next chapter, we will discuss principal component analysis (PCA) as an unsupervised machine learning algorithm for finding low-dimensional feature subspaces that capture the variability in the dataset.

## CHAPTER 26

Principal Component Analysis (PCA)

Principal component analysis (PCA) is an essential algorithm in machine learning. It is a mathematical method for evaluating the principal components of a dataset. The principal components are a set of vectors in high-dimensional space that capture the variance (i.e., spread) or variability of the feature space.

The goal of computing principal components is to find a low-dimensional feature sub-space that captures as much information as possible from the original higher-dimensional features of the dataset.

PCA is particularly useful for simplifying data visualization of high-dimensional features by reducing the dimensions of the dataset to a lower sub-space. For example, since we can easily visualize relationships on a 2-D plane using scatter diagrams, it will be useful to condense an n-dimensional space into two dimensions that retain as much information as possible in the n-dimensional dataset. This technique is popularly called dimensionality reduction.

How Are Principal Components Computed

The mathematical details for computing principal components are somewhat involved. This section will instead provide a conceptual but solid overview of this process.

The first step is to find the covariance matrix of the dataset. The covariance matrix captures the linear relationship between variables or features in the dataset. In a covariance matrix, an increasingly positive number represents a growing relationship, while the converse is represented by an increasingly negative number. Numbers around zero indicate a non-linear relationship between the variables. The covariance matrix is a square matrix (that means it has the same rows and columns). Hence, given a dataset with m rows and p columns, the covariance matrix will be a $m \times p$ matrix.

© Ekaba Bisong 2019
E. Bisong, *Building Machine Learning and Deep Learning Models on Google Cloud Platform*,
https://doi.org/10.1007/978-1-4842-4470-8_26

The next step is to find the eigenvectors of the covariance matrix dataset. In linear algebra theory, eigenvectors are non-zero vectors that merely stretch by a scalar factor, but do not change direction when acted upon by a linear transformation. We find the eigenvectors using a linear algebra technique called the singular value decomposition or SVD for short (see Figure 26-1). This advanced mathematical concept is beyond the scope of this book.

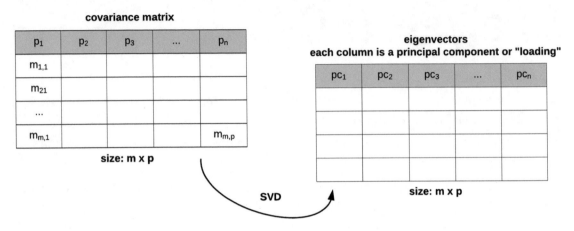

Figure 26-1. *Decompose the covariance matrix using SVD to get the eigenvector matrix*

The critical point to note at this junction is that the SVD also outputs a square matrix $(p \times p)$, and each column of the matrix is an eigenvector of the original dataset. This output is the same across different software packages that compute the eigenvectors because the covariance matrix satisfies a mathematical property of being symmetric and positive semi-definite (the non-math inclined can conveniently ignore this point). We have as many eigenvectors as they are attributes or features in the dataset.

Without delving into mathematical theory, we can conclude that the eigenvectors are the principal components or loadings of the feature space. Again remember that the principal components capture the most significant variance in the dataset by projecting the data onto a vector called the first principal component. Other principal components are perpendicular to each other and capture the variance not explained by the first principal component. The principal components are arranged in order of importance in the eigenvector matrix, with the first principal component in the first column, the second principal component in the second column, and so on.

Dimensionality Reduction with PCA

To reduce the dimensions of the original dataset using PCA, we multiply the desired number of components or loadings from the eigenvector matrix, A, by the design matrix X. Suppose the design matrix (or the original dataset) has m rows (or observations) and p columns (or features), if we want to reduce the dimensions of the original dataset to two dimensions, we will multiply the original dataset X by the first two columns of the eigenvector matrix, $A_{reduced}$. The result will be a reduced matrix of m rows and 2 columns.

If X is a $m \times p$ matrix and $A_{reduced}$ is a $p \times 2$ matrix,

$$T_{reduced} = X_{m \times p} \times A_{p \times 2}$$

Observe that the result $T_{reduced}$ is a $m \times 2$ matrix. Hence, T is a 2-D representation of the original dataset X as shown in Figure 26-2.

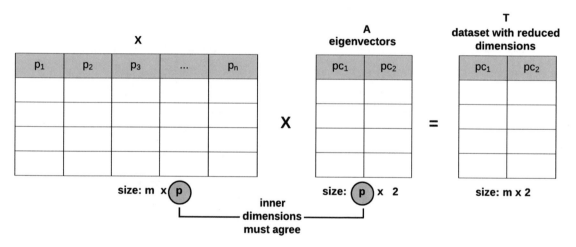

Figure 26-2. *Reducing the dimension of the original dataset*

In plotting the reduced dataset, the principal components are ranked in order of importance with the first principal component more prominent than the second and so on. Figure 26-3 illustrates a plot of the first two principal components.

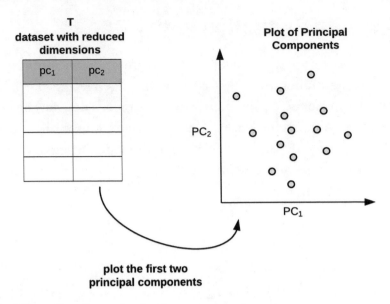

Figure 26-3. *Visualize the principal components*

Key Considerations for Performing PCA

It is vital to perform mean normalization and feature scaling on the variables of features of the original dataset before implementing PCA. This is because unscaled features can have stretched and narrow distance n-dimensional space, and this has a huge consequence when finding the principal components that explain the variance of the dataset (see Figure 26-4).

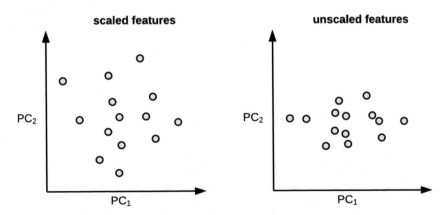

Figure 26-4. *Right: An illustration of PCA with scaled features. Left: An illustration of PCA with unscaled features.*

Again mean normalization ensures that every attribute or feature of the dataset has a zero mean, while feature scaling ensures all the features are within the same numeric range.

Finally, PCA is susceptible to vary wildly due to slight perturbations or changes in the dataset.

PCA with Scikit-learn

In this section, PCA is implemented using Scikit-learn.

```
# import packages
from sklearn.decomposition import PCA
from sklearn import datasets

from sklearn.preprocessing import Normalizer

# load dataset
data = datasets.load_iris()

# separate features and target
X = data.data

# normalize the dataset
scaler = Normalizer().fit(X)
normalize_X = scaler.transform(X)

# create the model.
pca = PCA(n_components=3)

# fit the model on the training set
pca.fit(normalize_X)

# examine the principal components percentage of variance explained
pca.explained_variance_ratio_

# print the principal components
pca_dataset = pca.components_
pca_dataset
'Output':
```

```
array([[ 0.18359702,   0.49546167, -0.76887947, -0.36004754],
       [ 0.60210709, -0.64966313, -0.05931229, -0.46031175],
       [-0.2436305 ,  0.28528504,  0.49319469, -0.78486663]])
```

In this chapter, we explained PCA giving a high-level overview of how it works to find a low-dimensional sub-space of a dataset. More so, we showed how PCA is implemented with Scikit-learn. This chapter concludes Part 4. In the next part, we introduce another scheme of learning methods called deep learning that builds on the machine learning neural network algorithm for learning complex representations.

PART V

Introducing Deep Learning

CHAPTER 27

What Is Deep Learning?

Deep learning is a class of machine learning algorithms called neural networks. Neural networks are mathematical models inspired by the structure of the brain. Deep learning enables the neural network algorithm to perform very well in building prediction models around complex problems such as computer vision and language modeling. Self-driving cars and automatic speech translation, to mention just a few, are examples of technologies that have resulted from advances in deep learning.

The Representation Challenge

Learning is a non-trivial task. The brain's ability to learn complex tasks is not yet fully understood by research communities in neurological science, psychology, and other brain-related fields. What we consider trivial, and to some others natural, are a system of complex and intricate processes that have set us apart from other life forms as intelligent beings.

Examples of complex tasks performed by the human brain include the ability to recognize faces at a millionth of a second (probably much faster), the uncanny aptitude for learning and understanding deep linguistic representations, and forming symbols for intelligent communications. Also, the adept skills to compose and perform masterful musical pieces are examples of the marvel of natural intelligence.

The challenge of AI research and engineering is to build machines that can understand and decompose the structural patterns inherent in complex problems in order to mimic natural intelligence. Deep learning as an AI technique approaches the representation problem by learning the underlying fundamental structure inherent in the dataset. Deep learning is also called representation learning.

© Ekaba Bisong 2019
E. Bisong, *Building Machine Learning and Deep Learning Models on Google Cloud Platform*,
https://doi.org/10.1007/978-1-4842-4470-8_27

Inspiration from the Brain

Scientists often look to nature for inspiration when performing incredible feats. Notably, the birds inspired the airplane. In that vein, there is no better type to study as an antitype of intelligence as the human brain.

We can view the brain as a society of intelligent agents that are networked together and communicate by passing information via electrical signals from one agent to another. These agents are known as neurons. Our principal interest here is to have a glimpse of what neurons are, what their components are, and how they pass information around to create intelligence.

A neuron is an autonomous agent in the brain and is a central part of the nervous system. Neurons are responsible for receiving and transmitting information to other cells within the body based on external or internal stimuli. Neurons react by firing electrical impulses generated at the stimuli source to the brain and other cells for the appropriate response. The intricate and coordinated workings of neurons are central to human intelligence.

The following are the three most essential components of neurons that are of primary interest to us:

- The axon

- The dendrite

- The synapse

The axon is a long tail connected to the nucleus of the neuron as seen in Figure 27-1. The axon is responsible for transmitting electrical signals from the nucleus to other neuron cells through the axon terminals. The dendrite, on the other hand, receives information as electrical impulses from other neuron cells through the synapses to the nucleus of a neuron cell.

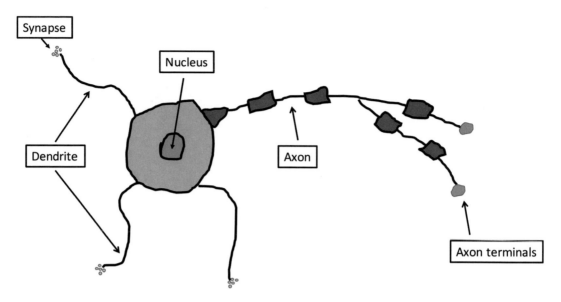

Figure 27-1. *A neuron*

By mimicking these three biological components of a neuron, scientists developed the core design and structure of an artificial neural network (ANN) that allows us to build machines that can learn. We will discuss the ANN in more detail in the next chapter. There is much hope that if we can mimic the capabilities of the brain from a science and engineering perspective, we can build machines that can learn hierarchical features from complex domain use cases.

This chapter introduces the field of deep learning as an engineering impersonation of how the brain learns to build artificial neural networks. In the next chapter, we'll go deeper to discuss the neural network algorithm.

## CHAPTER 28

Neural Network Foundations

Building on the inspiration of the biological neuron, the artificial neural network (ANN) is a society of connectionist agents that learn and transfer information from one artificial neuron to the other. As data transfers between neurons, a hierarchy of representations or a hierarchy of features is learned, hence the name deep representation learning or deep learning.

The Architecture

An artificial neural network is composed of

- An input layer
- Hidden layer(s)
- An output layer

© Ekaba Bisong 2019
E. Bisong, *Building Machine Learning and Deep Learning Models on Google Cloud Platform*,
https://doi.org/10.1007/978-1-4842-4470-8_28

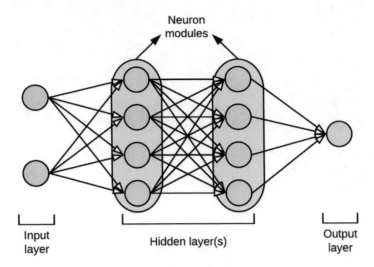

Figure 28-1. *Neural network architecture*

The input layer receives information from the features of the dataset, after which some computation takes place, and information that captures the learned patterns of the data is propagated across the hidden layer(s) with hopes to improve the learned patterns.

The hidden layer(s) is where the workhorse of deep learning occurs. The hidden layer(s) can consist of multiple neuron modules as shown in Figure 28-1. Each hidden network layer learns a more sophisticated set of feature representations. The decision on the number of neurons in a layer (network width) and the number of hidden layers (network depth) which forms the network topology is a design choice when training deep learning networks. The techniques for training a deep neural network are discussed in the next chapter.

## CHAPTER 29

Training a Neural Network

This chapter gives an overview of the techniques for training a deep neural network. Here, we briefly discuss

- How learned information flows through a neural network

- The role of the cost function at the output layer of the network

- One-hot encoding and the softmax activation function for determining class membership at the output layer of a classification problem

- The backpropagation algorithm for improving the learned parameters of the network

- Activation functions that enable the neural network to learn non-linear patterns

In this chapter, as we discuss the methods involved in training a neural network, we will use the example of a classification problem with two possible outputs. In designing a neural network, the number of neurons in the input layer is typically the number of features of the dataset, while the number of neurons in the output layer is the number of classes in the target variable that the neural network is learning to classify.

As illustrated in Figure 29-1, the dataset features are the inputs to the neural network, while the classes in the target variable determine the number of output neurons. In this example, the network learns two classes, 0 and 1.

© Ekaba Bisong 2019

E. Bisong, *Building Machine Learning and Deep Learning Models on Google Cloud Platform*, https://doi.org/10.1007/978-1-4842-4470-8_29

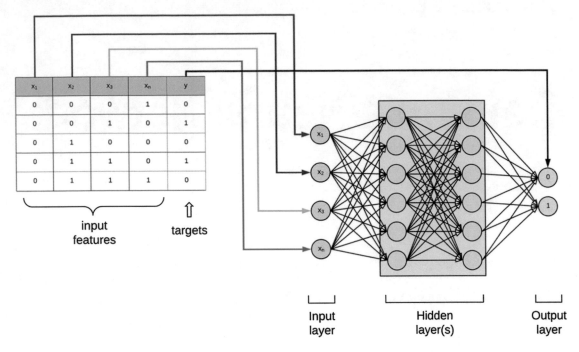

Figure 29-1. *Defining a neural network from a dataset*

A weight (also called parameter) is assigned to every neuron. The weights of neurons in a neural layer are multiplied by their inputs and then passed through an activation function (to be discussed in this chapter) for which the outputs are the inputs to the neurons in the next neural layer of the network (see Figure 29-2). This procedure is repeated as information of what the neural network is trying to learn moves from one layer of the network to another. Every neuron layer also has a bias neuron (typically set to 1) that controls the weighted sum. This is similar to the bias term in the logistic regression model.

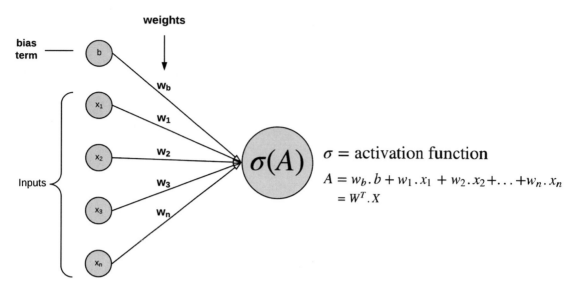

Figure 29-2. *Information flowing from a previous neural layer to a neuron in the next layer*

The weights are initialized as random values that are later adjusted as the network begins to learn using the backpropagation algorithm (to be discussed in this chapter). In summary, the outputs (or activations) of the neurons in the neural network layers are determined by the sum of the weight times the outputs plus the bias term of the neurons in the previous layer acted upon by a non-linear *activation function* (see Figure 29-2). This move is called the feedforward learning algorithm.

However, the output of the feedforward pass through the network may most likely result in an incorrect classification. The errors made from the feedforward procedure are later adjusted using the backpropagation algorithm (to be discussed). To evaluate the performance of the neural network, we define a cost function or loss function (similar to other machine learning algorithms) that captures the quality of the prediction made by the network.

The goal of the neural network is to minimize the cost function. Two commonly used cost functions are the squared error cost function for regression problems and the softmax cross-entropy cost function for classification problems.

Cost Function or Loss Function

The squared error cost function (also known as the mean squared error) finds the sum of the squared difference between the estimated target and the actual target for a real-valued problem, while the cross-entropy cost function finds the difference between the predicted class from the probability estimates of the actual class label in a classification problem.

Regardless of the cost function used, when the error loss is small, we say that the cost is minimized. In Figure 29-3, the correct output of the example passed into the network is **2.3**. After the output values are evaluated from the feedforward training, the squared error cost function is used to assess the quality of the network's output.

Remember that the MSE finds the average cost over all the data samples in the training dataset. In the example illustrated in Figure 29-3, we used just one data sample to demonstrate how the cost function works.

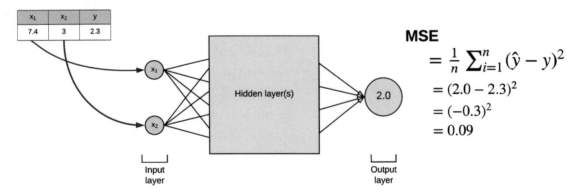

MSE

$$= \frac{1}{n} \sum_{i=1}^{n} (\hat{y} - y)^2$$
$$= (2.0 - 2.3)^2$$
$$= (-0.3)^2$$
$$= 0.09$$

Figure 29-3. *MSE estimate of the neural network*

One-Hot Encoding

In a classification problem, one-hot encoding is the process of transforming the class labels of the target variable into a matrix of binary variables. The one-hot encoder assigns 1 when the output belongs to a particular class and 0 otherwise. An illustration of one-hot encoding is shown in Figure 29-4.

original dataset

x_1	x_2	y
0	0	0
0	1	0
1	0	0
1	1	1

dataset with one-hot encoded label

x_1	x_2	y_0	y_1
0	0	1	0
0	1	1	0
1	0	1	0
1	1	0	1

one-hot encoding

Figure 29-4. *One-hot encoding*

In the final layer of the neural network, just before the output layer, an activation function called the softmax (same as discussed under "Logistic Regression") is applied to transform the activations to the probability that the example belongs to one of the output classes.

The purpose of applying one-hot encoding to the labels of the dataset is to represent the output as a vector of distinct classes with the probability that an example in the training dataset belongs to any one of the output categories.

The Backpropagation Algorithm

Backpropagation is the process by which we train the neural network to improve its prediction accuracy. To train the neural network, we need to find a mechanism for adjusting the weights of the network; this in turn affects the value of the activations within each neuron and consequently updates the value of the predicted output layer. The first time we run the feedforward algorithm, the activations at the output layer are most likely incorrect with a high error estimate or cost function.

The goal of backpropagation is to repeatedly go back and adjust the weights of each preceding neural layer and perform the feedforward algorithm again until we minimize the error made by the network at the output layer (see Figure 29-5).

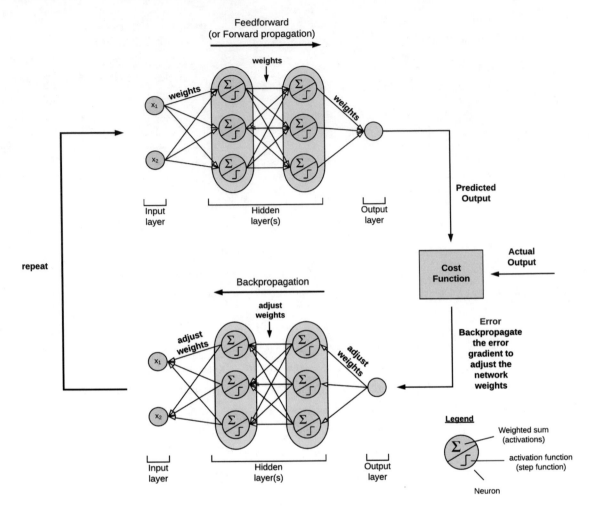

Figure 29-5. *Backpropagation*

The backpropagation algorithm works by computing the cost function at the output layer by comparing the predicted output of the neural network with the actual outputs from the dataset. It then employs gradient descent (earlier discussed in Chapter 16) to calculate the gradient of the cost function using the weights of the neurons at each successive layer and update the weights propagating back through the network.

Activation Functions

Up till now, we have mentioned activation functions. Now let's go a bit deeper into what activation functions are and why do we have them.

Activation functions act on the weighted sum in the neuron (which is nothing more than the weighted sum of weights and their added bias) by passing it through a non-linear function to decide if that neuron should fire (propagate) its information or not to the succeeding neural layers.

In other words, the activation function determines if a particular neuron has the information to result in a correct prediction at the output layer for an observation in the training dataset. Activation functions are analogous to how neurons communicate and transfer information in the brain, by firing when the activation goes above a particular threshold value.

These activation functions are also called non-linearities because they inject non-linear capabilities to our network and can learn a mapping from inputs to output for a dataset whose fundamental structure is non-linear. An illustration of passing the weighted sum of weights and biases through an activation function is shown in Figure 29-6.

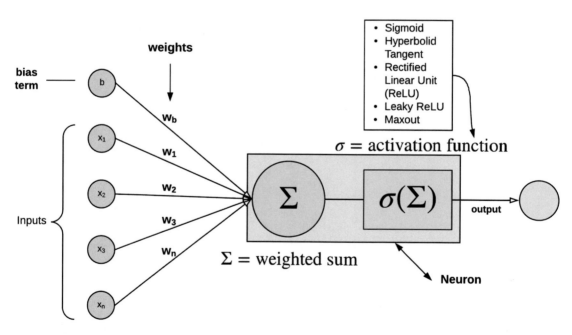

Figure 29-6. *Activation function*

The following are examples of activation functions used in a neural network:

- Sigmoid

- Hyperbolic tangent (tanh)

- Rectified linear unit (ReLU)

- Leaky ReLU

- Maxout

Let's briefly examine them.

Sigmoid

The sigmoid function illustrated in Figure 29-7 is a non-linear function that brings (or squashes) the activations to fall within a range of 0 and 1. This brings large negative and positive numbers to 0 and 1, respectively. The neurons typically begin firing when the function output is above a threshold of 0.5.

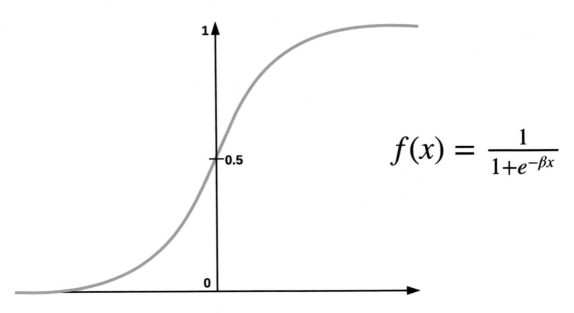

$$f(x) = \frac{1}{1+e^{-\beta x}}$$

Figure 29-7. *Sigmoid activation function*

However, a significant drawback of the sigmoid function is its susceptibility to a phenomenon called exploding and vanishing gradients. In the process of optimizing the weights of the network during backpropagation, the gradients can become disproportionately small or large with their activations concentrated at either 0 or 1.

When this happens, we say that the gradients have saturated. Hence, further multiplication via backpropagation causes the gradient to either vanish or explode; and as a result, the affected neurons become dead and transfer no information across the network, thus negatively affecting training.

Another drawback is that the outputs of the function are not zero-centered. As a consequence, during backpropagation, the gradients can either become all positive or all negative. This has a negative effect in minimizing the function objective (i.e., the cost function).

Hyperbolic Tangent (tanh)

The hyperbolic tangent illustrated in Figure 29-8 improves on the sigmoid function by bordering its output within a range of −1 and 1. So, while it still suffers from the exploding and vanishing gradient problem, its outputs are now zero-centered. From the formula, the reader will observe that tanh is merely a scaled sigmoid function.

Tanh Activation Function

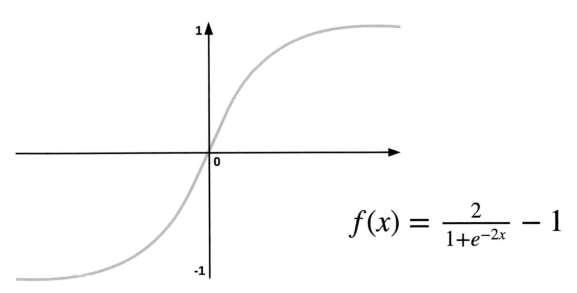

$$f(x) = \frac{2}{1+e^{-2x}} - 1$$

Figure 29-8. *The hyperbolic tangent activation function*

Rectified Linear Unit (ReLU)

The rectified linear unit or ReLU activation function is illustrated in Figure 29-9 and works by setting the activation to 0 for values, x, less than 0 and a linear slope of 1 when values, x, are greater than 0.

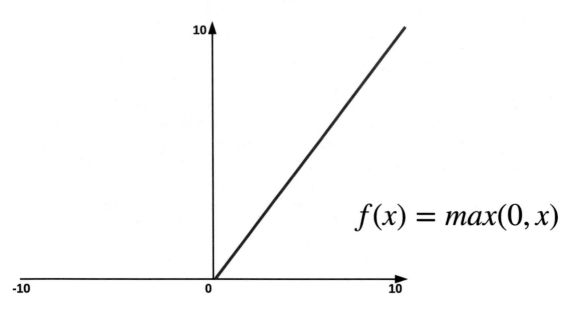

$$f(x) = max(0, x)$$

Figure 29-9. *ReLU activation function*

ReLU offers a vast improvement on the tanh and sigmoid activation functions by greatly mitigating the vanishing and exploding gradient problem. However, some gradients can still die out during backpropagation with a large learning rate. However, with a well-defined learning rate, we should not have a problem.

Leaky ReLU

Leaky ReLU is another activation function that is proposed to solve the case of some neurons completely dying out in ReLU by avoiding zero gradients. Leaky ReLU is illustrated in Figure 29-10. The function works by setting the activation to a small negative slope when the value $x < 0$.

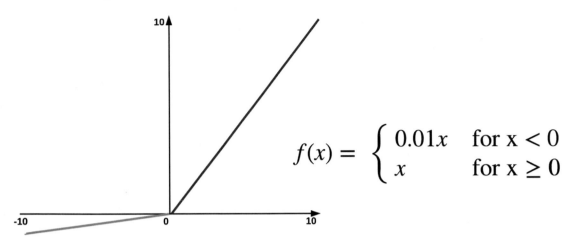

$$f(x) = \begin{cases} 0.01x & \text{for x} < 0 \\ x & \text{for x} \geq 0 \end{cases}$$

Figure 29-10. *Leaky ReLU activation function*

Maxout

The Maxout activation function generalizes the ReLU and leaky ReLU functions and hence takes advantage of the efficiency of ReLU while avoiding its pitfalls of some neurons dying out. In any case, a trade-off needs to be made, because Maxout increases the parameter size of each neuron during training.

As a rule of thumb, different types of activation functions are not mixed in the same network. Also, ReLU is typically used for the hidden layers, and the softmax activation is used for classification problems at the output layer since this layer returns a probability of membership of a particular class.

This chapter provided an overview on how to train a predictive model using neural networks. This chapter ends Part 5 on introducing deep learning. The chapters in Part 6 will cover deep learning algorithms and their implementation with TensorFlow and Keras.

PART VI

Deep Learning in Practice

## CHAPTER 30

TensorFlow 2.0 and Keras

TensorFlow (TF) is a specialized numerical computation library for deep learning. It is the preferred tool by numerous deep learning researchers and industry practitioners for developing deep learning models and architectures as well as for serving learned models into production servers and software products. This chapter is focused on TensorFlow 2.0.

Navigating Through the TensorFlow API

Understanding the different levels of the TF API hierarchy is critical to working effectively with TF. The task of building a TF deep learning model may be addressed via different TF API levels. An understanding of the API hierarchy provides clarity on implementing neural network models with TF as well as navigating the TF ecosystem. The TF API hierarchy is primarily composed of three API levels, the high-level API, the mid-level API which provides components for building neural network models, and the low-level API. A diagrammatic representation of this is shown in Figure 30-1.

© Ekaba Bisong 2019
E. Bisong, *Building Machine Learning and Deep Learning Models on Google Cloud Platform*,
https://doi.org/10.1007/978-1-4842-4470-8_30

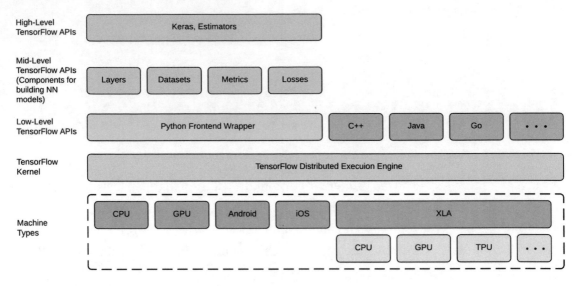

Figure 30-1. *TensorFlow API hierarchy*

The Low-Level TensorFlow APIs

The low-level API gives the tools for building network graphs from the ground up using mathematical operations. This API level affords the greatest level of flexibility to tweak and tune the model as desired. Moreover, the higher-level APIs implement low-level operations under the hood.

The Mid-Level TensorFlow APIs

TensorFlow provides a set of reusable packages for simplifying the process involved in creating neural network models. Some examples of these functions include the layers **(tf.keras.layers)**, Datasets **(tf.data)**, metrics **(tf.keras.metrics)**, loss **(tf.keras.losses)**, and FeatureColumns **(tf.feature_column)** packages.

Layers

The layers package **(tf.keras.layers)** provides a handy set of functions to simplify the construction of layers in a neural network architecture. For example, consider the convolutional network architecture in Figure 30-2 and how the layers API simplifies the creation of the network layers.

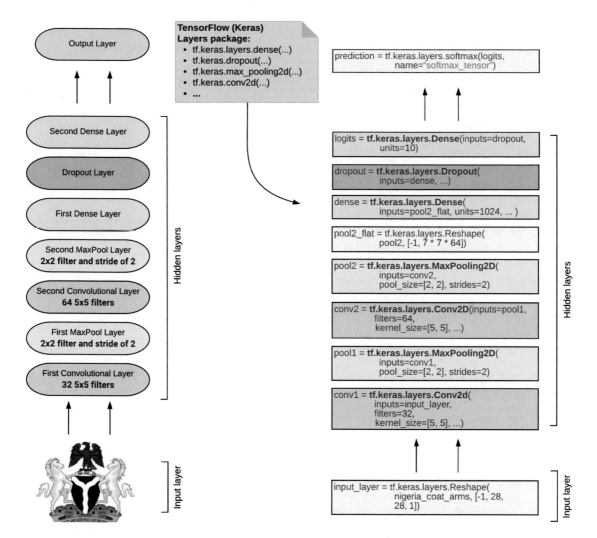

Figure 30-2. *Using the layers API to simplify creating the layers of a neural network*

Datasets

The Dataset package **(tf.data)** provides a convenient set of high-level functions for creating complex dataset input pipelines. The goal of the Dataset package is to have a fast, flexible, and easy-to-use interface for fetching data from various data sources, performing data transform operations on them before passing them as inputs to the learning model. The Dataset API provides a more efficient means of fetching records from a dataset. The major classes of the Dataset API are illustrated in Figure 30-3.

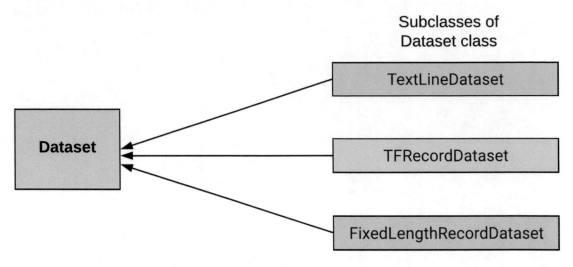

Figure 30-3. *Dataset API class hierarchy*

From the illustration in Figure 30-3, the subclasses perform the following functions:

- TextLineDataset: This class is used for reading lines from text files.

- TFRecordDataset: This class is responsible for reading records from TFRecord files. A TFRecord file is a TensorFlow binary storage format. It is faster and easier to work with data stored as TFRecord files as opposed to raw data files. Working with TFRecord also makes the data input pipeline more easily aligned for applying vital transformations such as shuffling and returning data in batches.

- FixedLengthRecordDataset: This class is responsible for reading records of fixed sizes from binary files.

FeatureColumns

FeatureColumns **tf.feature_column** is a TensorFlow functionality for describing the features of the dataset that will be fed into a high-level Keras or Estimator models for training and validation. FeatureColumns makes it easy to prepare data for modeling by carrying out tasks such as the conversion of categorical features of the dataset into a one-hot encoded vector.

The **feature_column** API is broadly divided into two categories; they are the categorical and dense columns. The categories and subsequent functions are illustrated in Figure 30-4.

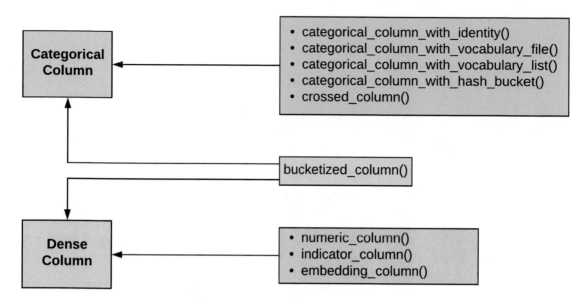

Figure 30-4. *Function calls of the Feature Column API*

Let's go through each API function briefly in Table 30-1.

Table 30-1. *tf.feature_column API Functions*

Function name	Description
Numeric column – **tf.feature_column. numeric_column()**	This is a high-level wrapper for numeric features in the dataset.
Indicator column – **tf.feature_column. indicator_column()**	The indicator column takes as input a categorical column and transforms it into a one-hot encoded vector.
Embedding column – **tf.feature_column. embedding_column()**	The embedding column function transforms a categorical column with multiple levels or classes into a lower-dimensional numeric representation that captures the relationships between the categories. Using embeddings mitigates the problem of a large sparse vector (an array with mostly zeros) created via one-hot encoding for a dataset feature with lots of different classes.

(*continued*)

Table 30-1. (*continued*)

Function name	Description
Categorical column with identity – **tf.feature_ column.categorical_ column_with_identity()**	This function creates a one-hot encoded output of a categorical column containing identities, e.g, ['0', '1', '2', '3'].
Categorical column with vocabulary list – **tf.feature_column. categorical_ column_ with_vocabulary_list()**	This function creates a one-hot encoded output of a categorical column with strings. It maps each string to an integer based on a vocabulary list. However, if the vocabulary list is long, it is best to create a file containing the vocabulary and use the function **tf.feature_ column.categorical_ column_with_vocabulary_file()**.
Categorical column with hash bucket – **tf.feature_column. categorical_ column_ with_hash_buckets()**	This function specifies the number of categories by using the hash of the inputs. It is used when it is not possible to create a vocabulary for the number of categories due to memory considerations.
Crossed column – **tf.feature_columns. crossed_column()**	The function gives the ability to combine multiple input features into a single input feature.
Bucketized column – **tf.feature_column. bucketized_column()**	The function splits a column of numerical inputs into buckets to form new classes based on a specified set of numerical ranges.

The High-Level TensorFlow APIs

The high-level API provides simplified API calls that encapsulate lots of the details that are typically involved in creating a deep learning TensorFlow model. These high-level abstractions make it easier to develop powerful deep learning models quickly with fewer lines of code.

Estimator API

The Estimator API is a high-level TensorFlow functionality that is aimed at reducing the complexity involved in building machine learning models by exposing methods that abstract common models and processes. There are two ways of working with Estimators, and they include

- **Using the premade Estimators**: The premade Estimators are black box models made available by the TensorFlow team for building common machine learning/deep learning architectures such as linear regression/classification, Random forest regression/ classification and deep neural networks for regression and classification. An illustration of the premade Estimators as subclasses of the Estimator class is shown in Figure 30-5.

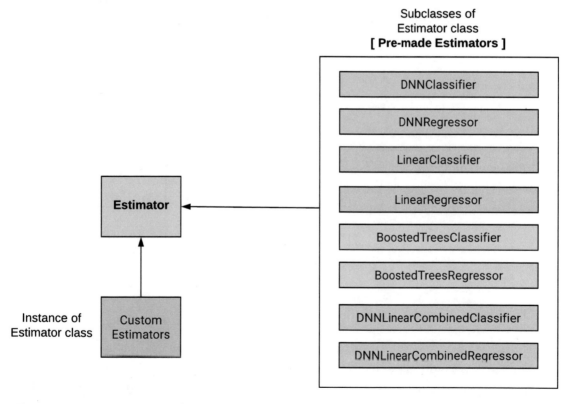

Figure 30-5. *Estimator class API hierarchy*

- **Creating a custom Estimator**: It is also possible to use the low-level TensorFlow methods to create a custom black box model for easy reusability. To do this, you must put your code in a method called the **model_fn**. The model function will include code that defines operations such as the labels or predictions, loss function, the training operations, and the operations for evaluation.

The Estimator class exposes four major methods, namely, the **fit()**, **evaluate()**, **predict()**, and **export_savedmodel()** methods. The **fit()** method is called to train the data by running a loop of training operations. The **evaluate()** method is called to evaluate the model performance by looping through a set of evaluation operations. The **predict()** method uses the trained model to make predictions, while the **export_savedmodel()** method is used for exporting the trained model to a specified directory. For both the premade and custom Estimators, we must write a method to build the data input pipeline into the model. This pipeline is built for both the training and evaluation data inputs. This is further illustrated in Figure 30-6.

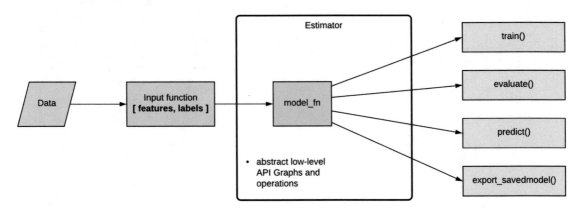

Figure 30-6. *Estimator data input pipeline*

Keras API

Keras provides a high-level specification for developing deep neural network models. The Keras API was initially separate from TensorFlow and only provided an interface for model building with TensorFlow as one of the frameworks running at the backend. However, in TensorFlow 2.0, Keras is an integral part of the TensorFlow codebase as preferred high-level API.

The Keras API version internal to TensorFlow is available from the 'tf.keras' package, whereas the broader Keras API blueprint that is not tied to a specific backend will remain available from the 'keras' package. In summary, when working with the 'keras' package, the backend can run with either TensorFlow, Microsoft CNTK, or Theano. On the other hand, working with 'tf.keras' provides a TensorFlow only version which is tightly integrated and compatible with all of the functionality of the core TensorFlow library.

In this book, we will focus on **'tf.Keras'** as a high-level API of TensorFlow.

The Anatomy of a Keras Program

The Keras **'Model'** forms the core of a Keras program. A 'Model' is first constructed, then it is compiled. Next, the compiled model is trained and evaluated using their respective training and evaluation datasets. Upon successful evaluation using the relevant metrics, the model is then used for making predictions on previously unseen data samples. Figure 30-7 shows the program flow for modeling with Keras.

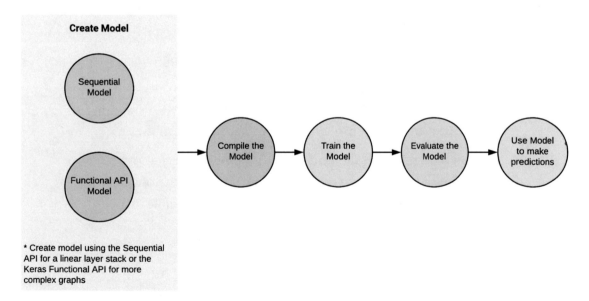

Figure 30-7.   *The anatomy of a Keras program*

As shown in Figure 30-7, the Keras 'Model' can be constructed using the Sequential API 'tf.keras.Sequential' or the Keras Functional API which defines a model instance 'tf. keras.Model'. The Sequential model is the simplest method for creating a linear stack of neural network layers. The Functional model is used if a more complex graph is desired. Keras is the de facto API for building neural network architectures with TensorFlow.

From here on, the code examples in this book will use the Sequential API, Functional API, and Model subclassing methods for building neural network architectures with Keras. In doing this, the reader can play around with the various examples as samples to get a feel of how they work.

TensorBoard

TensorBoard is an interactive visualization tool that comes bundled with TensorFlow. The goal of TensorBoard is to gain a visual insight into how the computational graph is constructed and executed. This information provides greater visibility for understanding, optimizing, and debugging deep learning models.

TensorBoard has a variety of visualization dashboard, such as

- Scalar dashboard: This dashboard captures metrics that change with time, such as the loss of a model or other model evaluation metrics such as accuracy, precision, recall, f1, and so on.

- Histogram dashboard: This dashboard shows the histogram distribution for a Tensor as it has changed over time.

- Distribution dashboard: This dashboard is similar to the histogram dashboard. However, it displays the histogram as a distribution.

- Graph explorer: This dashboard gives a graphical overview of the TensorFlow computational graph and how information flows from one node to the other. This dashboard provides invaluable insights into the network architecture.

- Image dashboard: This dashboard displays images saved using the method **tf.summary.image**.

- Audio dashboard: This dashboard provides audio clips saved using the method **tf.summary.audio**.

- Embedding projector: The dashboard makes it easy to visualize high-dimensional datasets after they have been transformed using **Embeddings**. The visualization uses principal component analysis (PCA) and another technique called t-distributed Stochastic Neighbor Embedding (t-SNE). Embedding is a technique for capturing the latent variables in a high-dimensional dataset by converting the data units into real numbers that capture their relationship. This technique is broadly similar to how PCA reduces data dimensionality. Embeddings are also useful for converting sparse matrices (matrices made up of mostly zeros) into a dense representation.

- Text dashboard: This dashboard is for displaying textual information.

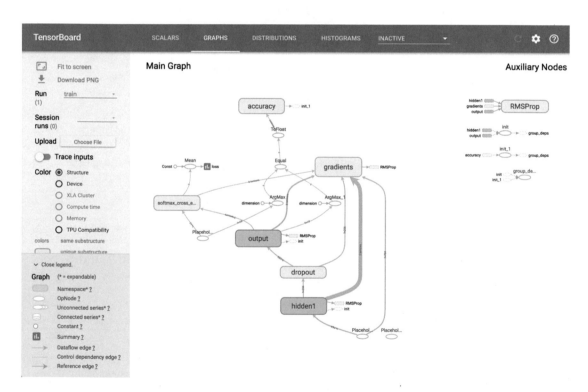

Figure 30-8. *TensorBoard*

Features in TensorFlow 2.0

TensorFlow 2.0 comes with new features for building machine learning models. Some of these new features include

- A more pythonic feel to model design and debugging with eager execution as the de facto execution mode.

- Eager execution enables instant evaluation of TensorFlow operations. This is opposed to previous versions of Tensorflow where we first construct a computational graph and then execute it in a session.

- Using tf.function to transform a Python method into high-performance TensorFlow graphs.

- Using Keras as the core high-level API for model design.

- Using FeatureColumns to parse data as input into Keras models.

- The ease of training on distributed architectures and devices.

To install and work with TensorFlow 2.0 on Google Colab, run

```
!pip install -q tensorflow==2.0.0-beta0
```

The GCP Deep Learning VM has images with TensorFlow 2.0 pre-configured.

A Simple TensorFlow Program

Let's start by building a simple TF program. Here, we will build a graph to find the roots of the quadratic expression $x^2 + 3x - 4 = 0$.

```
# import tensorflow
import tensorflow as tf

# Quadratic expression: x**2 + 3x - 4 = 0.
a = tf.constant(1.0)
b = tf.constant(3.0)
c = tf.constant(-4.0)
```

```
print(a)
print(b)
print(c)
```

```
'Output':
tf.Tensor(1.0, shape=(), dtype=float32)
tf.Tensor(3.0, shape=(), dtype=float32)
tf.Tensor(-4.0, shape=(), dtype=float32)
```

tf.constant() is a Tensor for storing a constant type. Now let's calculate the roots of the expression.

```
x1 = (-b + tf.math.sqrt(b**2 - (4*a*c))) / 2**a
x2 = (-b - tf.math.sqrt(b**2 - (4*a*c))) / 2**a

roots = (x1, x2)
print(roots)
```

```
'Output':
(<tf.Tensor: id=163, shape=(), dtype=float32, numpy=1.0>, <tf.Tensor:
id=175, shape=(), dtype=float32, numpy=-4.0>)
```

TensorFlow 2.0 is eager-first; this implies that operations are executed immediately after they are defined, just like regular python code.

Building Efficient Input Pipelines with the Dataset API

The Dataset API **'tf.data'** offers an efficient mechanism for building robust input pipelines for passing data into a TensorFlow program. This section uses the Boston housing dataset to illustrate working with the Dataset API methods for building data input pipelines in TensorFlow.

```
# import packages
import tensorflow as tf
from tensorflow.keras.datasets import boston_housing

# load dataset and split in train and test sets
(X_train, y_train), (X_test, y_test) = boston_housing.load_data()
```

```
# construct data input pipelines
dataset = tf.data.Dataset.from_tensor_slices((X_train, y_train))
dataset = dataset.shuffle(buffer_size=1000)
dataset = dataset.batch(5)

# retrieve first data batch from dataset
for features, labels in dataset:
    print('Features:', features)
    print('Shape of Features:', features.shape)

    print('Labels:', labels)
    print('Shape of Labels:', labels.shape)
    break

'Output':
Features: tf.Tensor(
[[8.19900e-02 0.00000e+00 1.39200e+01 0.00000e+00 4.37000e-01 6.00900e+00
  4.23000e+01 5.50270e+00 4.00000e+00 2.89000e+02 1.60000e+01 3.96900e+02
  1.04000e+01]
 [8.82900e-02 1.25000e+01 7.87000e+00 0.00000e+00 5.24000e-01 6.01200e+00
  6.66000e+01 5.56050e+00 5.00000e+00 3.11000e+02 1.52000e+01 3.95600e+02
  1.24300e+01]
 [2.90900e-01 0.00000e+00 2.18900e+01 0.00000e+00 6.24000e-01 6.17400e+00
  9.36000e+01 1.61190e+00 4.00000e+00 4.37000e+02 2.12000e+01 3.88080e+02
  2.41600e+01]
 [5.87205e+00 0.00000e+00 1.81000e+01 0.00000e+00 6.93000e-01 6.40500e+00
  9.60000e+01 1.67680e+00 2.40000e+01 6.66000e+02 2.02000e+01 3.96900e+02
  1.93700e+01]
 [1.71710e-01 2.50000e+01 5.13000e+00 0.00000e+00 4.53000e-01 5.96600e+00
  9.34000e+01 6.81850e+00 8.00000e+00 2.84000e+02 1.97000e+01 3.78080e+02
  1.44400e+01]], shape=(5, 13), dtype=float64)
Shape of Features: (5, 13)
Labels: tf.Tensor([21.7 22.9 14.  12.5 16. ], shape=(5,), dtype=float64)
Shape of Labels: (5,)
```

From the preceding code listing, take note of the following:

- The method **'tf.data.Dataset.from_tensor_slices()'** is used to create a Dataset whose elements are Tensor slices.

- The Dataset method **'shuffle()'** shuffles the Dataset at each epoch.

- The Dataset method **'batch()'** is used to set the size of each mini-batch of the Dataset. In the preceding example, each Dataset batch contains five observations.

Linear Regression with TensorFlow

In this section, we use TensorFlow to implement a linear regression machine learning model. In the following example, we use the Boston house-prices dataset from the **Keras dataset package** to build a linear regression model with TensorFlow 2.0.

```
# import packages
import numpy as np
import tensorflow as tf
from tensorflow.keras.datasets import boston_housing
from tensorflow.keras import Model
from sklearn.preprocessing import StandardScaler

# load dataset and split in train and test sets
(X_train, y_train), (X_test, y_test) = boston_housing.load_data()

# standardize the dataset
scaler_X_train = StandardScaler().fit(X_train)
scaler_X_test = StandardScaler().fit(X_test)
X_train = scaler_X_train.transform(X_train)
X_test = scaler_X_test.transform(X_test)

# reshape y-data to become column vector
y_train = np.reshape(y_train, [-1, 1])
y_test = np.reshape(y_test, [-1, 1])

# build the linear model
class LinearRegressionModel(Model):
```

```python
  def __init__(self):
    super(LinearRegressionModel, self).__init__()
    # initialize weight and bias variables
    self.weight = tf.Variable(
        initial_value = tf. random.normal(
            [13, 1], dtype=tf.float64),
        trainable=True)
    self.bias = tf.Variable(initial_value = tf.constant(
        1.0, shape=[], dtype=tf.float64), trainable=True)

  def call(self, inputs):
    return tf.add(tf.matmul(inputs, self.weight), self.bias)

model = LinearRegressionModel()

# parameters
batch_size = 32
learning_rate = 0.01

# use tf.data to batch and shuffle the dataset
train_ds = tf.data.Dataset.from_tensor_slices(
    (X_train, y_train)).shuffle(len(X_train)).batch(batch_size)
test_ds = tf.data.Dataset.from_tensor_slices((X_test, y_test)).batch(batch_size)

loss_object = tf.keras.losses.MeanSquaredError()

optimizer = tf.keras.optimizers.SGD(learning_rate=0.01)

train_loss = tf.keras.metrics.Mean(name='train_loss')
train_rmse = tf.keras.metrics.RootMeanSquaredError(name='train_rmse')

test_loss = tf.keras.metrics.Mean(name='test_loss')
test_rmse = tf.keras.metrics.RootMeanSquaredError(name='test_rmse')

# use tf.GradientTape to train the model
@tf.function
def train_step(inputs, labels):
  with tf.GradientTape() as tape:
    predictions = model(inputs)
    loss = loss_object(labels, predictions)
```

```
  gradients = tape.gradient(loss, model.trainable_variables)
  optimizer.apply_gradients(zip(gradients, model.trainable_variables))

  train_loss(loss)
  train_rmse(labels, predictions)

@tf.function
def test_step(inputs, labels):
  predictions = model(inputs)
  t_loss = loss_object(labels, predictions)

  test_loss(t_loss)
  test_rmse(labels, predictions)

num_epochs = 1000

for epoch in range(num_epochs):
  for train_inputs, train_labels in train_ds:
    train_step(train_inputs, train_labels)

  for test_inputs, test_labels in test_ds:
    test_step(test_inputs, test_labels)

  template = 'Epoch {}, Loss: {}, RMSE: {}, Test Loss: {}, Test RMSE: {}'

  if ((epoch+1) % 100 == 0):
    print (template.format(epoch+1,
                           train_loss.result(),
                           train_rmse.result(),
                           test_loss.result(),
                           test_rmse.result()))
```

'Output':
Epoch 100, Loss: 23.531124114990234, RMSE: 4.862841606140137, Test Loss: 21.077274322509766, Test RMSE: 4.591667175292969
Epoch 200, Loss: 23.51316261291504, RMSE: 4.860987663269043, Test Loss: 21.067768096923828, Test RMSE: 4.590633869171143
Epoch 300, Loss: 23.496540069580078, RMSE: 4.859271049499512, Test Loss: 21.058971405029297, Test RMSE: 4.589677333831787

```
Epoch 400, Loss: 23.481115341186523, RMSE: 4.857677459716797, Test Loss:
21.050806045532227, Test RMSE: 4.588788986206055
Epoch 500, Loss: 23.466760635375977, RMSE: 4.856194019317627, Test Loss:
21.043209075927734, Test RMSE: 4.587962627410889
Epoch 600, Loss: 23.453369140625, RMSE: 4.8548102378845215, Test Loss:
21.036123275756836, Test RMSE: 4.587191581726074
Epoch 700, Loss: 23.440847396850586, RMSE: 4.853515625, Test Loss:
21.029495239257812, Test RMSE: 4.586470603942871
Epoch 800, Loss: 23.429113388061523, RMSE: 4.852302074432373, Test Loss:
21.02336311340332, Test RMSE: 4.585799694061279
Epoch 900, Loss: 23.4180965423584, RMSE: 4.851161956787109, Test Loss:
21.017648696899414, Test RMSE: 4.585177898406982
Epoch 1000, Loss: 23.407730102539062, RMSE: 4.8500895500183105, Test Loss:
21.012271881103516, Test RMSE: 4.584592819213867
```

Here are a few points and methods to take note of in the preceding code listing for linear regression with TensorFlow:

- Note that transformation to standardize the feature dataset is performed after splitting the data into train and test sets. This action is performed in this manner to prevent information from the training data to pollute the test data which must remain unseen by the model.

- The class named **'LinearRegressionModel'** builds a Keras model by subclassing the **'tf.keras.Model'** class. The linear regression model is created as a layer of the neural network in the **'__init__'** method, and it is defined as a forward pass in the **'call'** method. In Chapter 31 on Keras, we will see how to use simpler routines with the Keras Functional API.

- The **'tf.data.Dataset.from_tensor_slices'** method uses the **'.minimize()'** method to update the loss function.

- The squared error loss function is defined with **'tf.keras.losses. MeanSquaredError()'**.

- The gradient descent optimization algorithm is defined using **'tf.keras.optimizers.SGD()'** with the learning rate set as a parameter to the method.

- The method to capture the loss and root mean squared error estimates is defined using **'tf.keras.metrics.Mean(name='train_loss')'** and **'tf.keras.metrics.RootMeanSquaredError()'** functions, respectively.

- The @tf.function is a python decorator to transform a method into high-performance TensorFlow graphs.

- The method **'train_step'** uses the **'tf.GradientTape()'** method to record operations for automatic differentiation. These gradients are later used to minimize the cost function by calling the **'apply_gradients()'** method of the optimization algorithm.

- The method **'test_step'** uses the trained model to obtain predictions on test data.

Classification with TensorFlow

In this example, we'll use the Iris flower dataset to build a multivariable logistic regression machine learning classifier with TensorFlow 2.0. The dataset is gotten from the Scikit-learn dataset package.

```
# import packages
import numpy as np
import tensorflow as tf
from sklearn import datasets
from tensorflow.keras import Model
from sklearn.model_selection import train_test_split
from sklearn.preprocessing import OneHotEncoder

 # load dataset
data = datasets.load_iris()

# separate features and target
X = data.data
y = data.target

# apply one-hot encoding to targets
one_hot_encoder = OneHotEncoder(categories='auto')
```

```python
encode_categorical = y.reshape(len(y), 1)
y = one_hot_encoder.fit_transform(encode_categorical).toarray()

# split in train and test sets
X_train, X_test, y_train, y_test = train_test_split(X, y, shuffle=True)

# build the linear model
class LogisticRegressionModel(Model):
  def __init__(self):
    super(LogisticRegressionModel, self).__init__()
    # initialize weight and bias variables
    self.weight = tf.Variable(
        initial_value = tf.random.normal(
            [4, 3], dtype=tf.float64),
        trainable=True)
    self.bias = tf.Variable(initial_value = tf.random.normal(
        [3], dtype=tf.float64), trainable=True)

  def call(self, inputs):
    return tf.add(tf.matmul(inputs, self.weight), self.bias)

model = LogisticRegressionModel()

# parameters
batch_size = 32
learning_rate = 0.1

# use tf.data to batch and shuffle the dataset
train_ds = tf.data.Dataset.from_tensor_slices(
    (X_train, y_train)).shuffle(len(X_train)).batch(batch_size)
test_ds = tf.data.Dataset.from_tensor_slices((X_test, y_test)).batch(batch_size)

optimizer = tf.keras.optimizers.SGD(learning_rate=learning_rate)

train_loss = tf.keras.metrics.Mean(name='train_loss')
train_accuracy = tf.keras.metrics.Accuracy(name='train_accuracy')

test_loss = tf.keras.metrics.Mean(name='test_loss')
test_accuracy = tf.keras.metrics.Accuracy(name='test_accuracy')
```

```python
# use tf.GradientTape to train the model
@tf.function
def train_step(inputs, labels):
  with tf.GradientTape() as tape:
    predictions = model(inputs)
    loss = tf.reduce_mean(tf.nn.softmax_cross_entropy_with_logits(labels,
    predictions))
  gradients = tape.gradient(loss, model.trainable_variables)
  optimizer.apply_gradients(zip(gradients, model.trainable_variables))

  train_loss(loss)
  train_accuracy(tf.argmax(labels,1), tf.argmax(predictions,1))

@tf.function
def test_step(inputs, labels):
  predictions = model(inputs)
  t_loss = tf.reduce_mean(tf.nn.softmax_cross_entropy_with_logits(labels,
  predictions))

  test_loss(t_loss)
  test_accuracy(tf.argmax(labels,1), tf.argmax(predictions,1))

num_epochs = 1000

for epoch in range(num_epochs):
  for train_inputs, train_labels in train_ds:
    train_step(train_inputs, train_labels)

  for test_inputs, test_labels in test_ds:
    test_step(test_inputs, test_labels)

  template = 'Epoch {}, Loss: {}, Accuracy: {}, Test Loss: {}, Test Accuracy: {}'

  if ((epoch+1) % 100 == 0):
    print (template.format(epoch+1,
                           train_loss.result(),
                           train_accuracy.result()*100,
                           test_loss.result(),
                           test_accuracy.result()*100))
```

```
'Output':
Epoch 100, Loss: 0.3510790765285492, Accuracy: 89.63029479980469, Test
Loss: 0.44924452900886536, Test Accuracy: 84.37885284423828
Epoch 200, Loss: 0.3282322287559509, Accuracy: 91.29582214355469, Test
Loss: 0.43276602029800415, Test Accuracy: 85.73675537109375
Epoch 300, Loss: 0.3093726634979248, Accuracy: 92.46343231201172, Test
Loss: 0.41915151476860046, Test Accuracy: 86.6886978149414
Epoch 400, Loss: 0.29340484738349915, Accuracy: 93.3273696899414, Test
Loss: 0.40762627124786377, Test Accuracy: 87.43070220947266
Epoch 500, Loss: 0.2796294391155243, Accuracy: 93.99247741699219, Test
Loss: 0.3976936936378479, Test Accuracy: 88.27145385742188
Epoch 600, Loss: 0.2675718069076538, Accuracy: 94.52030944824219, Test
Loss: 0.38901543617248535, Test Accuracy: 88.93867492675781
Epoch 700, Loss: 0.25689396262168884, Accuracy: 94.94937896728516, Test
Loss: 0.38134896755218506, Test Accuracy: 89.48106384277344
Epoch 800, Loss: 0.24734711647033691, Accuracy: 95.3050537109375, Test
Loss: 0.3745149075984955, Test Accuracy: 89.9306640625
Epoch 900, Loss: 0.23874221742153168, Accuracy: 95.60466766357422, Test
Loss: 0.3683767020702362, Test Accuracy: 90.30940246582031
Epoch 1000, Loss: 0.23093272745609283, Accuracy: 95.86051177978516, Test
Loss: 0.3628271818161011, Test Accuracy: 90.63280487060547
```

From the preceding code, listing is similar to the example on linear regression with TensorFlow 2.0. However, take note of the following procedures:

- The target variable **'y'** is converted to a one-hot encoded matrix by using the **'OneHotEncoder'** function from Scikit-learn. There exists a TensorFlow method named **'tf.one_hot'** for performing the same function, even easier! The reader is encouraged to Experiment with this.

- Observe how the **'tf.reduce_mean'** and the **'tf.nn.softmax_cross_entropy_with_logits'** methods are used to implement the loss for optimizing the logistic model.

- The Stochastic Gradient Descent optimization algorithm **'tf.keras.optimizers.SGD()'** is used to train the logistic model.

- Observe how the **'weight'** and **'bias'** variables are updated by the gradient descent optimizer within the **'train_step'** method using **'tf. GradientTape()'** to capture and compute the derivatives from the trainable model variables.

- The **'tf.keras.metrics.Accuracy'** method is used to evaluate the accuracy of the model.

Visualizing with TensorBoard

In this section, we will go through visualizing TensorFlow graphs and statistics with TensorBoard. The following code improves on the previous code to build a linear regression model by adding methods to visualize the graph and other variable statistics in TensorBoard using the **'tf.summary'** method calls. The TensorBoard output (illustrated in Figure 30-9) is displayed within the notebook.

```
# import packages
import datetime
import numpy as np
import tensorflow as tf
from tensorflow.keras.datasets import boston_housing
from tensorflow.keras import Model
from sklearn.preprocessing import StandardScaler

# load the TensorBoard notebook extension
%load_ext tensorboard

# load dataset and split in train and test sets
(X_train, y_train), (X_test, y_test) = boston_housing.load_data()

# standardize the dataset
scaler_X_train = StandardScaler().fit(X_train)
scaler_X_test = StandardScaler().fit(X_test)
X_train = scaler_X_train.transform(X_train)
X_test = scaler_X_test.transform(X_test)

# reshape y-data to become column vector
y_train = np.reshape(y_train, [-1, 1])
```

```
y_test = np.reshape(y_test, [-1, 1])

# build the linear model
class LinearRegressionModel(Model):
  def __init__(self):
    super(LinearRegressionModel, self).__init__()
    # initialize weight and bias variables
    self.weight = tf.Variable(
        initial_value = tf. random.normal(
            [13, 1], dtype=tf.float64),
        trainable=True)
    self.bias = tf.Variable(initial_value = tf.constant(
        1.0, shape=[], dtype=tf.float64), trainable=True)

  def call(self, inputs):
    return tf.add(tf.matmul(inputs, self.weight), self.bias)

model = LinearRegressionModel()

# parameters
batch_size = 32
learning_rate = 0.01

# use tf.data to batch and shuffle the dataset
train_ds = tf.data.Dataset.from_tensor_slices(
    (X_train, y_train)).shuffle(len(X_train)).batch(batch_size)
test_ds = tf.data.Dataset.from_tensor_slices((X_test, y_test)).batch(batch_size)

loss_object = tf.keras.losses.MeanSquaredError()
optimizer = tf.keras.optimizers.SGD(learning_rate=0.01)

train_loss = tf.keras.metrics.Mean(name='train_loss')
train_rmse = tf.keras.metrics.RootMeanSquaredError(name='train_rmse')

test_loss = tf.keras.metrics.Mean(name='test_loss')
test_rmse = tf.keras.metrics.RootMeanSquaredError(name='test_rmse')
```

```python
# use tf.GradientTape to train the model
@tf.function
def train_step(inputs, labels):
  with tf.GradientTape() as tape:
    predictions = model(inputs)
    loss = loss_object(labels, predictions)
  gradients = tape.gradient(loss, model.trainable_variables)
  optimizer.apply_gradients(zip(gradients, model.trainable_variables))

  train_loss(loss)
  train_rmse(labels, predictions)

@tf.function
def test_step(inputs, labels):
  predictions = model(inputs)
  t_loss = loss_object(labels, predictions)

  test_loss(t_loss)
  test_rmse(labels, predictions)

# Clear any logs from previous runs
!rm -rf ./logs/

# set up summary writers to write the summaries to disk in a different logs
directory
current_time = datetime.datetime.now().strftime("%Y%m%d-%H%M%S")
train_log_dir = 'logs/gradient_tape/' + current_time + '/train'
test_log_dir = 'logs/gradient_tape/' + current_time + '/test'
train_summary_writer = tf.summary.create_file_writer(train_log_dir)
test_summary_writer = tf.summary.create_file_writer(test_log_dir)

num_epochs = 1000

for epoch in range(num_epochs):
  for train_inputs, train_labels in train_ds:
    train_step(train_inputs, train_labels)
  with train_summary_writer.as_default():
    tf.summary.scalar('loss', train_loss.result(), step=epoch)
    tf.summary.scalar('rmse', train_rmse.result(), step=epoch)
```

```
for test_inputs, test_labels in test_ds:
  test_step(test_inputs, test_labels)
with test_summary_writer.as_default():
  tf.summary.scalar('loss', test_loss.result(), step=epoch)
  tf.summary.scalar('rmse', test_rmse.result(), step=epoch)

template = 'Epoch {}, Loss: {}, RMSE: {}, Test Loss: {}, Test RMSE: {}'

if ((epoch+1) % 100 == 0):
  print (template.format(epoch+1,
                         train_loss.result(),
                         train_rmse.result(),
                         test_loss.result(),
                         test_rmse.result()))

# Reset metrics every epoch
train_loss.reset_states()
test_loss.reset_states()
train_rmse.reset_states()
test_rmse.reset_states()
```

```
'Output':
Epoch 100, Loss: 22.03757667541504, RMSE: 4.726028919219971, Test Loss:
29.092111587524414, Test RMSE: 4.577760696411133
Epoch 200, Loss: 21.973844528198242, RMSE: 4.719051837921143, Test Loss:
29.113895416259766, Test RMSE: 4.585252285003662
Epoch 300, Loss: 21.970674514770508, RMSE: 4.7187066078186035, Test Loss:
29.13644790649414, Test RMSE: 4.587917327880859
Epoch 400, Loss: 21.970500946044922, RMSE: 4.718687534332275, Test Loss:
29.1422119140625, Test RMSE: 4.588583469390869
Epoch 500, Loss: 21.970489501953125, RMSE: 4.718685626983643, Test Loss:
29.14352035522461, Test RMSE: 4.588735103607178
Epoch 600, Loss: 21.970487594604492, RMSE: 4.718685626983643, Test Loss:
29.143817901611328, Test RMSE: 4.58876895904541
Epoch 700, Loss: 21.970487594604492, RMSE: 4.718685626983643, Test Loss:
29.143882751464844, Test RMSE: 4.588776111602783
```

Epoch 800, Loss: 21.970487594604492, RMSE: 4.718685626983643, Test Loss: 29.14389419555664, Test RMSE: 4.588778018951416

Epoch 900, Loss: 21.970487594604492, RMSE: 4.718685626983643, Test Loss: 29.143898010253906, Test RMSE: 4.588778495788574

Epoch 1000, Loss: 21.970487594604492, RMSE: 4.718685626983643, Test Loss: 29.143898010253906, Test RMSE: 4.588778495788574

```
# launch tensorboard
%tensorboard --logdir logs/gradient_tape
```

From the preceding code listing, take note of the following steps:

- The **'tf.summary.create_file_writer'** method creates summary writers to write the summaries to disk.

- The **'tf.summary.scalar'** method is used to capture scalar metrics for TensorBoard.

- The magic command '%tensorboard' is used to launch TensorBoard by pointing to the appropriate log directory.

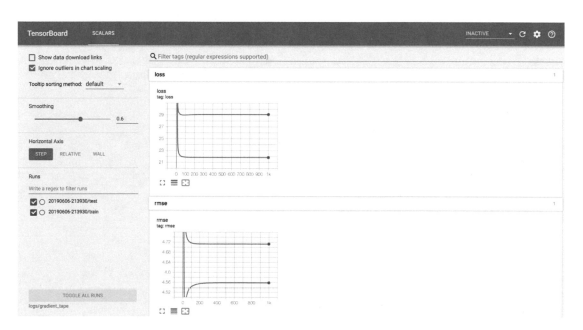

Figure 30-9. *TensorBoard visualization dashboard for linear regression metrics*

Running TensorFlow with GPUs

GPU is short for graphics processing unit. It is a specialized processor designed for carrying out complex computations on large memory blocks. GPUs provide more efficient processing for building deep learning models.

TensorFlow can leverage processing on multiple GPUs to speed up computation especially when training a complex network architecture. To take advantage of parallel processing, a replica of the network architecture resides on each GPU machine and trains a subset of the data. However, for synchronous updates, the model parameters from each tower (or GPU machines) are stored and updated on a CPU. It turns out that CPUs are generally good at mean or averaging processing. A diagram of this operation is shown in Figure 30-10.

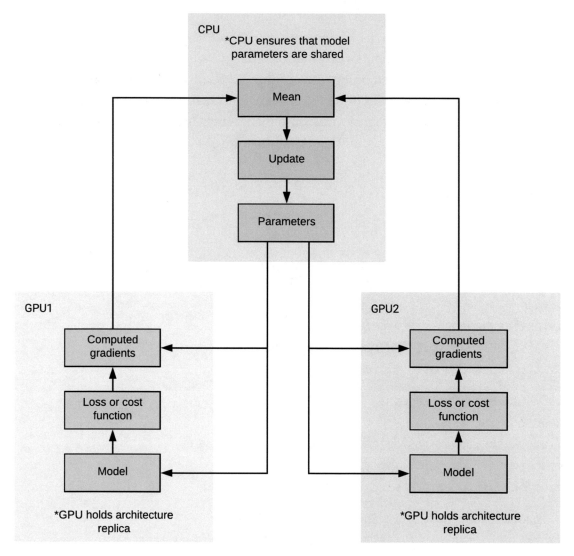

Figure 30-10. *Framework for training on multiple GPUs*

TensorFlow 2.0 performs distributed training across multiple machines (i.e., CPUs, GPUs, or TPUs) using the **'tf.distribute.Strategy'** API. To use GPUs on Google Colab, first change the runtime type to GPU and install the TensorFlow with GPU library by running the following code in the notebook cell:

```
!pip install -q tf-nightly-gpu-2.0-preview
```

The following code block uses GPUs for model training. In this example we train a simple regression model on the Boston housing dataset. The method **'tf.distribute. MirroredStrategy()'** implements a distribution strategy called MirroredStrategy. This strategy supports distributed training with multiple GPUs on a single machine. The code is similar to the previous code for linear regression with TensorFlow 2.0. However, minimal changes are added to make the components such as variables, layers, models, optimizers, metrics, summaries, and checkpoints strategy-aware using the strategy scope().

```
# import TensorFlow 2.0 with GPU
!pip install -q tf-nightly-gpu-2.0-preview

# confirm tensorflow can see GPU
import tensorflow as tf

device_name = tf.test.gpu_device_name()
if device_name != '/device:GPU:0':
  raise SystemError('GPU device not found')
print('Found GPU at: {}'.format(device_name))

# import other packages
import numpy as np
from tensorflow.keras.datasets import boston_housing
from tensorflow.keras import Model
from sklearn.preprocessing import StandardScaler

# load dataset and split in train and test sets
(X_train, y_train), (X_test, y_test) = boston_housing.load_data()

# standardize the dataset
scaler_X_train = StandardScaler().fit(X_train)
scaler_X_test = StandardScaler().fit(X_test)
X_train = scaler_X_train.transform(X_train)
X_test = scaler_X_test.transform(X_test)

# reshape y-data to become column vector
y_train = np.reshape(y_train, [-1, 1])
y_test = np.reshape(y_test, [-1, 1])
```

```python
# build the linear model
class LinearRegressionModel(Model):
  def __init__(self):
    super(LinearRegressionModel, self).__init__()
    # initialize weight and bias variables
    self.weight = tf.Variable(
        initial_value = tf. random.normal(
            [13, 1], dtype=tf.float64),
        trainable=True)
    self.bias = tf.Variable(initial_value = tf.constant(
        1.0, shape=[], dtype=tf.float64), trainable=True)

  def call(self, inputs):
    return tf.add(tf.matmul(inputs, self.weight), self.bias)

# create a strategy to distribute the variables and the graph
strategy = tf.distribute.MirroredStrategy()

# print number of machines with GPUs
print ('Number of devices: {}'.format(strategy.num_replicas_in_sync))

# parameters
batch_size_per_replica = 32
global_batch_size = batch_size_per_replica * strategy.num_replicas_in_sync

learning_rate = 0.01

# create the distributed datasets inside a strategy.scope:
with strategy.scope():
  train_ds = tf.data.Dataset.from_tensor_slices(
      (X_train, y_train)).shuffle(len(X_train)).batch(global_batch_size)
  train_dist_ds = strategy.experimental_distribute_dataset(train_ds)

  test_ds = tf.data.Dataset.from_tensor_slices((X_test, y_test)).
  batch(global_batch_size)
  test_dist_ds = strategy.experimental_distribute_dataset(test_ds)
```

```
# define the loss function
with strategy.scope():
  # Set reduction to `none` so we can do the reduction afterwards and
  divide by
  # global batch size.
  loss_object = tf.keras.losses.MeanSquaredError(
      reduction=tf.keras.losses.Reduction.NONE)

  def compute_loss(labels, predictions):
    per_example_loss = loss_object(labels, predictions)
    return tf.reduce_sum(per_example_loss) * (1. / global_batch_size)

# define metrics to track loss and rmse
with strategy.scope():
  test_loss = tf.keras.metrics.Mean(name='test_loss')

  train_rmse = tf.keras.metrics.RootMeanSquaredError(
      name='train_rmse')
  test_rmse = tf.keras.metrics.RootMeanSquaredError(
      name='test_rmse')

# model and optimizer must be created under `strategy.scope`.
with strategy.scope():
  model = LinearRegressionModel()
  optimizer = tf.keras.optimizers.SGD(learning_rate=0.01)

with strategy.scope():
  def train_step(inputs, labels):
    with tf.GradientTape() as tape:
      predictions = model(inputs)
      loss = compute_loss(labels, predictions)

    gradients = tape.gradient(loss, model.trainable_variables)
    optimizer.apply_gradients(zip(gradients, model.trainable_variables))

    train_rmse.update_state(labels, predictions)
    return loss
```

```python
def test_step(inputs, labels):
    predictions = model(inputs)
    t_loss = loss_object(labels, predictions)

    test_loss.update_state(t_loss)
    test_rmse.update_state(labels, predictions)

num_epochs = 1000

with strategy.scope():
    # `experimental_run_v2` replicates the provided computation and runs it
    # with the distributed input.
    @tf.function
    def distributed_train_step(inputs, labels):
        per_replica_losses = strategy.experimental_run_v2(train_step,
                                            args=(inputs, labels))
        return strategy.reduce(tf.distribute.ReduceOp.SUM, per_replica_losses,
                            axis=None)

    @tf.function
    def distributed_test_step(inputs, labels):
        return strategy.experimental_run_v2(test_step, args=(inputs, labels))

    for epoch in range(num_epochs):
        # Train loop
        total_loss = 0.0
        num_batches = 0
        for train_inputs, train_labels in train_dist_ds:
            total_loss += distributed_train_step(train_inputs, train_labels)
            num_batches += 1
        train_loss = total_loss / num_batches

        # Test loop
        for test_inputs, test_labels in test_dist_ds:
            distributed_test_step(test_inputs, test_labels)

        if (epoch+1) % 100 == 0:
            template = ("Epoch {}, Loss: {}, RMSE: {}, Test Loss: {}, "
                        "Test RMSE: {}")
```

```
print (template.format(epoch+1, train_loss,
                       train_rmse.result(), test_loss.result(),
                       test_rmse.result()))

test_loss.reset_states()
train_rmse.reset_states()
test_rmse.reset_states()
```

'Output:'
Epoch 100, Loss: 21.673020569627965, RMSE: 4.724063396453857, Test Loss: 20.915191650390625, Test RMSE: 4.573312759399414
Epoch 200, Loss: 21.594741116702117, RMSE: 4.715524196624756, Test Loss: 20.994861602783203, Test RMSE: 4.582014560699463
Epoch 300, Loss: 21.590902259189097, RMSE: 4.7151055335998535, Test Loss: 21.02731704711914, Test RMSE: 4.585555076599121
Epoch 400, Loss: 21.59074064145569, RMSE: 4.715087413787842, Test Loss: 21.03565216064453, Test RMSE: 4.5864644050598145
Epoch 500, Loss: 21.590740279510765, RMSE: 4.715087413787842, Test Loss: 21.037595748901367, Test RMSE: 4.586676120758057
Epoch 600, Loss: 21.590742194311133, RMSE: 4.715087890625, Test Loss: 21.03803825378418, Test RMSE: 4.586724281311035
Epoch 700, Loss: 21.59074262401866, RMSE: 4.715087890625, Test Loss: 21.03813934326172, Test RMSE: 4.586735248565674
Epoch 800, Loss: 21.59074272223048, RMSE: 4.715087413787842, Test Loss: 21.038162231445312, Test RMSE: 4.586737632751465
Epoch 900, Loss: 21.59074286927267, RMSE: 4.715087413787842, Test Loss: 21.03816795349121, Test RMSE: 4.586737632751465
Epoch 1000, Loss: 21.590742907190307, RMSE: 4.715087413787842, Test Loss: 21.03816795349121, Test RMSE: 4.586738109588623

Please note the following from the preceding code block:

- When writing a custom training loop, sum the per example losses and divide the sum by the global batch size. In the code tf.reduce_sum(per_example_loss) ∗ (1. / global_batch_size). This needs to be done because after calculation on each replica, the gradients are synced across the replicas by summing them. When using tf.keras.losses classes, the loss reduction needs to be explicitly specified to be one of NONE or SUM.

TensorFlow High-Level APIs: Using Estimators

In this section, we will use the high-level TensorFlow Estimator API for modeling with premade Estimators. Estimators provide another high-level API for building TensorFlow models for execution on CPUs, GPUs, or TPUs with minimal code modification.

The following steps are typically followed when working with premade Estimators:

1. Write the **'input_fn'** to handle the data pipeline.

2. Define the type of data attributes into the model using feature columns **'tf.feature_column'**.

3. Instantiate one of the premade Estimators by passing in the feature columns and other relevant attributes.

4. Use the **'train()'**, **'evaluate()'**, and **'predict()'** methods to train and evaluate the model on evaluation dataset and use the model to make prediction/inference.

Let's see a simple example of working with a TensorFlow premade Estimator again using the Boston housing dataset.

Note   Reset session before running the following cells and change runtime type to None.

```
# import packages
import datetime
import numpy as np
import pandas as pd
import tensorflow as tf
from tensorflow.keras.datasets import boston_housing
from tensorflow.keras import Model
from sklearn.preprocessing import StandardScaler

# load dataset and split in train and test sets
(X_train, y_train), (X_test, y_test) = boston_housing.load_data()
```

```
# standardize the dataset
scaler_X_train = StandardScaler().fit(X_train)
scaler_X_test = StandardScaler().fit(X_test)
X_train = scaler_X_train.transform(X_train)
X_test = scaler_X_test.transform(X_test)

# reshape y-data to become column vector
y_train = np.reshape(y_train, [-1, 1])
y_test = np.reshape(y_test, [-1, 1])

# parameters
batch_size = 32
learning_rate = 0.01

# create an input_fn
def input_fn(features, labels, batch_size=30, training=True):
  dataset = tf.data.Dataset.from_tensor_slices((features, labels))
  if training:
      dataset = dataset.shuffle(buffer_size=1000)
      dataset = dataset.repeat()
  return dataset.batch(batch_size)

# use feature columns to define the attributes to the model
feature_columns = []
columns_names = []
for i in range(X_train.shape[1]):
  feature_columns.append(tf.feature_column.numeric_column(key=str(i)))
  columns_names.append(str(i))

# instantiate a LinearRegressor Estimator
estimator = tf.estimator.DNNRegressor(
    feature_columns=feature_columns,
    hidden_units=[20]
)

# convert feature datasets to dictionary
X_train_pd = pd.DataFrame(X_train)
X_train_pd.columns = columns_names
```

```
X_test_pd = pd.DataFrame(X_test)
X_test_pd.columns = columns_names

# train model
estimator.train(input_fn=lambda:input_fn(dict(X_train_pd), y_train),
steps=2000)

# evaluate model
metrics = estimator.evaluate(input_fn=lambda:input_fn(dict(X_test_pd),
y_test, training=False))

# print model metrics
metrics
```

Neural Networks with Keras

In this section, we will use the Sequential and Functional Keras API to build a simple neural network model. A Sequential API is the most commonly used method to build deep neural network models by stacking one layer on another. The Functional API offers more flexibility to build more complex neural network architectures. Both API methods are relatively easy to construct in Keras as we will see in the examples.

Subclassing a model as we did in the preceding examples provides even more flexibility for building and inspecting complex models. However, the code is more verbose and may be prone to errors. This technique should be used when it makes the most sense to, depending on the problem use case. We used them previously to serve as an illustration.

The following examples will use the Iris Dataset to build a neural network with one hidden layer as illustrated in Figure 30-11.

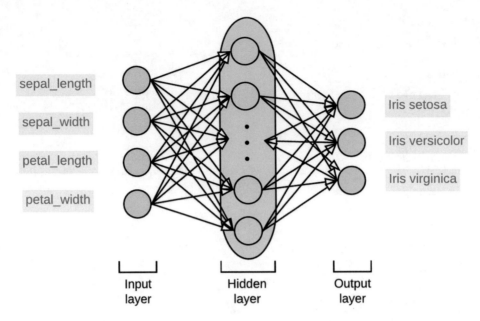

Figure 30-11. *Iris dataset – neural network architecture*

Using the Keras Sequential API

This code segment will construct a neural network model with the Sequential API using the method **'tf.keras.Sequential()'** to stack layers on each other. The model creates a hidden layer with 32 neurons and an output layer with 3 output units because the Iris target contains 3 classes.

```
!pip install -q tensorflow==2.0.0-beta0

# import packages
import tensorflow as tf
import pandas as pd
from sklearn.preprocessing import OneHotEncoder

# dataset url
train_data_url = "https://storage.googleapis.com/download.tensorflow.org/
data/iris_training.csv"
test_data_url = "https://storage.googleapis.com/download.tensorflow.org/
data/iris_test.csv"
```

```python
# define column names
columns = ['sepal_length', 'sepal_width', 'petal_length', 'petal_width', 'species']

# download and load the csv files
train_data = pd.read_csv(tf.keras.utils.get_file('iris_train.csv', train_
data_url),

                                skiprows=1, header=None, names=columns)

test_data = pd.read_csv(tf.keras.utils.get_file('iris_test.csv', test_data_url),
                                skiprows=1, header=None, names=columns)

# separate the features and targets
(X_train, y_train) = (train_data.iloc[:,0:-1], train_data.iloc[:,-1])
(X_test, y_test) = (test_data.iloc[:,0:-1], test_data.iloc[:,-1])

# apply one-hot encoding to targets
y_train=tf.keras.utils.to_categorical(y_train)
y_test=tf.keras.utils.to_categorical(y_test)

# create the sequential model
def model_fn():
    model = tf.keras.Sequential()
    # Add a densely-connected layer with 32 units to the model:
    model.add(tf.keras.layers.Dense(32, activation='sigmoid', input_dim=4))
    # Add a softmax layer with 3 output units:
    model.add(tf.keras.layers.Dense(3, activation='softmax'))

    # compile the model
    model.compile(optimizer=tf.keras.optimizers.SGD(),
                    loss='categorical_crossentropy',
                    metrics=['accuracy'])
    return model

# parameters
batch_size=50
```

```
# use tf.data to batch and shuffle the dataset
train_ds = tf.data.Dataset.from_tensor_slices(
    (X_train.values, y_train)).shuffle(len(X_train)).repeat().batch(batch_size)
test_ds = tf.data.Dataset.from_tensor_slices((X_test.values, y_test)).
batch(batch_size)

# build train model
model = model_fn()

# print train model summary
model.summary()

# train the model
history = model.fit(train_ds,steps_per_epoch=5000)

# evaluate the model
score = model.evaluate(test_ds)

print('Test loss: {:.2f} \nTest accuracy: {:.2f}%'.format(score[0],
score[1]*100))

'Output':
Test loss: 0.22
Test accuracy: 96.67%
```

Using the Keras Functional API

The general code pattern for the Functional API is structurally the same as the Sequential version. The only change here is in how the network model is constructed. We also demonstrated the Keras feature for printing the graph of the model in this example. The output is illustrated in Figure 30-12.

```
!pip install -q tensorflow==2.0.0-beta0

# import packages
import tensorflow as tf
import pandas as pd
from sklearn.preprocessing import OneHotEncoder
```

```python
# dataset url
train_data_url = "https://storage.googleapis.com/download.tensorflow.org/
data/iris_training.csv"
test_data_url = "https://storage.googleapis.com/download.tensorflow.org/
data/iris_test.csv"

# define column names
columns = ['sepal_length', 'sepal_width', 'petal_length', 'petal_width', 'species']

# download and load the csv files
train_data = pd.read_csv(tf.keras.utils.get_file('iris_train.csv',
train_data_url),
                                  skiprows=1, header=None, names=columns)

test_data = pd.read_csv(tf.keras.utils.get_file('iris_test.csv', test_data_url),
                                  skiprows=1, header=None, names=columns)

# separate the features and targets
(X_train, y_train) = (train_data.iloc[:,0:-1], train_data.iloc[:,-1])
(X_test, y_test) = (test_data.iloc[:,0:-1], test_data.iloc[:,-1])

# apply one-hot encoding to targets
y_train=tf.keras.utils.to_categorical(y_train)
y_test=tf.keras.utils.to_categorical(y_test)

# create the functional model
def model_fn():
    # Model input
    model_input = tf.keras.layers.Input(shape=(4,))
    # Adds a densely-connected layer with 32 units to the model:
    x = tf.keras.layers.Dense(32, activation='relu')(model_input)
    # Add a softmax layer with 3 output units:
    predictions = tf.keras.layers.Dense(3, activation='softmax')(x)

    # the model
    model = tf.keras.Model(inputs=model_input,
                           outputs=predictions,
                           name='iris_model')
```

```
    # compile the model
    model.compile(optimizer='sgd',
                    loss='categorical_crossentropy',
                    metrics=['accuracy'])
    return model

# parameters
batch_size=50

# use tf.data to batch and shuffle the dataset
train_ds = tf.data.Dataset.from_tensor_slices(
    (X_train.values, y_train)).shuffle(len(X_train)).repeat().batch(batch_size)
test_ds = tf.data.Dataset.from_tensor_slices((X_test.values, y_test)).
batch(batch_size)

# build train model
model = model_fn()

# print train model summary
model.summary()

# plot the model as a graph
tf.keras.utils.plot_model(model, 'keras_iris_model.png', show_shapes=True)

# train the model
history = model.fit(train_ds, steps_per_epoch=5000)

# evaluate the model
score = model.evaluate(test_ds)

print('Test loss: {:.2f} \nTest accuracy: {:.2f}%'.format(score[0],
score[1]*100))


'Output':
Test loss: 0.07
Test accuracy: 96.67%
```

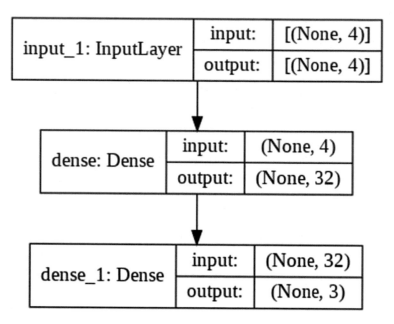

Figure 30-12. *The graph of the model – produced with Keras*

Model Visualization with Keras

With Keras, it is quite easy and straightforward to plot the metrics of the model to have a better graphical perspective as to how the model is performing for every training epoch. This view is also useful for dealing with issues of bias or variance of the model.

A callback function of the 'model.fit()' method returns the loss and evaluation score for each epoch. This information is stored in a variable and plotted.

In this example, we use the same Iris dataset model to illustrate visualization with Keras. The plots of the loss and accuracy of the model at each epoch are shown in Figure 30-13 and Figure 30-14, respectively.

```
!pip install -q tensorflow==2.0.0-beta0

# import packages
import tensorflow as tf
import pandas as pd
import matplotlib.pyplot as plt
from sklearn.preprocessing import OneHotEncoder
```

```python
# dataset url
train_data_url = "https://storage.googleapis.com/download.tensorflow.org/
data/iris_training.csv"
test_data_url = "https://storage.googleapis.com/download.tensorflow.org/
data/iris_test.csv"

# define column names
columns = ['sepal_length', 'sepal_width', 'petal_length', 'petal_width',
'species']

# download and load the csv files
train_data = pd.read_csv(tf.keras.utils.get_file('iris_train.csv', train_
data_url),

                        skiprows=1, header=None, names=columns)

test_data = pd.read_csv(tf.keras.utils.get_file('iris_test.csv', test_data_url),
                        skiprows=1, header=None, names=columns)

# separate the features and targets
(X_train, y_train) = (train_data.iloc[:,0:-1], train_data.iloc[:,-1])
(X_test, y_test) = (test_data.iloc[:,0:-1], test_data.iloc[:,-1])

# apply one-hot encoding to targets
y_train=tf.keras.utils.to_categorical(y_train)
y_test=tf.keras.utils.to_categorical(y_test)

# create the functional model
def model_fn():
    # Model input
    model_input = tf.keras.layers.Input(shape=(4,))
    # Adds a densely-connected layer with 32 units to the model:
    x = tf.keras.layers.Dense(32, activation='relu')(model_input)
    # Add a softmax layer with 3 output units:
    predictions = tf.keras.layers.Dense(3, activation='softmax')(x)

    # the model
    model = tf.keras.Model(inputs=model_input,
                           outputs=predictions,
                           name='iris_model')
```

```
    # compile the model
    model.compile(optimizer='sgd',
                  loss='categorical_crossentropy',
                  metrics=['accuracy'])
    return model

# parameters
batch_size=50

# use tf.data to batch and shuffle the dataset
train_ds = tf.data.Dataset.from_tensor_slices(
    (X_train.values, y_train)).shuffle(len(X_train)).repeat().batch(batch_size)
test_ds = tf.data.Dataset.from_tensor_slices((X_test.values, y_test)).
batch(batch_size)

# build train model
model = model_fn()

# print train model summary
model.summary()

# train the model
history = model.fit(train_ds, epochs=10,
                    steps_per_epoch=100,
                    validation_data=test_ds)

# list metrics returned from callback function
history.history.keys()

# plot loss metric
plt.figure(1)
plt.plot(history.history['loss'], '--')
plt.plot(history.history['val_loss'], '--')
plt.title('Model loss per epoch: Training')
plt.ylabel('loss')
plt.xlabel('epoch')
plt.legend(['train', 'evaluation'])
plt.show()
```

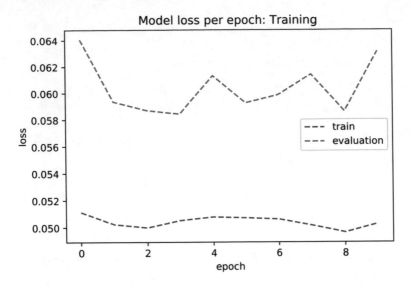

Figure 30-13. *Model loss per epoch*

```
# plot accuracy metric
plt.figure(2)
plt.plot(history.history['accuracy'], '--')
plt.plot(history.history['val_accuracy'], '--')
plt.title('Model accuracy per epoch: Training')
plt.ylabel('loss')
plt.xlabel('epoch')
plt.legend(['train', 'evaluation'])
plt.show()
```

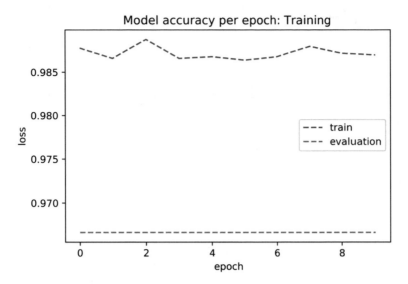

Figure 30-14. *Model accuracy per epoch*

TensorBoard with Keras

To visualize models with TensorBoard, attach a TensorBoard callback **'tf.keras. callbacks.TensorBoard()'** to the **'model.fit()'** method before training the model. The model graph, scalars, histograms, and other metrics are stored as event files in the log directory.

For this example, we modify the Iris model to use TensorBoard. The TensorBoard output is shown in Figure 30-15.

```
!pip install -q tensorflow==2.0.0-beta0

# import packages
import tensorflow as tf
import pandas as pd
from sklearn.preprocessing import OneHotEncoder

# load the TensorBoard notebook extension
%load_ext tensorboard

# dataset url
train_data_url = "https://storage.googleapis.com/download.tensorflow.org/
data/iris_training.csv"
```

```
test_data_url = "https://storage.googleapis.com/download.tensorflow.org/
data/iris_test.csv"

# define column names
columns = ['sepal_length', 'sepal_width', 'petal_length', 'petal_width',
'species']

# download and load the csv files
train_data = pd.read_csv(tf.keras.utils.get_file('iris_train.csv',
train_data_url),
                                  skiprows=1, header=None, names=columns)

test_data = pd.read_csv(tf.keras.utils.get_file('iris_test.csv', test_data_url),
                              skiprows=1, header=None, names=columns)

# separate the features and targets
(X_train, y_train) = (train_data.iloc[:,0:-1], train_data.iloc[:,-1])
(X_test, y_test) = (test_data.iloc[:,0:-1], test_data.iloc[:,-1])

# apply one-hot encoding to targets
y_train=tf.keras.utils.to_categorical(y_train)
y_test=tf.keras.utils.to_categorical(y_test)

# create the functional model
def model_fn():
    # Model input
    model_input = tf.keras.layers.Input(shape=(4,))
    # Adds a densely-connected layer with 32 units to the model:
    x = tf.keras.layers.Dense(32, activation='relu')(model_input)
    # Add a softmax layer with 3 output units:
    predictions = tf.keras.layers.Dense(3, activation='softmax')(x)

    # the model
    model = tf.keras.Model(inputs=model_input,
                           outputs=predictions,
                           name='iris_model')
```

```python
    # compile the model
    model.compile(optimizer='sgd',
                  loss='categorical_crossentropy',
                  metrics=['accuracy'])
    return model

# parameters
batch_size=50

# use tf.data to batch and shuffle the dataset
train_ds = tf.data.Dataset.from_tensor_slices(
    (X_train.values, y_train)).shuffle(len(X_train)).repeat().batch(batch_size)
test_ds = tf.data.Dataset.from_tensor_slices((X_test.values, y_test)).
batch(batch_size)

# build train model
model = model_fn()

# print train model summary
model.summary()

# tensorboard
tensorboard = tf.keras.callbacks.TensorBoard(log_dir='./tmp/logs_iris_keras',
                                             histogram_freq=0, write_
                                             graph=True,
                                             write_images=True)

# assign callback
callbacks = [tensorboard]

# train the model
history = model.fit(train_ds, epochs=10,
                    steps_per_epoch=100,
                    validation_data=test_ds,
                    callbacks=callbacks)
```

```
# evaluate the model
score = model.evaluate(test_ds)
print('Test loss: {:.2f} \nTest accuracy: {:.2f}%'.format(score[0],
score[1]*100))

# execute the command to run TensorBoard
%tensorboard --logdir tmp/logs_iris_keras
```

Figure 30-15. *TensorBoard output of Iris model*

Checkpointing to Select Best Models

Checkpointing makes it possible to save the weights of the neural network model when there is an increase in the validation accuracy metric. This is achieved in Keras using the **'tf.keras.callbacks.ModelCheckpoint()'**. The saved weights can then be loaded back into the model and used to make predictions. Using the Iris dataset, we'll build a model that saves the weights to file only when there is an improvement in the validation set performance. For completeness sake as we have done in the previous segments, we will produce this example within a complete code listing.

```
!pip install -q tensorflow==2.0.0-beta0
```

```
# import packages
import tensorflow as tf
import pandas as pd
from sklearn.preprocessing import OneHotEncoder

# dataset url
train_data_url = "https://storage.googleapis.com/download.tensorflow.org/
data/iris_training.csv"
test_data_url = "https://storage.googleapis.com/download.tensorflow.org/
data/iris_test.csv"

# define column names
columns = ['sepal_length', 'sepal_width', 'petal_length', 'petal_width', 'species']

# download and load the csv files
train_data = pd.read_csv(tf.keras.utils.get_file('iris_train.csv', train_
data_url),
                                  skiprows=1, header=None, names=columns)

test_data = pd.read_csv(tf.keras.utils.get_file('iris_test.csv', test_data_url),
                                  skiprows=1, header=None, names=columns)

# separate the features and targets
(X_train, y_train) = (train_data.iloc[:,0:-1], train_data.iloc[:,-1])
(X_test, y_test) = (test_data.iloc[:,0:-1], test_data.iloc[:,-1])

# apply one-hot encoding to targets
y_train=tf.keras.utils.to_categorical(y_train)
y_test=tf.keras.utils.to_categorical(y_test)

# create the functional model
def model_fn():
    # Model input
    model_input = tf.keras.layers.Input(shape=(4,))
    # Adds a densely-connected layer with 32 units to the model:
    x = tf.keras.layers.Dense(32, activation='relu')(model_input)
    # Add a softmax layer with 3 output units:
    predictions = tf.keras.layers.Dense(3, activation='softmax')(x)
```

```
    # the model
    model = tf.keras.Model(inputs=model_input,
                           outputs=predictions,
                           name='iris_model')

    # compile the model
    model.compile(optimizer='sgd',
                  loss='categorical_crossentropy',
                  metrics=['accuracy'])
    return model

# parameters
batch_size=50

# use tf.data to batch and shuffle the dataset
train_ds = tf.data.Dataset.from_tensor_slices(
    (X_train.values, y_train)).shuffle(len(X_train)).repeat().batch(batch_size)
test_ds = tf.data.Dataset.from_tensor_slices((X_test.values, y_test)).
batch(batch_size)

# build train model
model = model_fn()

# print train model summary
model.summary()

# checkpointing
checkpoint = tf.keras.callbacks.ModelCheckpoint(
    './tmp/iris_weights.h5',
    monitor='val_accuracy',
    verbose=1,
    save_best_only=True,
    mode='max')

# assign callback
callbacks = [checkpoint]
```

```
# train the model
history = model.fit(train_ds, epochs=10,
                    steps_per_epoch=100,
                    validation_data=test_ds,
                    callbacks=callbacks)

# build evaluation model and upload saved weights
eval_model = model_fn()
eval_model.load_weights('./tmp/iris_weights.h5')

# evaluate the model
score = eval_model.evaluate(test_ds)
print('Test loss: {:.2f} \nTest accuracy: {:.2f}%'.format(score[0],
score[1]*100))
```

This chapter covered the foundation of working with TensorFlow 2.0 and its exciting features for developing machine learning models. Some of these new features include a more pythonic feel to model design and debugging, using tf.function to transform a Python method into high-performance TensorFlow graphs, using Keras as the core high-level API for model design, using FeatureColumns to parse data as input into Keras models, and the ease of training on distributed architectures and devices. The chapter also covered the principles of building models using the high-level Estimator API.

In the next chapters, we will take a deeper dive into deep neural networks and how they are implemented in TensorFlow with Keras. In TensorFlow 2.0, Keras is the de facto method for developing neural network.

CHAPTER 31

The Multilayer Perceptron (MLP)

The multilayer perceptron (MLP) is the fundamental example of a deep neural network. The architecture of a MLP consists of multiple hidden layers to capture more complex relationships that exist in the training dataset. Another name for the MLP is the deep feedforward neural network (DFN). An illustration of an MLP is shown in Figure 31-1.

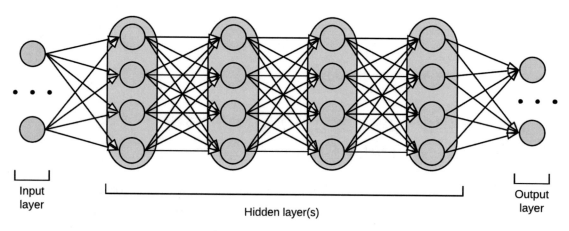

Input layer   Hidden layer(s)   Output layer

Figure 31-1. *Deep feedforward neural network*

The Concept of Hierarchies

The more the number of hidden layers in a neural network, the deeper the network becomes. Deep networks are able to learn more sophisticated representations of the inputs. The concept of hierarchical representation is when each layer learns a set of features that describe the input and hierarchically pass that information across the hidden layers. Initially, the hidden layers closer to the input layer learn a simple set

© Ekaba Bisong 2019
E. Bisong, *Building Machine Learning and Deep Learning Models on Google Cloud Platform,*
https://doi.org/10.1007/978-1-4842-4470-8_31

of features, which then grow to increasingly complex features as information flows to deeper layers of the network, to capture the mapping between the inputs and the target. See Figure 31-2.

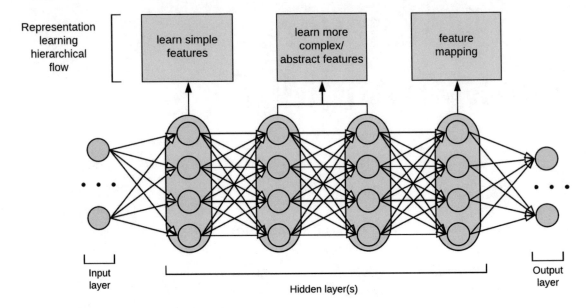

Figure 31-2. *Hierarchical learning*

Choosing the Number of Hidden Layers: Bias/Variance Trade-Off

From experience, increasing the number of hidden layers may improve the representational quality of the network; however, arbitrarily increasing the number of hidden layers in your network design can have detrimental effects on the overall network performance with respect to generalizing to unseen observations. This is because the neural network will learn more closely the irreducible errors inherent in the training dataset and will fail to generalize to new examples.

Appropriate caution should be taken when selecting the number of hidden layers to avoid overfitting. Regularization techniques for neural networks such as Tikhonov regularization, Dropout, or early stopping are different methods of mitigating overfitting. Regularization for neural networks will be covered in more detail in a later section.

Empirically, one hidden layer will produce good results for simple learning problems, but if the number of output classes increases or there exists a high degree of non-linearities among the data features, then it is recommended to add more layers while taking care to ensure that the model performs well on test data. Choosing the number of neurons in a hidden layer and the number of hidden layers is usually a case of a trial-and-error heuristics and presents the case of applying hyper-parameter tuning to improve the network performance. Using a grid search for hyper-parameter tuning is a good way to approximate an optimal neural network architecture that performs well on test data.

Multilayer Perceptron (MLP) with Keras

In this section, we examine a motivating example by building an MLP model with Keras. In doing so, we'll go through the following steps:

- Import and transform the dataset.

- Build and compile the model.

- Train the data using **'Model.fit()'**.

- Evaluate the model using **'Model.evaluate()'**.

- Predict on unseen data using **'Model.predict()'**.

The dataset used for this example is the Fashion-MNIST database of fashion articles. This dataset contains 60,000 28 x 28 pixel grayscale images of ten clothing items (the target classes). This dataset is downloaded from the **'tf.keras.datasets'** package. The following code example will build a simple MLP neural network for the computer to classify an image of a clothing item into its appropriate class. The network architecture has the following layers:

- A dense hidden layer with 250 neurons

- A second hidden layer with 64 neurons

- A third hidden layer with 32 neurons

- An output layer with 10 output classes

```
# install tensorflow 2.0
!pip install -q tensorflow==2.0.0-beta0

# import packages
```

```python
import tensorflow as tf
import numpy as np

# import dataset
(x_train, y_train), (x_test, y_test) = tf.keras.datasets.fashion_mnist.
load_data()

# flatten the 28*28 pixel images into one long 784 pixel vector
x_train = np.reshape(x_train, (-1, 784)).astype('float32')
x_test = np.reshape(x_test, (-1, 784)).astype('float32')

# scale dataset from 0 -> 255 to 0 -> 1
x_train /= 255
x_test /= 255

# one-hot encode targets
y_train = tf.keras.utils.to_categorical(y_train)
y_test = tf.keras.utils.to_categorical(y_test)

# create the model
def model_fn():
    model = tf.keras.Sequential()
    # Adds a densely-connected layer with 256 units to the model:
    model.add(tf.keras.layers.Dense(256, activation='relu', input_dim=784))
    # Add Dense layer with 64 units
    model.add(tf.keras.layers.Dense(64, activation='relu'))
    # Add another densely-connected layer with 32 units:
    model.add(tf.keras.layers.Dense(32, activation='relu'))
    # Add a softmax layer with 10 output units:
    model.add(tf.keras.layers.Dense(10, activation='softmax'))

    # compile the model
    model.compile(optimizer=tf.keras.optimizers.SGD(0.01),
                  loss='categorical_crossentropy',
                  metrics=['accuracy'])
    return model

# build model
model = model_fn()
```

```
# use tf.data to batch and shuffle the dataset
train_ds = tf.data.Dataset.from_tensor_slices(
    (x_train, y_train)).shuffle(len(x_train)).repeat().batch(32)
test_ds = tf.data.Dataset.from_tensor_slices((x_test, y_test)).batch(32)

# train the model
model.fit(train_ds, epochs=10,
          steps_per_epoch=2000)

# evaluate the model
score = model.evaluate(test_ds)
print('Test loss: {:.2f} \nTest accuracy: {:.2f}%'.format(score[0],
score[1]*100))

'Ouput:'
Test loss: 0.35
Test accuracy: 87.36%
```

Observe the following from the preceding code:

- A Keras Sequential Model is built by calling the **'tf.keras. Sequential()'** method from which layers are then added to the model.

- After constructing the model layers, the model is compiled by calling the method '.compile()'.

- The model is trained by calling the **'fit()'** method which receives the training features and targets from the **'tf.data.Dataset'** pipeline.

- The method '.evaluate()' is used to get the final metric estimate and the loss score of the model after training.

In this chapter, we introduced the multilayer perceptron network and how it achieves good performance on complex learning problems by stacking layers of neurons together to form a deep representational hierarchy. By doing this, the network learns what features are relevant and also learns what weights of the network will best approximate the target function.

In the next chapter, we will discuss on other considerations for training deep neural network.

CHAPTER 32

Other Considerations for Training the Network

In this chapter, we will cover some other important techniques to consider when training a deep neural network.

Weight Initialization

Weight initialization is a technique for assigning initial values to the weights (parameters) of the neural network before training (see Figure 32-1). Proper weight initialization may mitigate the effects of vanishing and exploding gradients when training the network. It may also speed up the training process. Two commonly used methods for weight initializations are the Xavier and the He techniques. We will not go into the technical explanation of these initialization strategies. However, they are implemented in the standard deep learning framework libraries such as TensorFlow and Keras. In TensorFlow 2.0, the dense layer in **'tf.keras.layers.Dense()'** has the Glorot uniform initializer, also called Xavier uniform initializer as its default kernel initializer.

© Ekaba Bisong 2019

E. Bisong, *Building Machine Learning and Deep Learning Models on Google Cloud Platform*, https://doi.org/10.1007/978-1-4842-4470-8_32

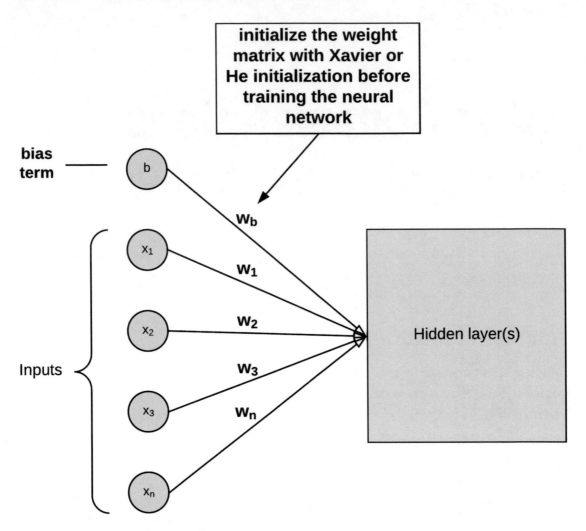

Figure 32-1. *Weight initialization*

Batch Normalization

The technique of batch normalization involves normalizing the data (to have zero mean and unit variance), as well as scaling and shifting the data batch at each layer of the neural network during the training phase. Batch normalization occurs after the affine transformation of the input matrix and their weights, but before passing the transformation into the activation function (see Figure 32-2).

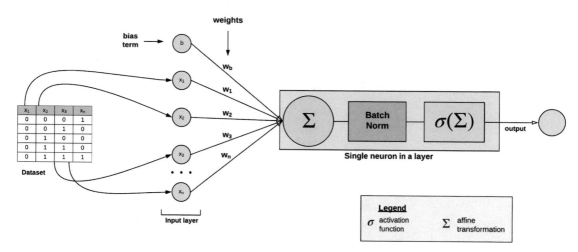

Figure 32-2. *Batch normalization, also known as batch norm*

The neural network learns the parameters for scaling and shifting the data at every layer during training. Also at the training phase, the score of the running mean and standard deviation of the data is maintained so that it can be used to normalize the test data before evaluation.

Batch normalization also mitigates the exploding and vanishing gradient problem irrespective of weight initialization. However, due to the added computational step at each layer, the network may be a bit slower. A batch normalization layer is added to a TensorFlow 2.0 network model by calling the method **'tf.keras.layers. BatchNormalization()'** as shown in the following code listing.

```
# create the model
def model_fn():
    model = tf.keras.Sequential()
    # Adds a densely-connected layer with 256 units to the model:
    model.add(tf.keras.layers.Dense(256, activation='relu', input_dim=784))
    # Add Dense layer with 64 units
    model.add(tf.keras.layers.Dense(64, activation='relu'))
    # Add a Batch Normalization layer
    model.add(tf.keras.layers.BatchNormalization())
    # Add another densely-connected layer with 32 units:
    model.add(tf.keras.layers.Dense(32, activation='relu'))
    # Add a softmax layer with 10 output units:
    model.add(tf.keras.layers.Dense(10, activation='softmax'))
```

```
# compile the model
model.compile(optimizer=tf.keras.optimizers.SGD(0.01),
              loss='categorical_crossentropy',
              metrics=['accuracy'])
return model
```

Gradient Clipping

Gradient clipping is another technique for hemming the problem of vanishing and exploding gradients mostly seen in recurrent networks due to training via backpropagation across a large number of deep recurrent layers. Gradient clipping involves trimming the computed gradients so that they remain within a specific range; in doing so, the gradients are prevented from saturating as the network trains across multiple deep layers.

Gradient clipping is implemented in TensorFlow 2.0 by adjusting the **'clipnorm'** or **'clipvalue'** parameters of the selected optimizer from the **'tf.keras.optimizers'** package. **'clipnorm'** clips the gradients by norm, while **'clipvalue'** clips the gradients by value.

This chapter introduces some important techniques that are employed to improve the performance of a neural network by further mitigating the issue of vanishing and exploding gradients. In the next chapter, we will see more optimization techniques for training deep neural network model.

## CHAPTER 33

More on Optimization Techniques

In this chapter, we'll go over some other optimization techniques for improving the ability of a neural network to learn complex patterns in a dataset.

Momentum

Momentum is a technique for improving the convergence speed of stochastic gradient descent (SGD) optimization. Remember that stochastic gradient works by learning the direction of steepest descent by evaluating a training example at each time step to optimize the weights of the network. Momentum improves on this by calculating the average of previous gradients in a process called exponentially smoothed averages. It then uses this computed average to continue to move in the direction of steepest descent. By doing so, it quickens the learning process. In computing this exponentially decayed average, a momentum hyper-parameter is introduced to control how the weight parameters are updated. Figure 33-1 shows an example of stochastic gradient descent with and without momentum as it converges in a function space. In TensorFlow 2.0, momentum is added to a SGD optimizer by adjusting the **'momentum'** parameter of the **SGD method, 'tf.keras.optimizers.SGD(momentum=[float >=0])'**. The momentum value must be a float value that is greater or equal to 0 that accelerates SGD in the relevant direction and dampens oscillations.

© Ekaba Bisong 2019
E. Bisong, *Building Machine Learning and Deep Learning Models on Google Cloud Platform*,
https://doi.org/10.1007/978-1-4842-4470-8_33

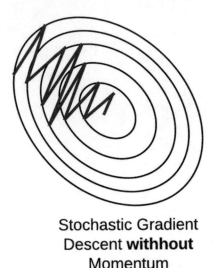

Stochastic Gradient
Descent **withhout**
Momentum

Stochastic Gradient
Descent **with**
Momentum

Figure 33-1. *SGD with and without momentum*

Variable Learning Rates

Remember that the learning rate controls how large a step the gradient descent algorithm makes when moving in the direction of steepest descent. If the learning rate is large, the algorithm takes larger steps in the direction of the steepest gradient, as is faster. However, the algorithm may overshoot the global minimum and fail to converge. But if the learning rate is set to a small number, closer to zero, the algorithm converges slowly, but it is more guaranteed to converge.

Variable learning rates are a set of techniques for adjusting the learning rate of the gradient descent algorithm at every time instance while training. These methods are also called learning rate scheduling. Examples of variable learning rates include

- Step decay: This method reduces the learning rate by a constant factor after a certain number of iterations.

- Exponential decay: The exponential decay adapts the learning rate following an exponential distribution.

- Decay proportion: This method reduces the learning rate by a ratio of 1 over the time instance, **t**. The learning rate decay can be adjusted by modifying the proportionality constant.

In TensorFlow 2.0, the **'decay'** parameter of the selected optimizer from the **'tf.keras. optimizers'** module allows time inverse decay of learning rate.

Adaptive Learning Rates

Adaptive learning rate, on the other hand, re-adjusts the learning rate in accordance with the training data. It basically uses a different learning rate for each parameter and adapts it during training. These techniques are based on the observation that each parameter results in a different type of gradient. The following list outlines types of adaptive learning rates in use and their method calls in TensorFlow 2.0:

- AdaGrad: **tf.keras.optimizers.Adagrad()**

- AdaDelta: **tf.keras.optimizers.Adadelta()**

- RMSProp: **tf.keras.optimizers.RMSprop()**

- Adaptive Moments, (Adam): **tf.keras.optimizers.Adam()**

However, AdaGrad performs poorly when used for training deep learning models due to its monotonic learning rate which could be too aggressive, and learning may stop early during training. As of now, there is no proven best optimization technique, so the choice of the optimization technique is down to the preference of the model designer.

This chapter surveys some other techniques for optimizing the weights of a deep neural network. These techniques have implementations in deep learning libraries such as Tensorflow and Keras and can be explored as hyper-parameters when designing a neural network solution for a particular learning use case.

In the next chapter, we will discuss techniques for applying regularization to a deep neural network to prevent overfitting.

## CHAPTER 34

Regularization for Deep Learning

Regularization is a technique for reducing the variance in the validation set, thus preventing the model from overfitting during training. In doing so, the model can better generalize to new examples. When training deep neural networks, a couple of strategies exist for use as a regularizer.

Dropout

Dropout is a regularization technique that prevents a deep neural network from overfitting by randomly discarding a number of neurons at every layer during training. In doing so, the neural network is not overly dominated by any one feature as it only makes use of a subset of neurons in each layer during training. In doing so, Dropout resembles an ensemble of neural networks as a similar but distinct neural network is trained at each layer. Dropout works by designating a probability that a neuron will be dropped in a layer. This probability value is called the Dropout rate. Figure 34-1 shows an example of a network with and without Dropout.

© Ekaba Bisong 2019
E. Bisong, *Building Machine Learning and Deep Learning Models on Google Cloud Platform*,
https://doi.org/10.1007/978-1-4842-4470-8_34

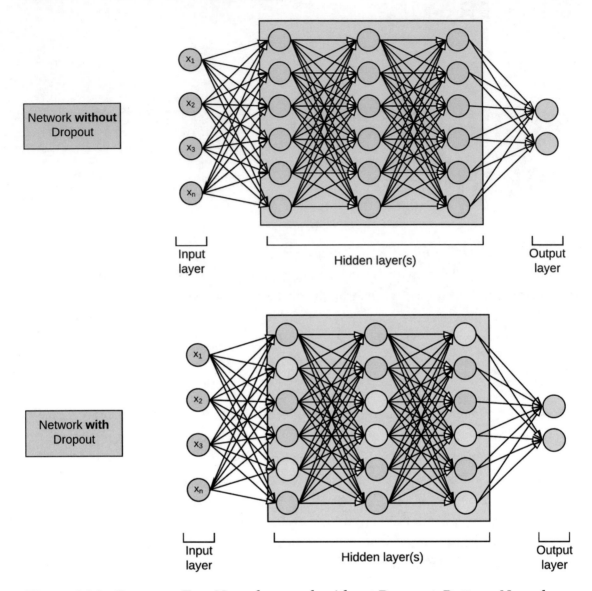

Figure 34-1. *Dropout. Top: Neural network without Dropout. Bottom: Neural network with Dropout.*

In TensorFlow 2.0 Dropout is added to a model with the method **'tf.keras.layers. Dropout()'**. The **'rate'** parameter of the method controls the fraction of the input units to drop. It is assigned a float value between 0 and 1. The following code listing shows an MLP Keras model with Dropout applied.

```
# create the model
def model_fn():
    model = tf.keras.Sequential()
    # Adds a densely-connected layer with 256 units to the model:
    model.add(tf.keras.layers.Dense(256, activation='relu', input_dim=784))
    # Add Dropout layer
    model.add(tf.keras.layers.Dropout(rate=0.2))
    # Add another densely-connected layer with 64 units:
    model.add(tf.keras.layers.Dense(64, activation='relu'))
    # Add a softmax layer with 10 output units:
    model.add(tf.keras.layers.Dense(10, activation='softmax'))

    # compile the model
    model.compile(optimizer=tf.train.AdamOptimizer(0.001),
                  loss='categorical_crossentropy',
                  metrics=['accuracy'])
    return model
```

Data Augmentation

Data augmentation is a method for artificially generating more training data points. This technique is precipitated on the observation that for an increasingly large training dataset mitigates the problem of overfitting. For some problems, it may be easy to artificially generate fake data, while for others it may not readily be the case. A classic example where we can use data augmentation is in the case of image classification. Here artificial images can easily be created by rotating or scaling the original images to create more variations of the dataset for a particular image class.

Noise Injection

The noise injection regularization method adds some Gaussian noise to the network inputs during training. Also, Gaussian noise can be added to the hidden units to mitigate overfitting. Yet still another form of injecting noise into the network is to add some Gaussian noise to the network weights. Noise injection can be considered as a form of data augmentation. The amount of noise added is a configurable hyper-parameter.

Too little noise has no effect, whereas too much noise makes the mapping function too challenging to learn.

In TensorFlow 2.0, noise injection can be added to the model as a form of data augmentation using the method **'tf.keras.layers.GaussianNoise()'**. The **'stddev'** parameter of the method controls the standard deviation of the noise distribution. The following code listing shows an MLP Keras model with Gaussian noise applied to the model.

```
# create the model
def model_fn():
    model = tf.keras.Sequential()
    # Adds a densely-connected layer with 256 units to the model:
    model.add(tf.keras.layers.Dense(256, activation='relu', input_dim=784))
    # Add Gaussian Noise
    model.add(tf.keras.layers.GaussianNoise(stddev=1.0))
    # Add another densely-connected layer with 64 units:
    model.add(tf.keras.layers.Dense(64, activation='relu'))
    # Add a softmax layer with 10 output units:
    model.add(tf.keras.layers.Dense(10, activation='softmax'))

    # compile the model
    model.compile(optimizer=tf.keras.optimizers.RMSprop(),
                  loss='categorical_crossentropy',
                  metrics=['accuracy'])
    return model
```

Early Stopping

Early stopping involves storing the model parameters each time there is an improvement in the loss (or error) estimate on the validation dataset. At the end of the training phase, the stored model parameters are used rather than the last known parameter before termination.

The technique of early stopping is based on the observation that for a sufficiently complex classifier, as the training phase progresses, the error estimate on the training data continues to decrease, whereas the validation data will see an increase in the model error measure. This is illustrated in Figure 34-2.

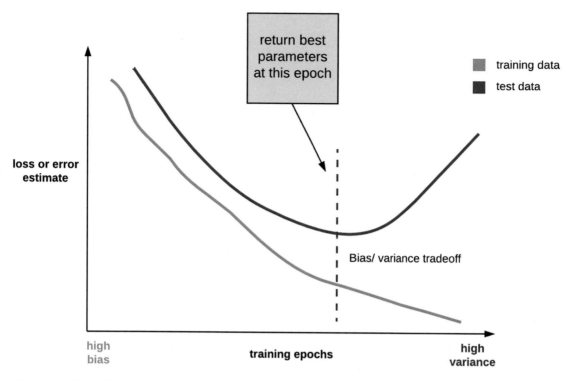

Figure 34-2. *Early stopping*

In TensorFlow 2.0, early stopping can be applied to stop training when there is no improvement in the validation accuracy or loss by applying the **'tf. keras.callbacks. EarlyStopping()'** method as a callback when training the model. For completeness sake, we will produce a complete code listing with early stopping applied to the MLP Fashion-MNIST model.

```
# install tensorflow 2.0
!pip install -q tensorflow==2.0.0-beta0

# import packages
import tensorflow as tf
import numpy as np

# import dataset
(x_train, y_train), (x_test, y_test) = tf.keras.datasets.fashion_mnist.
load_data()

# flatten the 28*28 pixel images into one long 784 pixel vector
```

```python
x_train = np.reshape(x_train, (-1, 784)).astype('float32')
x_test = np.reshape(x_test, (-1, 784)).astype('float32')

# scale dataset from 0 -> 255 to 0 -> 1
x_train /= 255
x_test /= 255

# one-hot encode targets
y_train = tf.keras.utils.to_categorical(y_train)
y_test = tf.keras.utils.to_categorical(y_test)

# create the model
def model_fn():
    model = tf.keras.Sequential()
    # Adds a densely-connected layer with 256 units to the model:
    model.add(tf.keras.layers.Dense(256, activation='relu', input_dim=784))
    # Add another densely-connected layer with 128 units:
    model.add(tf.keras.layers.Dense(128, activation='relu'))
    # Add another densely-connected layer with 64 units:
    model.add(tf.keras.layers.Dense(64, activation='relu'))
    # Add another densely-connected layer with 32 units:
    model.add(tf.keras.layers.Dense(32, activation='relu'))
    # Add a softmax layer with 10 output units:
    model.add(tf.keras.layers.Dense(10, activation='softmax'))

    # compile the model
    model.compile(optimizer=tf.keras.optimizers.RMSprop(),
                  loss='categorical_crossentropy',
                  metrics=['accuracy'])
    return model

# use tf.data to batch and shuffle the dataset
train_ds = tf.data.Dataset.from_tensor_slices(
    (x_train, y_train)).shuffle(len(x_train)).repeat().batch(32)
test_ds = tf.data.Dataset.from_tensor_slices((x_test, y_test)).batch(32)

# build model
model = model_fn()
```

```
# early stopping
checkpoint = tf.keras.callbacks.EarlyStopping(
    monitor='val_loss',
    mode='auto',
    patience=5)

# assign callback
callbacks = [checkpoint]

# train the model
history = model.fit(train_ds, epochs=10,
                    steps_per_epoch=100,
                    validation_data=test_ds,
                    callbacks=callbacks)

# evaluate the model
score = model.evaluate(test_ds)
print('Test loss: {:.2f} \nTest accuracy: {:.2f}%'.format(score[0],
score[1]*100))
```

With early stopping applied to the preceding code, the training will stop once there is no improvement to the loss on the validation dataset. The **'patience'** parameter in the EarlyStopping method represents the number of epochs with no improvement, after which training will be stopped.

This chapter surveys some techniques to tackle the problem of overfitting when training with a deep neural network. In the next chapter, we will discuss on convolutional neural networks for building predictive models for computer vision use cases such as image recognition with TensorFlow 2.0.

CHAPTER 35

Convolutional Neural Networks (CNN)

Convolutional neural networks (CNN) are a specific type of neural network systems that are particularly suited for computer vision problems such as image recognition. In such tasks, the dataset is represented as a 2-D grid of pixels. See Figure 35-1.

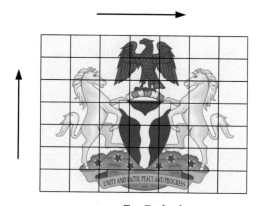

7 x 7 pixels

Figure 35-1. *2-D representation of an image*

An image is depicted in the computer as a matrix of pixel intensity values ranging from 0 to 255. A grayscale (or black and white) image consists of a single channel with 0 representing the black areas and 255 the white regions with the values in between for various shades of gray.

For example, the image in Figure 35-2 is a 10 x 10 grayscale image with its matrix representation.

© Ekaba Bisong 2019

E. Bisong, *Building Machine Learning and Deep Learning Models on Google Cloud Platform,*
https://doi.org/10.1007/978-1-4842-4470-8_35

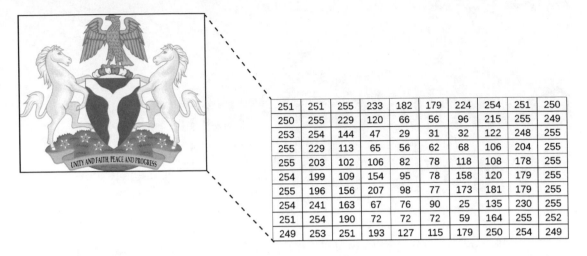

251	251	255	233	182	179	224	254	251	250
250	255	229	120	66	56	96	215	255	249
253	254	144	47	29	31	32	122	248	255
255	229	113	65	56	62	68	106	204	255
255	203	102	106	82	78	118	108	178	255
254	199	109	154	95	78	158	120	179	255
255	196	156	207	98	77	173	181	179	255
254	241	163	67	76	90	25	135	230	255
251	254	190	72	72	72	59	164	255	252
249	253	251	193	127	115	179	250	254	249

Figure 35-2. *Grayscale image with matrix representation*

On the other hand, a colored image consists of three channels, red, green, and blue, with each channel also containing pixel intensity values from 0 to 255. A colored image has a matrix shape of [height x width x channel]. In Figure 35-3, we have an image of shape [10 x 10 x 3] indicating a 10 x 10 matrix with three channels.

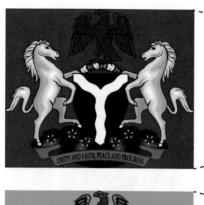

250	250	255	246	249	251	245	251	250	250
249	255	246	206	118	97	183	241	255	250
253	253	218	60	8	6	28	203	254	254
255	242	226	89	37	45	89	214	230	253
254	231	208	235	122	112	235	213	217	255
254	238	203	253	139	111	254	204	228	251
255	234	229	196	114	101	155	230	233	255
253	247	254	55	93	132	0	215	252	253
251	253	236	144	74	74	121	221	255	252
250	255	249	242	218	209	239	246	253	249

250	250	255	236	216	209	231	255	252	250
251	254	234	161	78	52	120	223	255	250
253	255	173	43	5	8	4	148	255	255
255	249	123	0	50	60	2	101	217	255
255	230	94	54	53	25	119	113	179	255
255	226	130	214	49	2	150	136	208	255
255	216	218	205	109	94	147	216	210	255
254	244	237	47	87	122	0	178	243	255
251	254	197	69	61	60	51	165	255	252
248	255	250	203	156	137	188	249	253	249

250	250	253	224	103	97	202	252	251	252
248	255	212	6	8	21	4	183	253	248
253	254	54	35	118	119	64	31	239	255
253	205	2	73	103	103	83	0	175	252
255	165	0	0	58	67	0	0	132	253
253	150	2	23	74	83	65	27	119	255
253	150	50	236	77	42	255	109	106	255
255	229	33	120	53	37	127	35	202	254
251	255	138	2	68	83	0	106	255	252
249	253	252	148	33	29	122	254	253	249

Figure 35-3. *Colored image with matrix representation*

Local Receptive Fields of the Visual Cortex

The core concept of convolutional neural networks is built on understanding the local receptive fields found in the neurons of the visual cortex – the part of the brain responsible for processing visual information.

A local receptive field is an area on the neuron that excites or activates that neuron to fire information to other neurons. When viewing an image, the neurons in the visual cortex react to a small or limited area of the overall image due to the presence of a small local receptive field.

Hence, the neurons in the visual cortex do not all sense the entire image at the same time, but they are activated by viewing a local area of the image via its local receptive field.

In Figure 35-4, the local receptive fields overlap to give a collective perspective on the entire image. Each neuron in the visual cortex reacts to a different type of visual information (e.g., lines with different orientations).

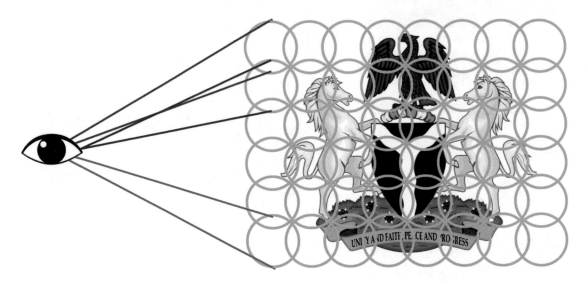

Figure 35-4. *Local receptive field*

Other neurons have large receptive fields that react to more complex visual patterns such as edges, regions, and so on. From here we get the idea that neurons with larger receptive field receive information from those with lower receptive fields as they progressively learn the visual information of the image.

Advantages of CNN over MLP

Suppose we have a 28 x 28 pixel set of image data, a feedforward neural network or multilayer perceptron will need 784 input weights plus a bias. By flattening an image as you would in MLP, we lose the spatial relationship of the pixels in the image.

CNN, on the other hand, can learn complex image features by preserving the spatial relationship between the image pixels. It does so by stacking convolutional layers whereby the neurons in the higher layers with a larger receptive field receive information from neurons in the lower layers having a smaller receptive field. CNN learns a hierarchy of increasingly complex features from the input data as it flows through the network.

In CNN, the neurons (or filters) in the convolutional layer are not all connected to the pixels in the input image as we have in the dense multilayer perceptron. Hence, a CNN is also called a sparse neural network.

A distinct advantage of CNN over MLP is the reduced number of weights needed for training the network.

Convolutional neural networks are composed of three fundamental types of layers:

- Convolutional layer

- Pooling layer

- Fully connected layer

The Convolutional Layer

The convolution layer is made up of filters and feature maps. A filter is passed over the input image pixels to capture a specific set of features in a process called convolution (see Figure 35-5). The output of a filter is called a feature map.

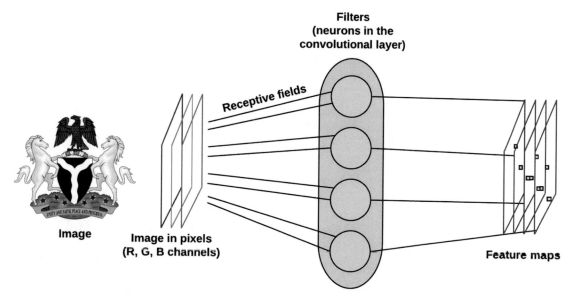

Figure 35-5. *The convolution process*

Convolution

Convolution is the process by which a function is applied to a matrix to extract specific information from the matrix. The function is implemented as a sliding window through the matrix, and it is more popularly called a convolutional filter or a kernel. Both terms are used interchangeably in the literature. The image in Figure 35-6 illustrates a filter sliding through a matrix to extract information from it.

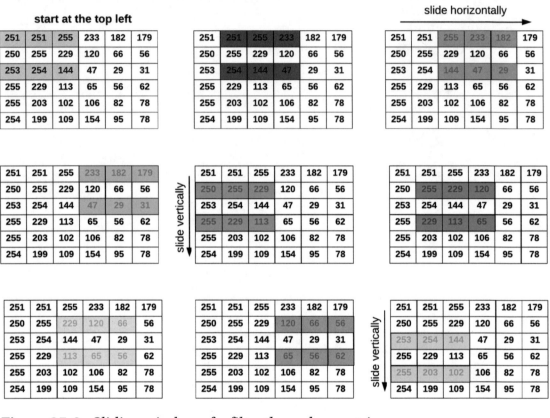

Figure 35-6. *Sliding window of a filter through a matrix*

Filters are neurons in the convolutional layer. They are assigned weights and are applied as a sliding window through the matrix. The output of a filter is a feature map. Filters which are basically neurons also have a non-linear activation function.

The inputs into a filter can be the matrix of the image pixels if the filter is at the input layer, or it can be the feature maps of a previous convolutional layer if the filter is applied at a deeper layer in the network.

Filters are assigned a fixed square block for its input size. This input size can also be seen as the local receptive field of the filter. A common input size for filters is a 3 x 3 square patch as illustrated in Figure 35-7; other standard sizes include a 5 x 5 or 7 x 7 filter for extracting features from images. It is also a best practice to use more filters at deeper layers of the network and fewer filter at the input layer.

-1	-1	-1
2	2	2
-1	-1	-1

A **3 x 3** Filter
function or
Kernel function

Figure 35-7. *An example of a 3 x 3 filter kernel*

Observe that each cell in the filter has an associated weight or value. These values are used to multiply its associated pixel intensities and then sum up their results, which are filled in the appropriate cell of the convolutional result. This procedure is illustrated in Figure 35-8.

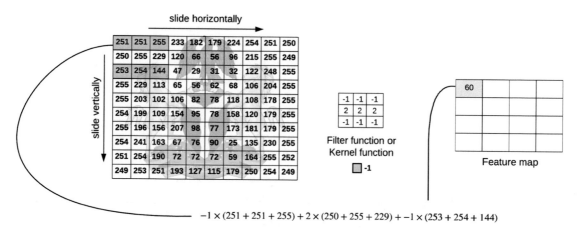

$$-1 \times (251 + 251 + 255) + 2 \times (250 + 255 + 229) + -1 \times (253 + 254 + 144)$$

Figure 35-8. *Sliding a convolutional filter across an image matrix to extract features*

The weights on the filter determine the filter operation and consequently the type of features that are extracted from the filter inputs. Different filters are responsible for edge detection, line detection, and so on. See Figure 35-9.

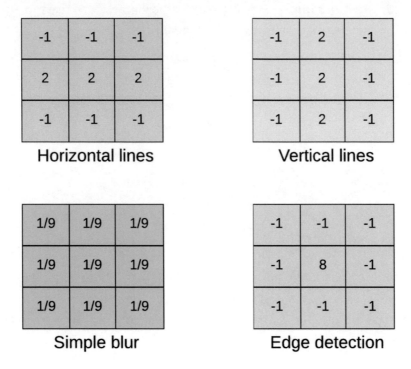

Figure 35-9. *Filter types*

Key considerations to make when designing a convolutional layer are

- The filter size

- The stride of the filter

- The padding for the layer input

The *stride* of the filter determines how many pixel steps the filter makes when moving from one image activation to another. It is typical to use a stride of 1, although this could be increased for large images. See Figure 35-10.

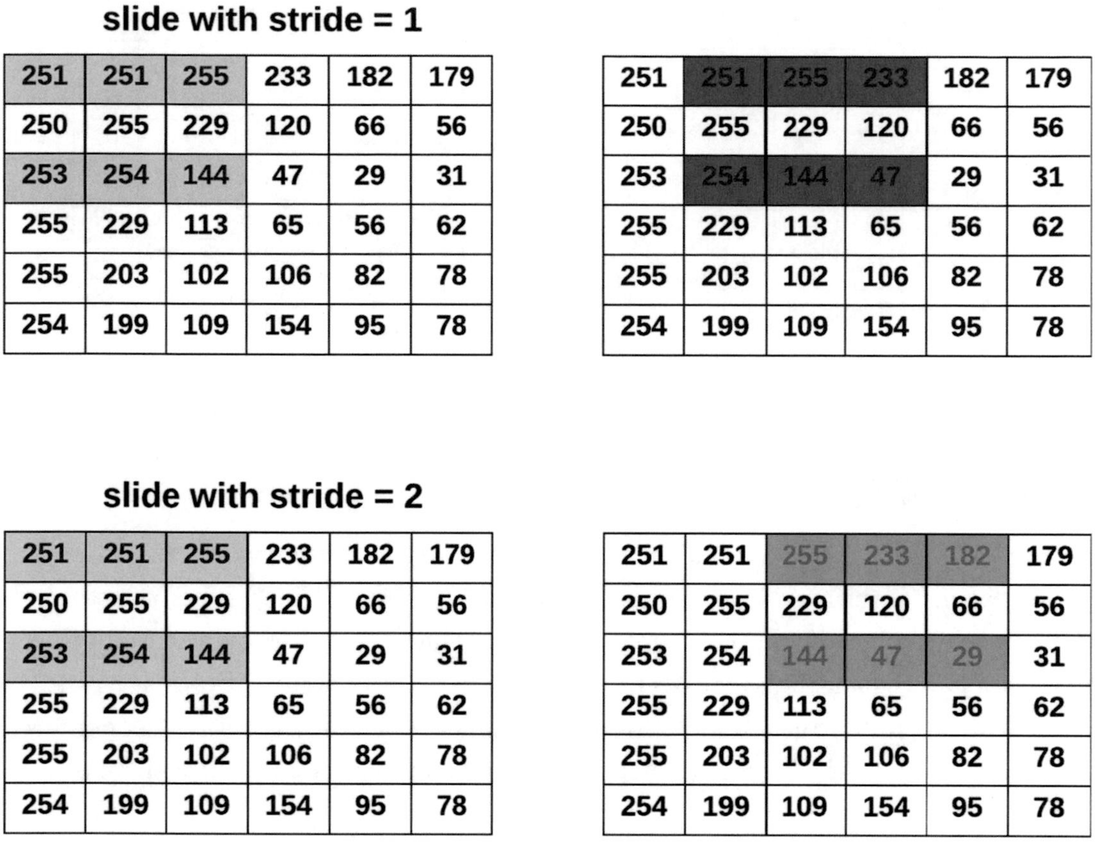

slide with stride = 1

251	251	255	233	182	179
250	255	229	120	66	56
253	254	144	47	29	31
255	229	113	65	56	62
255	203	102	106	82	78
254	199	109	154	95	78

251	251	255	233	182	179
250	255	229	120	66	56
253	254	144	47	29	31
255	229	113	65	56	62
255	203	102	106	82	78
254	199	109	154	95	78

slide with stride = 2

251	251	255	233	182	179
250	255	229	120	66	56
253	254	144	47	29	31
255	229	113	65	56	62
255	203	102	106	82	78
254	199	109	154	95	78

251	251	255	233	182	179
250	255	229	120	66	56
253	254	144	47	29	31
255	229	113	65	56	62
255	203	102	106	82	78
254	199	109	154	95	78

Figure 35-10. *An illustration of stride width*

Sometimes the choice of our filter size and the selected stride may not evenly divide up the size of the input to the filter. So to avoid losing pixel information since we don't slide past the edge of the image, a technique called *zero padding* is employed to pad the borders of the image pixels with a defined layer of zeros. This allows the filter to stride evenly through all the pixels in the image by including the zeros in the convolution. See Figure 35-11.

without zero padding

251	251	255	233	182	179
250	255	229	120	66	56
253	254	144	47	29	31
255	229	113	65	56	62
255	203	102	106	82	78
254	199	109	154	95	78

with zero padding

0	0	0	0	0	0	0	0
0	251	251	255	233	182	179	0
0	250	255	229	120	66	56	0
0	253	254	144	47	29	31	0
0	255	229	113	65	56	62	0
0	255	203	102	106	82	78	0
0	254	199	109	154	95	78	0
0	0	0	0	0	0	0	0

Figure 35-11.   An illustration of zero padding

Feature Maps

Feature maps are the outputs of a filter in a convolutional layer. Feature maps bring to the fore certain patterns of the input image such as horizontal lines, vertical lines, edges, and so on. These feature maps of the various neurons stacked together are what forms a convolutional neural layer and enable the layer to learn complex patterns and features of an image.

Moving deeper across the convolutional neural network, the inputs to a deeper convolutional layer are the feature maps of the previous layer. See Figure 35-12.

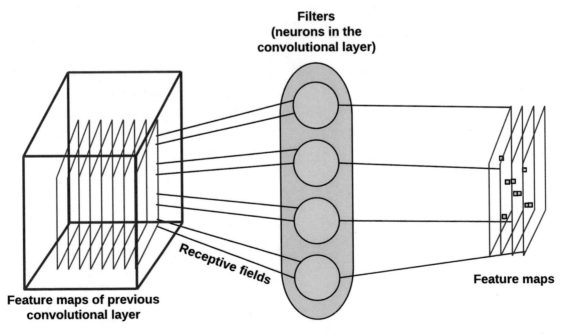

Figure 35-12.   Feature maps as inputs to a convolutional layer

432

The Pooling Layer

Pooling layers typically follow one or more convolutional layers. The goal of the pooling layer is to reduce or downsample the feature map of the convolutional layer. The pooling layer summarizes the image features learned in the previous network layers. By doing so, it also helps prevent the network from overfitting. Moreso, the reduction in the input size also bodes well for processing and memory costs when training the network.

The pooling layer can be seen as an aggregation function that consolidates learned features and extracts the essential features from previous layers. It does not conduct any multiplicative transformation on the input feature maps as seen in the convolutional layer.

The aggregation functions carried out by the pooling layer include max, sum, and average. The most frequently used aggregation function in practice is the max and is commonly called the MaxPool.

The aggregation functions of the pooling layer serve as the layers' filters. Just like the filters of the convolutional layer, they have a receptive field (although smaller in size than that of the convolutional layer) and a stride width. Howbeit, the filters which are the neurons of the pooling layer have no weight or biases. A typical size for the pooling filter is a 2 x 2 matrix as shown in Figure 35-13.

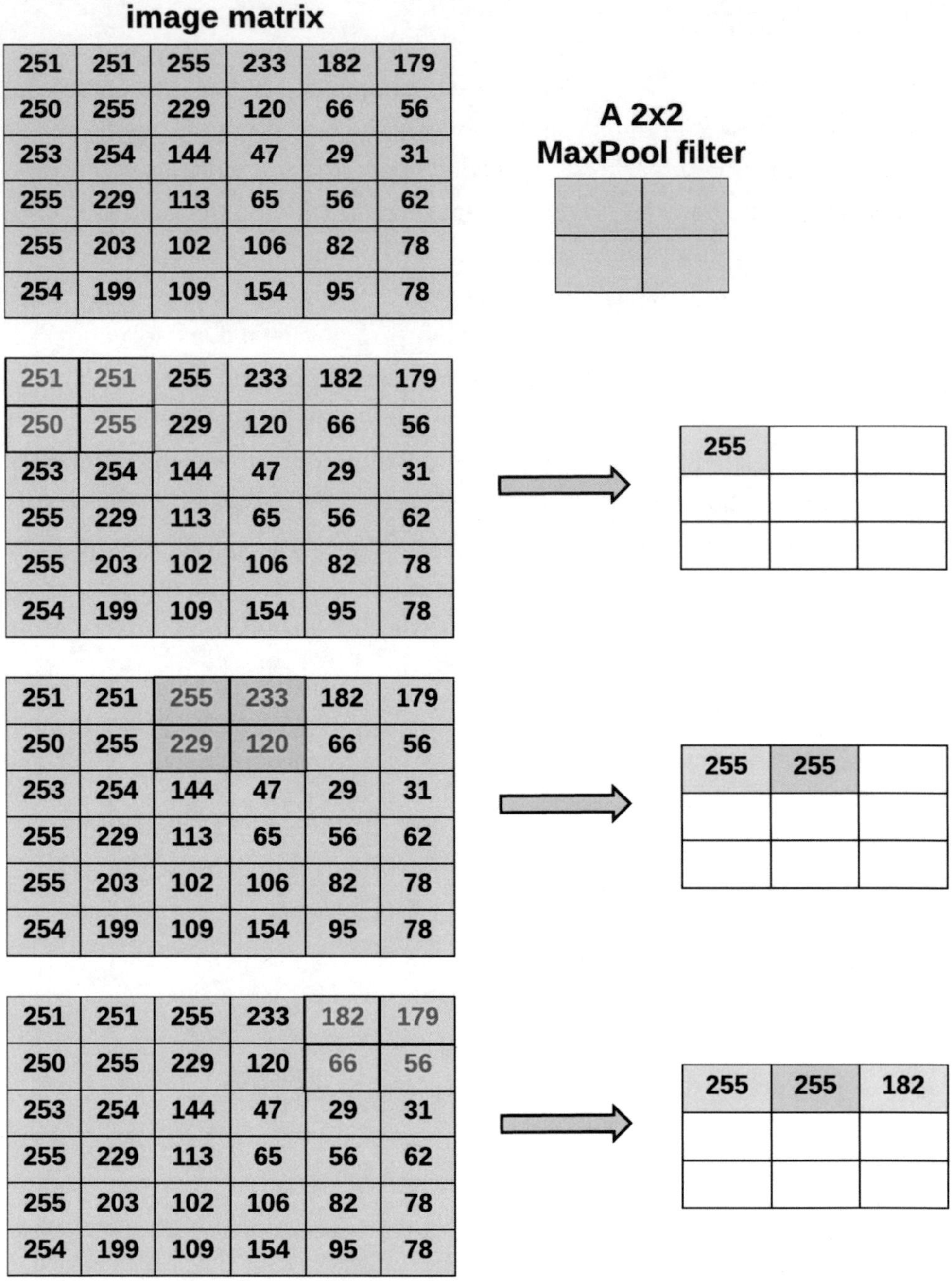

image matrix

251	251	255	233	182	179
250	255	229	120	66	56
253	254	144	47	29	31
255	229	113	65	56	62
255	203	102	106	82	78
254	199	109	154	95	78

A 2x2 MaxPool filter

251	251	255	233	182	179
250	255	229	120	66	56
253	254	144	47	29	31
255	229	113	65	56	62
255	203	102	106	82	78
254	199	109	154	95	78

255		

251	251	255	233	182	179
250	255	229	120	66	56
253	254	144	47	29	31
255	229	113	65	56	62
255	203	102	106	82	78
254	199	109	154	95	78

255	255	

251	251	255	233	182	179
250	255	229	120	66	56
253	254	144	47	29	31
255	229	113	65	56	62
255	203	102	106	82	78
254	199	109	154	95	78

255	255	182

Figure 35-13. *Example of pooling with MaxPooling*

434

The essential advantage of the pooling layer is its ability to inject location invariance into the network. Location invariance means that features can be detected by the network no matter where they are on the image.

The pooling layer applies its aggregation function to all the channels of the input image. For example, in an R, G, B image (i.e., an image with three channels, red, green, and blue), the MaxPool will be applied independently to all the three channels. Similarly, for feature maps with a particular depth, the pooling aggregation will be applied separately to each feature map. See Figure 35-14 as an example of applying pooling to the channel depth of its inputs.

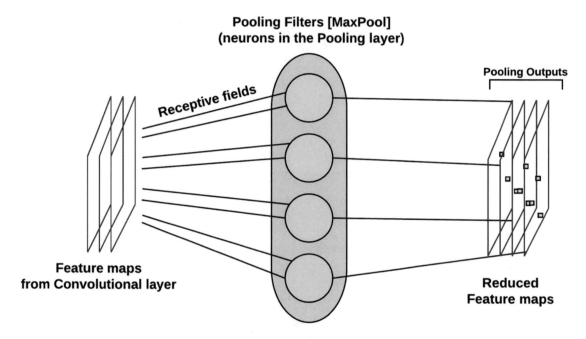

Figure 35-14. *Example of applying pooling to input with depth. Note that the filters in the pooling layer have no weights or biases*

The Fully Connected Network Layer

The fully connected network (FCN) layer is our regular feedforward neural network or multilayer perceptron. These layers typically have a non-linear activation function. In any case, the FCN is the final layer of the convolutional neural network. In this case, a softmax activation is used to output the probabilities that an input belongs to a particular class.

Before passing an input into the FCN, the image matrix will have to be flattened. For example, a 28 x 28 x 3 image matrix will become 2352 input weights plus a bias of 1 into the fully connected network.

In the case of our convolutional network, the feature maps of either the convolutional or pooling layer are flattened before passing into the FCN to compute the final network probabilities using the softmax function.

An Example CNN Architecture

We have discussed the building blocks of a convolutional neural network system. As you've seen, a CNN system is principally composed of convolution layers, pooling layers, and the fully connected layer. However, the way these layers are arranged and in what number are down to the preferred heuristics of the particular use case that a CNN is employed in solving.

An example CNN modeling pipeline is shown here:

1. The first layer following the input layer of images must be a convolutional layer for extracting image features. A 3 x 3 image filter is commonly used depending on the size of the input image.

2. Pooling layers typical follow a set of one or more convolutional layers. Typically, a 2 x 2 filter size is used in the pooling layer.

3. The fully connected layer must be the final layer of the CNN. It is also called the dense layer. It contains the softmax activation function to give the probabilities of class membership.

4. CNN may include one or more Dropout layers to prevent the network from overfitting.

Figure 35-15 is an example of a CNN architecture.

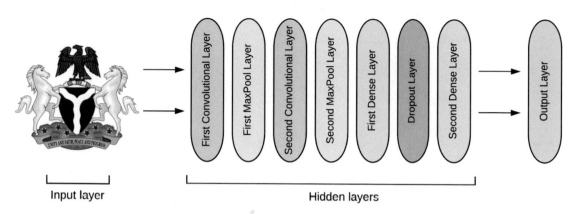

Figure 35-15. CNN architecture

CNN for Image Recognition with TensorFlow 2.0

In this example, we will build a convolutional neural network (CNN) to classify images from the CIFAR-10 dataset. CIFAR-10 is another standard image classification dataset to classify a colored 32 x 32 pixel image data into ten image classes, namely, airplane, automobile, bird, cat, deer, dog, frog, horse, ship, and truck. The focus of this section is exclusively on using TensorFlow 2.0 methods to build a CNN image classifier.

The CNN model architecture implemented loosely mirrors the Krizhevsky's architecture, also known as AlexNet. The network architecture has the following layers:

- Convolution layer: kernel_size = [5 x 5]

- Convolution layer: kernel_size = [5 x 5]

- Batch normalization layer

- Convolution layer: kernel_size = [5 x 5]

- Max pooling: pool size = [2 x 2]

- Convolution layer: kernel_size = [5 x 5]

- Convolution layer: kernel_size = [5 x 5]

- Batch normalization layer

- Max pooling: pool size = [2 x 2]

- Convolution layer: kernel_size = [5 x 5]

- Convolution layer: kernel_size = [5 x 5]

- Convolution layer: kernel_size = [5 x 5]

- Max pooling: pool size = [2 x 2]

- Dropout layer

- Dense layer: units = [512]

- Dense layer: units = [256]

- Dropout layer

- Dense layer: units = [10]

This CNN model has close to a million trainable variables as can be seen from the model summary when running **'model.summary()'**. Training on a CPU will take an inordinate amount of time (about 1 hour and 30 minutes). For this code example, we will train on a GPU instance. If running the code on Google Colab, change the runtime type to GPU and install TensorFlow 2.0 with GPU package. The graph of the model in Tensorboard is shown in Figure 35-16.

```
# import TensorFlow 2.0 with GPU
!pip install -q tf-nightly-gpu-2.0-preview

# import packages
import tensorflow as tf

# confirm tensorflow can see GPU
device_name = tf.test.gpu_device_name()
if device_name != '/device:GPU:0':
  raise SystemError('GPU device not found')
print('Found GPU at: {}'.format(device_name))

# load the TensorBoard notebook extension
%load_ext tensorboard

# import dataset
(x_train, y_train), (x_test, y_test) = tf.keras.datasets.cifar10.load_data()

# change datatype to float
```

```
x_train = x_train.astype('float32')
x_test = x_test.astype('float32')

# scale the dataset from 0 -> 255 to 0 -> 1
x_train /= 255
x_test /= 255

# one-hot encode targets
y_train = tf.keras.utils.to_categorical(y_train)
y_test = tf.keras.utils.to_categorical(y_test)

# parameters
batch_size = 100

# create dataset pipeline
train_ds = tf.data.Dataset.from_tensor_slices(
    (x_train, y_train)).shuffle(len(x_train)).repeat().batch(batch_size)
test_ds = tf.data.Dataset.from_tensor_slices((x_test, y_test)).batch(batch_size)

# create the model
def model_fn():
    model_input = tf.keras.layers.Input(shape=(32, 32, 3))
    x = tf.keras.layers.Conv2D(64, (5, 5), padding='same',
    activation='relu')(model_input)
    x = tf.keras.layers.Conv2D(64, (5, 5), padding='same', activation='relu')(x)
    x = tf.keras.layers.BatchNormalization()(x)
    x = tf.keras.layers.Conv2D(64, (5, 5), padding='same', activation='relu')(x)
    x = tf.keras.layers.MaxPooling2D(pool_size=(2, 2), strides=2,
    padding='same')(x)
    x = tf.keras.layers.Conv2D(64, (5, 5), padding='same', activation='relu')(x)
    x = tf.keras.layers.Conv2D(64, (5, 5), padding='same', activation='relu')(x)
    x = tf.keras.layers.BatchNormalization()(x)
    x = tf.keras.layers.MaxPooling2D(pool_size=(2, 2), strides=2, padding='same')(x)
    x = tf.keras.layers.Conv2D(32, (3, 3), padding='same', activation='relu')(x)
    x = tf.keras.layers.Conv2D(32, (3, 3), padding='same', activation='relu')(x)
    x = tf.keras.layers.Conv2D(32, (3, 3), padding='same', activation='relu')(x)
    x = tf.keras.layers.MaxPooling2D(pool_size=(2, 2), strides=2,
    padding='same')(x)
```

```python
    x = tf.keras.layers.Dropout(0.3)(x)
    x = tf.keras.layers.Flatten()(x)
    x = tf.keras.layers.Dense(512, activation='relu')(x)
    x = tf.keras.layers.Dense(256, activation='relu')(x)
    x = tf.keras.layers.Dropout(0.5)(x)
    output = tf.keras.layers.Dense(10, activation='softmax')(x)

    # the model
    model = tf.keras.Model(inputs=model_input, outputs=output)

    # compile the model
    model.compile(optimizer=tf.keras.optimizers.Nadam(),
                  loss='categorical_crossentropy',
                  metrics=['accuracy'])
    return model

# build the model
model = model_fn()

# print model summary
model.summary()

# tensorboard
tensorboard = tf.keras.callbacks.TensorBoard(log_dir='./tmp/logs_cifar10_
                keras',
                histogram_freq=0, write_graph=True,
                 write_images=True)

# assign callback
callbacks = [tensorboard]

# train the model
history = model.fit(train_ds, epochs=10,
                    steps_per_epoch=500,
                    callbacks=callbacks)

# evaluate the model
score = model.evaluate(test_ds)
print('Test loss: {:.2f} \nTest accuracy: {:.2f}%'.format(score[0], score[1]*100))
```

```
'Output:'
Test loss: 0.74
Test accuracy: 80.05%
```

```
# execute the command to run TensorBoard
%tensorboard --logdir tmp/logs_cifar10_keras
```

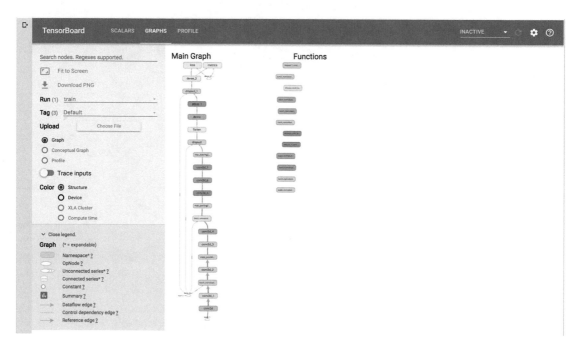

Figure 35-16. *Tensorboard output of CIFAR-10 model graph*

In this chapter, we discussed the convolutional neural network (CNN) as an example of a deep neural network. We went through the design details in architecting a CNN and implemented a CNN model with TensorFlow 2.0. In the next chapter, we will examine another type of deep neural network called the recurrent neural network.

CHAPTER 36

Recurrent Neural Networks (RNNs)

Recurrent neural networks (RNNs) are another specialized scheme of neural network architectures. RNNs are developed to solve learning problems where information about the past (i.e., past instants/events) is directly linked to making future predictions. Such sequential examples play up frequently in many real-world tasks such as language modeling where the previous words in the sentence are used to determine what the next word will be. Also in stock market prediction, the last hour/day/week stock prices define the future stock movement. RNNs are particularly tuned for time series or sequential tasks.

In a sequential problem, there is a looping or feedback framework that connects the output of one sequence to the input of the next sequence. RNNs are ideal for processing 1-D sequential data, unlike the grid-like 2-D image data in convolutional neural networks.

This feedback framework enables the network to incorporate information from past sequences or from time-dependent datasets when making a prediction.
In this section, we will cover the broad conceptual overview of recurrent neural networks and in particular the Long Short-Term Memory RNN variant (LSTM) which is the state-of-the-art technique for various sequential problems such as image captioning, stock market prediction, machine translation, and text classification.

The Recurrent Neuron

The first building block of the RNN is the recurrent neuron (see Figure 36-1). The neurons of the recurrent network are entirely different from those of other neural network architectures. The key difference here is that the recurrent neuron maintains a memory or a state from past computations. It does this by taking as input the output of the previous instant y_{t-1} in addition to its current input at a particular instant x_t.

© Ekaba Bisong 2019
E. Bisong, *Building Machine Learning and Deep Learning Models on Google Cloud Platform*,
https://doi.org/10.1007/978-1-4842-4470-8_36

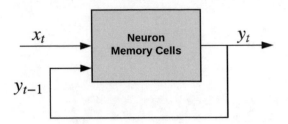

Figure 36-1. *A recurrent neuron*

In Figure 36-1, the recurrent neuron stands in contrast with neurons of the MLP and CNN architectures because instead of transferring a hierarchy of information across the network from one neuron to the other, data is looped back into the same neuron at every new time instant. A time instant can also mean a new sequence.

Hence, the recurrent neuron has two input weights, W_{x_t} and $W_{y_{t-1}}$, for the input at time x_t and for the input at time instant y_{t-1}. See Figure 36-2.

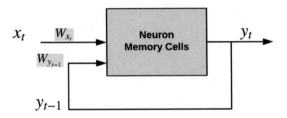

Figure 36-2. *Recurrent neuron with input weights*

Similar to other neurons, the recurrent neuron also injects non-linearity into the network by passing its weighted sums or affine transformations through a non-linear activation function.

Unfolding the Recurrent Computational Graph

A recurrent neural network is formalized as an unfolded computational graph. An unfolded computational graph shows the flow of information through the recurrent layer at every time instant in the sequence. Suppose we have a sequence of five time steps, we will unfold the recurrent neuron five times across the number of instants. The number of sequences constitutes the layers of the recurrent neural network architecture. See Figure 36-3.

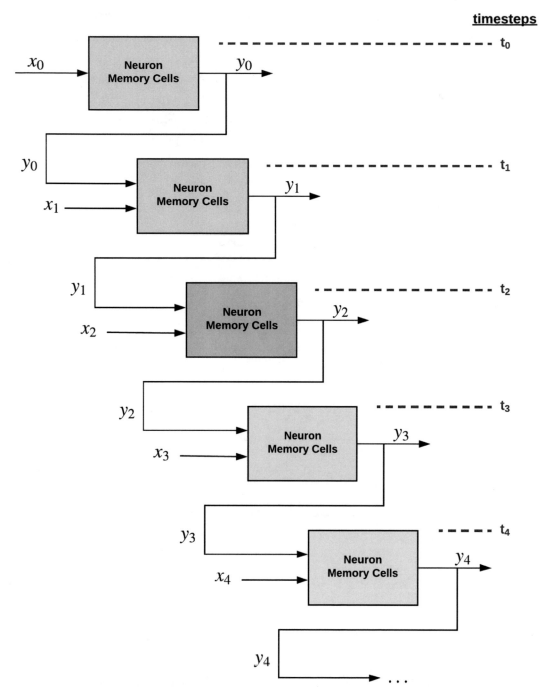

Figure 36-3. *Unfolding the recurrent neuron into a recurrent neural network*

From the unrolled graph of the recurrent neural network, we can observe how the input into the recurrent layer includes the output of the previous time step $t - 1$ in addition to the current input at time step t. This architecture of the recurrent neuron is central to how the recurrent neural network learns from past events or past sequences.

Up until now, we have seen that the recurrent neuron captures information from the past by storing memory or state in its memory cell. The recurrent neuron can have a much more complicated memory cell (such as the GRU or LSTM cell) than the basic RNN cell as illustrated in the images so far, where the output at time instant $t - 1$ holds the memory.

Basic Recurrent Neural Network

Earlier on, we mentioned that when a recurrent network is unfolded, we can see how information flows from one recurrent layer to the other. Further, we noted that the sequence length of the dataset determines the number of recurrent layers. Let's briefly illustrate this point in Figure 36-4. Suppose we have a time series dataset of ten layers, for each row sequence in the dataset, we will have ten layers in the recurrent network system.

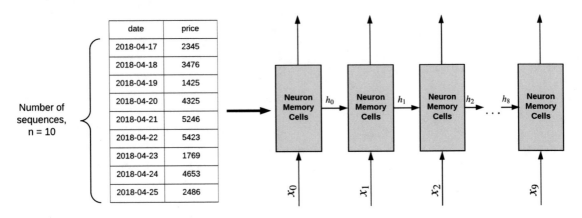

Figure 36-4. *Dataset to layers*

At this point, we must firmly draw attention to the fact that the recurrent layer does not comprise of just one neuron cell, but it is instead a set of neurons or neuron cells as shown in Figure 36-5. The choice of the number of neurons in a recurrent layer is a design decision when composing the network architecture.

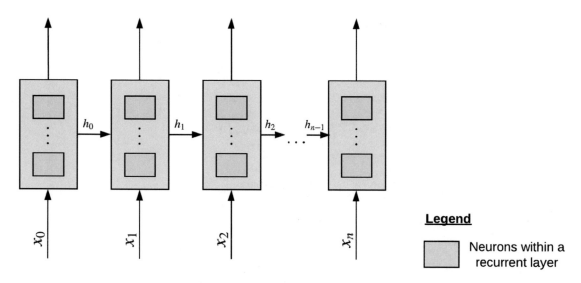

Figure 36-5. *Neurons in a recurrent layer*

Each neuron in a recurrent layer receives as input the output of the previous layer and its current input. Hence, the neurons each have two weight vectors. Again, just like other neurons, they perform an affine transformation of the inputs and pass it through a non-linear activation function (usually the hyperbolic tangent, tanh). Still, within the recurrent layer, the output of the neurons is moved to a dense or fully connected layer with a softmax activation function for outputting the class probabilities. This operation is illustrated in Figure 36-6.

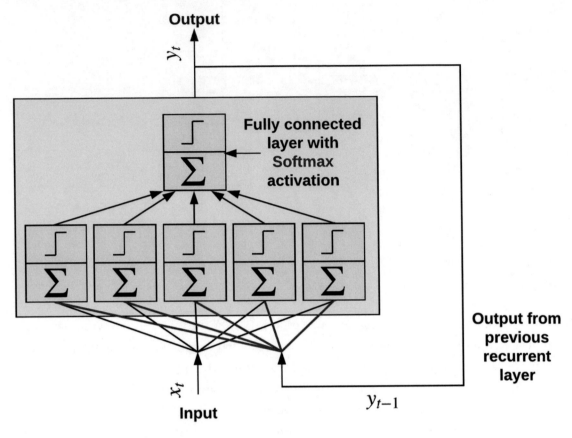

Figure 36-6. *Computations within a recurrent layer*

Recurrent Connection Schemes

There are two main schemes for forming recurrent connections from one recurrent layer to another. The first is to have recurrent connections between hidden units, and the other is recurrent connections between the hidden unit and the output of the previous layer. The different schemes are visually illustrated in Figure 36-7.

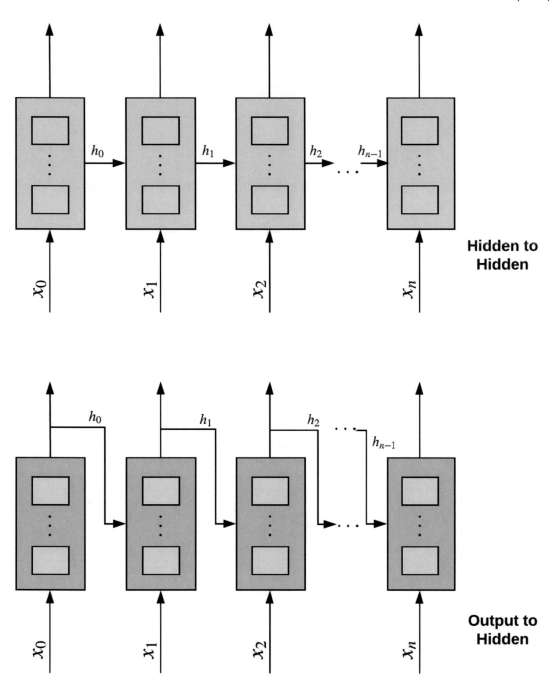

Figure 36-7. *Recurrent connection schemes*

The hidden-to-hidden recurrent configuration is found to be superior to the output-to-hidden form because it better captures the high-dimensional feature information about the past. In any case, the output-to-hidden recurrent form is less computationally expensive to train and can more easily be parallelized.

Sequence Mappings

Recurrent neural networks can represent sequence problems in a variety of ways. The flexibility of RNN mappings is that it operates on inputs and outputs of the network as sequences, thus freeing the network from the fixed sized input-output constraints found in other neural network architectures such as MLP and CNN.

Here are a few examples of variating sequence problems solved using RNNs:

1. An input to a sequence of output. This configuration is used for image captioning problems when an image is passed as an input to the network, and the output is a sequence of words. See Figure 36-8.

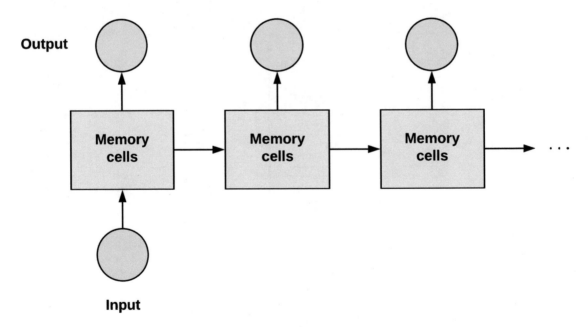

Figure 36-8. *An input to a sequence of output*

2. A sequence of inputs to an output. For example, in sentiment analysis, we need to pass in a sequence of words as input to the network, and the output is a class indicating either a positive or negative review or sentiment. See Figure 36-9.

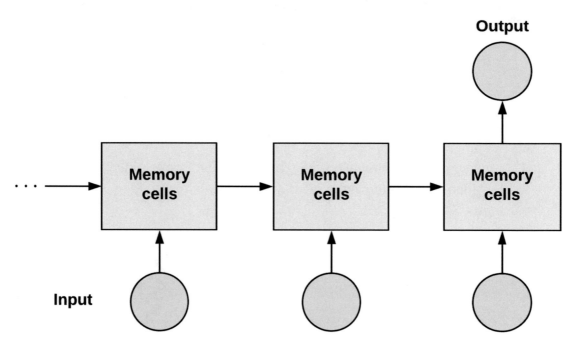

Figure 36-9. *A sequence of inputs to an output*

3. Sequence input to sequence output. This mapping operation is suited in application areas such as machine translation and speech recognition. It is more popularly called the encoder-decoder or sequence-to-sequence architecture. In this case, we may have a sequence of words in a particular language as input, and we want a sequence of words as output in another language. See Figure 36-10.

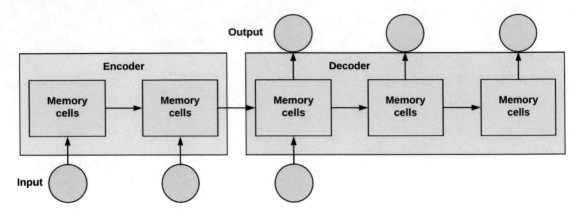

Figure 36-10. *Sequence input to sequence output*

4. Synced sequence input to output. This sort of framework is ideal for video classification in the event we want to label each video frame. See Figure 36-11.

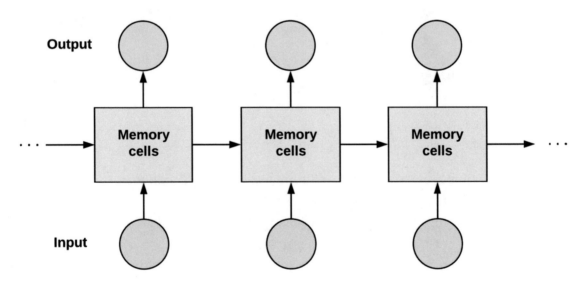

Figure 36-11. *Synced sequence input to output*

In the schemes illustrated in this sub-section, information flows from the hidden unit (or memory cell) of the recurrent layer at time instant $t - 1$ to the hidden unit at time instant t. As discussed earlier, this is because the transferred information is more feature-rich and contains more information from the past.

Training the Recurrent Network: Backpropagation Through Time

The recurrent neural network is trained in much the same way as other traditional neural networks by using the backpropagation algorithm. However, the backpropagation algorithm is modified into what is called backpropagation through time (BPTT).

Due to the architectural loop or recurrent structure of the recurrent network, vanilla backpropagation as is cannot work. Training a network using backpropagation involves calculating the error gradient, moving backward from the output layer through the hidden layers of the network and adjusting the network weights. However, this operation cannot work in the recurrent neuron because we have just one neural cell with recurrent connections to itself.

So, in order to train the recurrent network using backpropagation, we unroll the recurrent neuron across the time instants and apply backpropagation to the unrolled neurons at each time layer the same way it is done for a traditional feedforward neural network. This operation is further illustrated in Figure 36-12.

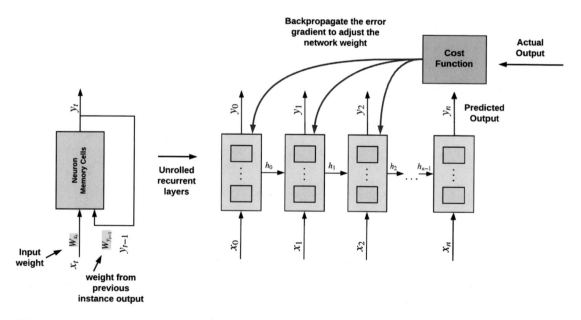

Figure 36-12. *Backpropagation through time*

A significant challenge of training the recurrent neural network is the vanishing and exploding gradient problem. When training a deep recurrent network for many layers of time instants, calculating the gradients of the weights of the neurons can become very volatile. When this happens, the value of the gradient can become extremely large tending to infinity, or they become tiny, all the way to zero. When this happens, the neurons become dead and cannot train or learn any new information further. This effect is called the exploding and vanishing gradient problem.

The exploding and vanishing gradient problem is most prevalent in recurrent neural networks because of the long-term dependencies or time instant of the unrolled recurrent neuron. A proposed alternative technique for mitigating this problem in recurrent networks (in addition to other discussed methods such as gradient clipping, batch normalization, and using a non-saturating activation function such as ReLu) is to discard early time instances or time instances in the distant past. This technique is called Truncated Backpropagation Through Time (truncated BPTT).

However, truncated BPTT suffers a significant drawback, and this is that some problems rely heavily on long-term dependencies to be able to make a prediction. A typical example is in language modeling where the long-term sequence of words in the past is vital in predicting the next word in the sequence.

The shortcoming of truncated BPTT and the need to deal with the problem of exploding and vanishing gradients led to the development of a memory cell called the Long Short-Term Memory or LSTM for short, which can store the long-term information of the problem in the memory cell of the recurrent network.

The Long Short-Term Memory (LSTM) Network

Long Short-Term Memory (LSTM) belongs to a class of RNN called gated recurrent unit. They are called *gated* because unlike the basic recurrent units, they contain extra components called gates that control the flow of information within the recurrent cell. This includes choosing what information to store in the cell and what information to discard or forget.

LSTM is very efficient for capturing the long-term dependencies across a large number of time instants. It does this by having a slightly more sophisticated cell than the basic recurrent units. The components of the LSTM are the

- Memory cell

- Input gate

- Forget gate

- Output gate

These extra components enable the RNN to remember and store important events from the distant past. The LSTM takes as input the previous cell state, c_{t-1}; the previous hidden state, h_{t-1}; and the current input, x_t. To keep in line with the simplicity of this book, we provide a high-level illustration of the LSTM cell showing how the extra components of the cell come together. In TensorFlow 2.0, LSTM layer is implemented in the method **'tf.keras.layers.LSTM()'**.

The illustration in Figure 36-13 is the LSTM memory cell. The components of the LSTM cell serve distinct functions in preserving long-term dependencies in sequence data. Let's go through them:

- The input gate: This gate is responsible for controlling what information gets stored in the long-term state or the memory cell, c. Working in tandem with the input gate is another gate that regulates the information flowing into the input gate. This gate analyzes the current input to the LSTM cell, x_t, and the previous short-term state, h_{t-1}.

- The forget gate: The role of this gate is to regulate how much of the information in the long-term state is persisted across time instants.

- The output gate: This gate controls how much information to output from the cell at a particular time instant. This gate controls the value of h_t (the short-term state) and y_t (the output at time t).

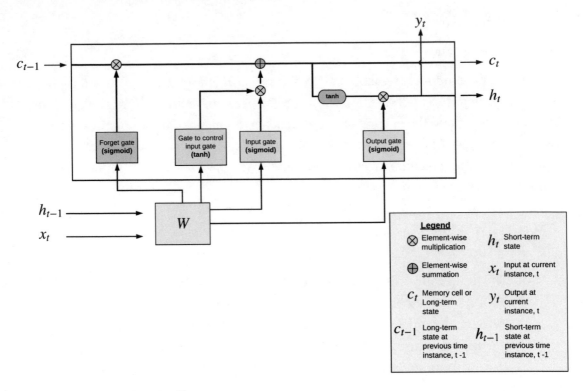

Figure 36-13. *LSTM cell*

It is important to note that the components of the LSTM cells are all fully connected neural networks. There exist other variants of recurrent networks with memory cells, two of such are the peephole connections and the gated recurrent units.

Peephole Connection

The peephole connection extends the LSTM network by also using information from the memory cell or long-term state of the previous time instant c_{t-1} as input to the LSTM gates. The goal of the peephole is to provide extra information into the LSTM unit by peeping at the stored long-term memory. This is further illustrated in Figure 36-14. In TensorFlow 2.0, the implementation of peephole connections to an LSTM layer is provided by the method **'tf.keras.experimental.PeepholeLSTMCell()'**.

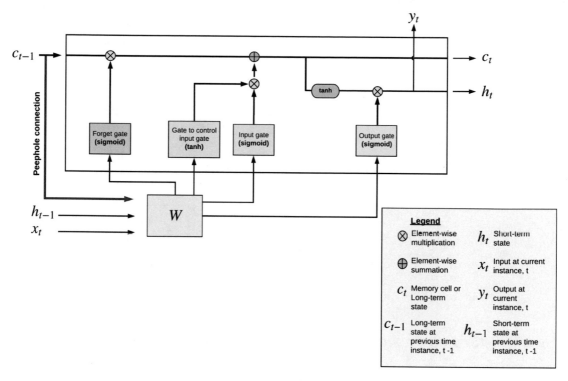

Figure 36-14. *Peephole connection*

Gated Recurrent Unit (GRU)

The gated recurrent unit (GRU) is a more recent recurrent neural network architecture than the LSTM, and it is also comparable simpler to implement with respect to the number of components within the unit and their operations. Despite its comparative simplicity, GRUs are high-performing recurrent architectures and, in most cases, even perform better than the LSTM in sequence modeling problems.

GRUs combine the forget and the input gates to decide on what information should be committed to the long-term memory or the memory cell and what information should be left out. Moreover, the GRU combines the cell (i.e., long-term state) and short-term states into a single state vector h_t. Also, the GRU removes the output gate and returns the state vector h_t at each time instant. This is further illustrated in Figure 36-15. In TensorFlow 2.0, the GRU layer is implemented in the method **'tf.keras.layers.GRU()'**.

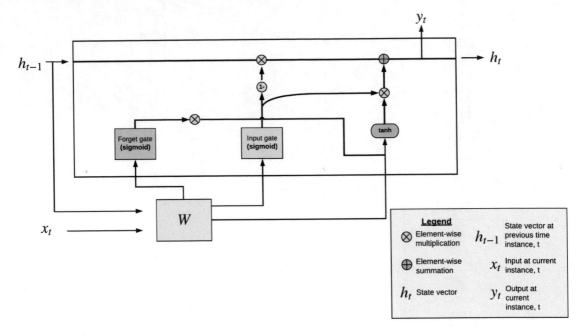

Figure 36-15. *Gated recurrent unit*

Recurrent Neural Networks Applied to Sequence Problems

Recurrent neural networks have many application areas for using LSTM models for sequence tasks. A couple of problems under this domain include sentiment analysis, machine translation, image captioning, video captioning, and voice recognition. As mentioned earlier, these problems can be modeled as a one-to-many model, a many-to-one model, or a many-to-many model. The section will survey a few LSTM architectures for tackling/modeling sequence problems:

- Long-term recurrent convolutional neural network, also known as CNN LSTM

- Encoder-Decoder LSTMs

- Bidirectional recurrent neural networks

Long-Term Recurrent Convolutional Network (LRCN)

The long-term recurrent convolutional network (LRCN) is a unique neural network architecture for generating descriptions of images and videos (which is seen as a sequence of images). These problems can be termed as visual time series modeling. The LRCN architecture combines the ability of the convolutional neural network (CNN) to extract image features together with a recurrent network for learning sequences or long-term dependencies. The LRCN passes visual inputs into a CNN to retrieve image features as outputs. These outputs are then passed into a recurrent LSTM network layer to generate the natural language descriptions. The recurrent layer can contain stacked LSTMs.

One core advantage of LRCN for modeling sequential vision problems such as image captioning and video captioning is that the network is not constrained to fixed lengths of inputs and outputs. Hence, it can be used to model sequential data with different lengths such as textual data and videos.

The following illustrations show how LRCN is applied to a variety of sequence problems:

1.   Image captioning: Image captioning can be seen as a one-to-many sequence problem. The input is an image and therefore a static input, and the output is a sequence of text that describes the objects in the image; this is a sequential output. The use of LRCN for image captioning is illustrated in Figure 36-16.

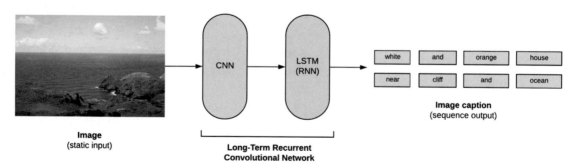

Figure 36-16. *Image captioning (photo by Daniel Llorente on Unsplash)*

2. Video captioning: Video can be seen as a sequence of images. Hence, in a video captioning problem, a sequence of images is passed as input to the LRCN model which in turn returns a sequence of outputs as a textual description for each video frame. Hence, video captioning can be seen as a many-to-many sequence problem. This approach is an example of an Encoder-Decoder LSTM where CNN is used as an image encoder that is initially trained for image classification. The final hidden layer, which is also called a bottleneck, is then passed as input to the RNN decode. It is typical to use an already pre-trained CNN on a large-scale image recognition task. A number of such models exist in the public domain. We will survey Encoder-Decoder LSTMs in more detail shortly. Video captioning is illustrated in Figure 36-17.

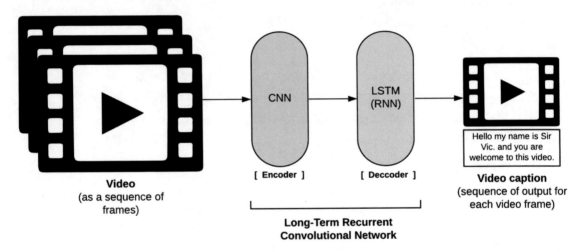

Figure 36-17. *Video captioning*

Encoder-Decoder LSTMs

Encoder-Decoder LSTM architecture handles a particular class of sequence problems that takes as input multiple time steps and also returns a multiple time step output. A major challenge of this sort of problems is that both the input and output sequences can have varied lengths.

The first part of the architecture, that is, the Encoder, is responsible for receiving and encoding the input sequence; the second part of the architecture, that is, the Decoder, takes in the output from the Encoder and then predicts the output sequence.

The sort of architecture is made for natural language processing problems where the output is a sequence of words. It is commonly used in machine translation, video captioning, and speech recognition. An illustration is already provided in Figure 36-10.

Bidirectional Recurrent Neural Networks

Bidirectional RNN is another particular type of recurrent neural network architecture that involves placing the recurrent layers beside each other where one layer works to learn the long-term dependencies from the past; this layer is called the forward LSTM. For the other layer, the input is reversed and fed into the network, so the network learns long-term dependencies from the future. This layer is called the backward LSTM. The bidirectional RNN is illustrated in Figure 36-18.

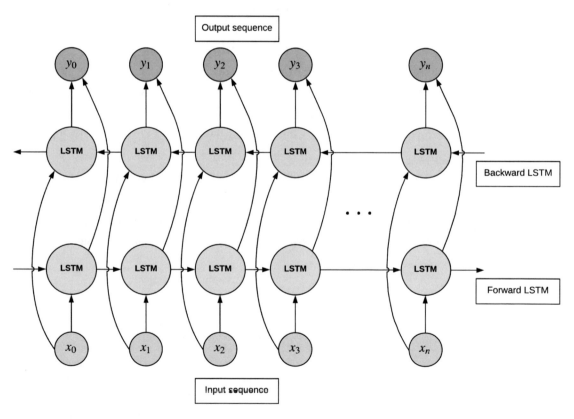

Figure 36-18. *Bidirectional LSTM*

When the outputs from these side-by-side networks are combined, it is easier to predict the next time step of a sequence having privy to the entire information gamut, because they process both information from the past and the future. Although this architecture was first designed for speech recognition tasks, it has performed impressively across a variety of other sequence prediction tasks. It is built to improve on the vanilla unidirectional LSTM which only has knowledge of the past.

This network is built on the understanding that some learning problems only make sense when a coherent set of information is present. For example, if a human interpreter is interpreting from one language to another, he first listens to a cohesive set of information in one language before interpreting to another language. This is because the context of an entire cohesive sentence gives the right basis for a correct interpretation.

RNN with TensorFlow 2.0: Univariate Timeseries

This section makes use of the Nigeria power consumption dataset to implement a univariate timeseries model with LSTM recurrent neural networks. The dataset for this example is the Nigeria power consumption data from January 1 to March 11 by Hipel and McLeod (1994), retrieved from DataMarket.

The dataset is preprocessed for timeseries modeling with RNNs by converting the data input and outputs into sequences using the method **'convert_to_ sequences'**. This method splits the dataset into rolling sequences consisting of 20 rows (or time steps) using a window of **1**}. In Figure 36-19, the example univariate dataset is converted into sequences of five time steps, where output sequence is one step ahead of the input sequence. Each sequence contains five rows (determined by the **time_steps** variable) and in this univariate case, 1 column.

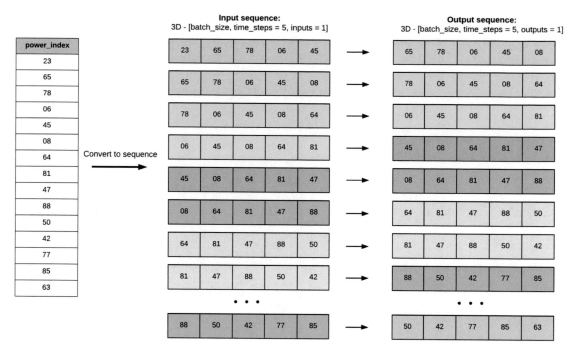

Figure 36-19. *Converting a univariate series into sequences for prediction with RNNs. Left: Sample univariate dataset. Center: Input sequence. Right: Output sequence*

When modeling using RNNs, it is important to scale the dataset to have values within the same range. The plot in Figure 36-20 shows predictions of the model along with the original targets and the lagging training instances. The next plots in Figure 36-21 and Figure 36-22 show the original series and the RNN generated series in both the scaled and normal values.

For increased training speed, the model will train on a GPU. If running the code on Google Colab, change the runtime type to GPU and install TensorFlow 2.0 with GPU package.

```
# import TensorFlow 2.0 with GPU
!pip install -q tf-nightly-gpu-2.0-preview

# import packages
import tensorflow as tf
import pandas as pd
import numpy as np
import matplotlib.pyplot as plt
```

```
from sklearn.model_selection import train_test_split
from sklearn.preprocessing import MinMaxScaler

# confirm tensorflow can see GPU
device_name = tf.test.gpu_device_name()
if device_name != '/device:GPU:0':
  raise SystemError('GPU device not found')
print('Found GPU at: {}'.format(device_name))

# data file path
file_path = "nigeria-power-consumption.csv"

# load data
parse_date = lambda dates: pd.datetime.strptime(dates, '%d-%m')
data = pd.read_csv(file_path, parse_dates=['Month'], index_col='Month',
                   date_parser=parse_date,
                   engine='python', skipfooter=2)

# print column name
data.columns

# change column names
data.rename(columns={'Nigeria power consumption': 'power-consumption'},
            inplace=True)

# split in training and evaluation set
data_train, data_eval = train_test_split(data, test_size=0.2,
shuffle=False)

# MinMaxScaler - center and scale the dataset
scaler = MinMaxScaler(feature_range=(0, 1))
data_train = scaler.fit_transform(data_train)
data_eval = scaler.fit_transform(data_eval)

# adjust univariate data for timeseries prediction
def convert_to_sequences(data, sequence, is_target=False):
    temp_df = []
    for i in range(len(data) - sequence):
        if is_target:
```

```
            temp_df.append(data[(i+1): (i+1) + sequence])
        else:
            temp_df.append(data[i: i + sequence])
    return np.array(temp_df)

# parameters
time_steps = 20
batch_size = 50

# create training and testing data
train_x = convert_to_sequences(data_train, time_steps, is_target=False)
train_y = convert_to_sequences(data_train, time_steps, is_target=True)

eval_x = convert_to_sequences(data_eval, time_steps, is_target=False)
eval_y = convert_to_sequences(data_eval, time_steps, is_target=True)

# build model
model = tf.keras.Sequential()
model.add(tf.keras.layers.LSTM(128, input_shape=train_x.shape[1:],
                                return_sequences=True))
model.add(tf.keras.layers.Dense(1))

# compile the model
model.compile(loss='mean_squared_error',
            optimizer='adam',
            metrics=['mse'])

# print model summary
model.summary()

# create dataset pipeline
train_ds = tf.data.Dataset.from_tensor_slices(
    (train_x, train_y)).shuffle(len(train_x)).repeat().batch(batch_size)
test_ds = tf.data.Dataset.from_tensor_slices((eval_x, eval_y)).batch(batch_
size)

# train the model
history = model.fit(train_ds, epochs=10,
                    steps_per_epoch=500)
```

```
# evaluate the model
loss, mse = model.evaluate(test_ds)

print('Test loss: {:.4f}'.format(loss))
print('Test mse: {:.4f}'.format(mse))

# predict
y_pred = model.predict(eval_x)

# plot predicted sequence
plt.title("Model Testing", fontsize=12)
plt.plot(eval_x[0,:,0], "b--", markersize=10, label="training instance")
plt.plot(eval_y[0,:,0], "g--", markersize=10, label="targets")
plt.plot(y_pred[0,:,0], "r--", markersize=10, label="model prediction")
plt.legend(loc="upper left")
plt.xlabel("Time")
plt.show()
```

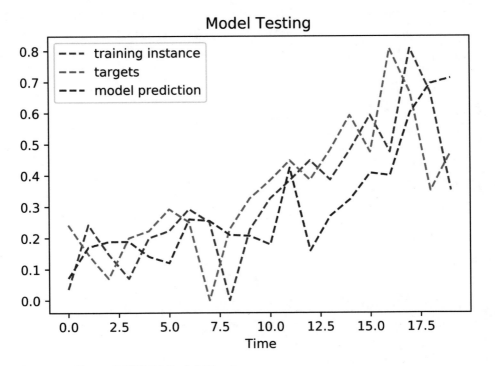

Figure 36-20. *Keras LSTM Model Testing*

```
# use model to predict sequences using training data as seed
rnn_data = list(data_train[:20])
for i in range(len(data_train) - time_steps):
    batch = np.array(rnn_data[-time_steps:]).reshape(1, time_steps, 1)
    y_pred = model.predict(batch)
    rnn_data.append(y_pred[0, -1, 0])

plt.title("RNN vs. Original series", fontsize=12)
plt.plot(data_train, "b--", markersize=10, label="Original series")
plt.plot(rnn_data, "g--", markersize=10, label="RNN generated series")
plt.legend(loc="upper left")
plt.xlabel("Time")
plt.show()
```

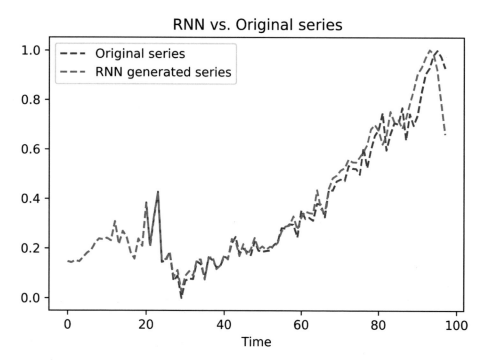

Figure 36-21. *Original series vs. RNN generated series – scaled data values*

```
# inverse to normal scale and plot
data_train_inverse = scaler.inverse_transform(data_train.reshape(-1, 1))
rnn_data_inverse = scaler.inverse_transform(np.array(rnn_data).reshape(-1, 1))
```

467

```
plt.title("RNN vs. Original series with normal scale", fontsize=12)
plt.plot(data_train_inverse, "b--", markersize=10, label="Original series")
plt.plot(rnn_data_inverse, "g--", markersize=10, label="RNN generated
series")
plt.legend(loc="upper left")
plt.xlabel("Time")
plt.show()
```

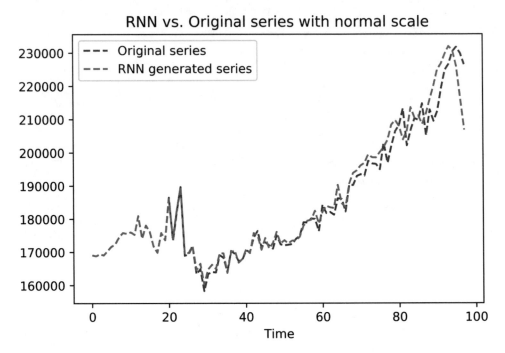

Figure 36-22. *Original series vs. RNN generated series – normal data values*

From the Keras LSTM code listing, the method **tf.keras.layers.LSTM()** is used to implement the LSTM recurrent layer. The attribute **return_sequences** is set to **True** to return the last output in the output sequence, or the full sequence.

RNN with TensorFlow 2.0: Multivariate Timeseries

The dataset for this example is the Dow Jones Index Data Set from the famous UCI Machine Learning Repository. In this stock dataset, each row contains the stock price record for a week including the percentage of return that stock has in the following week

percent_change_next_weeks_price(). For this example, the record for the previous week is used to predict the percent change in price for the next 2 weeks for Bank of America, BAC stock prices.

The method named **clean_dataset()** carries out some rudimentary cleanup of the dataset to make it suitable for modeling. The actions taken on this particular dataset involve removing the dollar sign from certain of the data columns, removing missing values, and rearranging the data columns so target attribute **percent_change_next_weeks_price** is the last column.

The method named **data_transform()** subselects the stock records belonging to 'Bank of America,' and the target attribute is adjusted so that the previous week record is used to predict the percent change in price for the next 2 weeks. Also, the dataset is split into training and testing sets. The method named **normalize_and_scale()** removes the non-numeric columns and scales the dataset attributes.

Again, the model will train on a GPU instance. The model will be a stacked GRU with multiple GRU layers. This stacking of RNN layers with memory cells makes the network more expressive and can learn more complex long-running sequences. If running the code on Google Colab, change the runtime type to GPU and install TensorFlow 2.0 with GPU package. The output plot in Figure 36-23 is the model predictions showing the targets and the lag training instances.

```
# import TensorFlow 2.0 with GPU
!pip install -q tf-nightly-gpu-2.0-preview
```

```
# import packages
import tensorflow as tf
import pandas as pd
import numpy as np
import matplotlib.pyplot as plt
from sklearn.model_selection import train_test_split
from sklearn.preprocessing import MinMaxScaler
```

```
# confirm tensorflow can see GPU
device_name = tf.test.gpu_device_name()
if device_name != '/device:GPU:0':
  raise SystemError('GPU device not found')
print('Found GPU at: {}'.format(device_name))
```

```
# data file path
file_path = "dow_jones_index.data"

# load data
data = pd.read_csv(file_path, parse_dates=['date'], index_col='date')

# print column name
data.columns

# print column datatypes
data.dtypes

# parameters
outputs = 1
stock ='BAC'   # Bank of America

def clean_dataset(data):
    # strip dollar sign from `object` type columns
    col = ['open', 'high', 'low', 'close', 'next_weeks_open', 'next_weeks_
    close']
    data[col] = data[col].replace({'\$': "}, regex=True)
    # drop NaN
    data.dropna(inplace=True)
    # rearrange columns
    columns = ['quarter', 'stock', 'open', 'high', 'low', 'close', 'volume',
        'percent_change_price', 'percent_change_volume_over_last_wk',
        'previous_weeks_volume', 'next_weeks_open', 'next_weeks_close',
        'days_to_next_dividend', 'percent_return_next_dividend',
        'percent_change_next_weeks_price']
    data = data[columns]
    return data

def data_transform(data):
    # select stock data belonging to Bank of America
    data = data[data.stock == stock]
    # adjust target(t) to depend on input (t-1)
    data.percent_change_next_weeks_price = data.percent_change_next_weeks_
    price.shift(-1)
```

```
    # remove nans as a result of the shifted values
    data = data.iloc[:-1,:]
    # split quarter 1 as training data and quarter 2 as testing data
    train_df = data[data.quarter == 1]
    test_df = data[data.quarter == 2]
    return (np.array(train_df), np.array(test_df))

def normalize_and_scale(train_df, test_df):
    # remove string columns and convert to float
    train_df = train_df[:,2:].astype(float,copy=False)
    test_df = test_df[:,2:].astype(float,copy=False)
    # MinMaxScaler - center and scale the dataset
    scaler = MinMaxScaler(feature_range=(0, 1))
    train_df_scale = scaler.fit_transform(train_df[:,2:])
    test_df_scale = scaler.fit_transform(test_df[:,2:])
    return (scaler, train_df_scale, test_df_scale)

# clean the dataset
data = clean_dataset(data)

# select Dow Jones stock and split into training and test sets
train_df, test_df = data_transform(data)

# scale the data
scaler, train_df_scaled, test_df_scaled = normalize_and_scale(train_df,
test_df)

# split train/ test
train_X, train_y = train_df_scaled[:, :-1], train_df_scaled[:, -1]
test_X, test_y = test_df_scaled[:, :-1], test_df_scaled[:, -1]

# reshape inputs to 3D array
train_X = train_X[:,None,:]
test_X = test_X[:,None,:]

# reshape outputs
train_y = np.reshape(train_y, (-1,outputs))
test_y = np.reshape(test_y, (-1,outputs))
```

```
# model parameters
batch_size = int(train_X.shape[0]/5)
length = train_X.shape[0]

# build model
model = tf.keras.Sequential()
model.add(tf.keras.layers.GRU(128, input_shape=train_X.shape[1:],
                                return_sequences=True))
model.add(tf.keras.layers.GRU(100, return_sequences=True))
model.add(tf.keras.layers.GRU(64))
model.add(tf.keras.layers.Dense(1))

# compile the model
model.compile(loss='mean_squared_error',
            optimizer='adam',
            metrics=['mse'])

# print model summary
model.summary()

# create dataset pipeline
train_ds = tf.data.Dataset.from_tensor_slices(
    (train_X, train_y)).shuffle(len(train_X)).repeat().batch(batch_size)
test_ds = tf.data.Dataset.from_tensor_slices((test_X, test_y)).batch(batch_size)

# train the model
history = model.fit(train_ds, epochs=10,
                    steps_per_epoch=500)

# evaluate the model
loss, mse = model.evaluate(test_ds)

print('Test loss: {:.4f}'.format(loss))
print('Test mse: {:.4f}'.format(mse))

# predict
y_pred = model.predict(test_X)

# plot
plt.figure(1)
```

```
plt.title("Keras - GRU RNN Model Testing for '{}' stock".format(stock),
fontsize=12)
plt.plot(test_y, "g--", markersize=10, label="targets")
plt.plot(y_pred, "r--", markersize=10, label="model prediction")
plt.legend()
plt.xlabel("Time")
plt.show()
# plt.savefig('gru-bac-model.png', dpi=800)
```

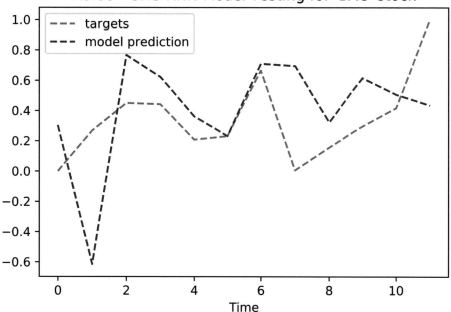

Figure 36-23. *GRU RNN Model Testing for Bank of America stock*

This chapter gave an overview of recurrent neural networks (RNNs) and its application in learning recurrent models for different types of sequence problems. The next chapter will discuss how we can use neural networks to reconstruct the inputs as some form of unsupervised learning using autoencoders.

CHAPTER 37

Autoencoders

Autoencoder is an unsupervised learning algorithm that uses neural networks to reconstruct the features of a dataset. Just like the unsupervised algorithms that we earlier discussed in the chapter on machine learning, autoencoders can be used to reduce the dimensionality of a dataset and to extract relevant features. Moreso, peculiar to autoencoders is the ability to generate more examples of the dataset after learning an internal representation (also called coding) that reconstructs the features of the inputs to the neural network.

An autoencoder receives as input the features of the dataset. These features are passed through a set of encoders, which are the hidden layers of a neural network to create an internal representation called codings. The learned coding is then used to reconstruct the output through a set of decoders, which are also hidden neural network layers. The autoencoder cannot merely do a trivial memorization of the inputs, because a constraint is placed on the encoders by reducing the input dimension to force the network to learn an efficient set of representation from which the decoders use to reconstruct the inputs.

Autoencoders with restricted Encoders and Decoders are called **undercomplete**. A reconstruction error term is used to evaluate the performance of an autoencoder by testing how well the output corresponds with the input. Of course, just like other neural networks, the neurons of the Encoders and Decoders have non-linear activation functions for learning complex patterns. An example of a simple autoencoder network architecture is shown in Figure 37-1.

© Ekaba Bisong 2019

E. Bisong, *Building Machine Learning and Deep Learning Models on Google Cloud Platform*, https://doi.org/10.1007/978-1-4842-4470-8_37

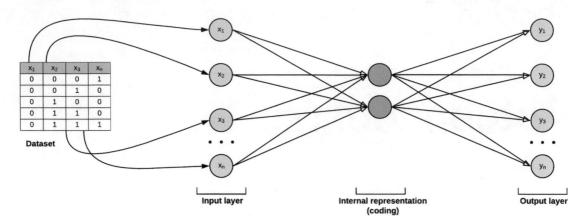

Figure 37-1. *A simple autoencoder architecture*

Stacked Autoencoders

Stacked autoencoder is when the simple autoencoder architecture as shown in Figure 37-1 is enhanced with multiple hidden layers. Just like other deep neural network architectures with hidden layers, the hidden layers of an autoencoder enable the network to learn more complex patterns of the input dataset.

The hidden layers of a stacked or deep autoencoder are added symmetrically at both the Encoder and Decoder part of the network as shown in Figure 22-2. The neurons of the hidden layers are restricted to be less than that of the input layer. This formulation places a restriction on the network, so it doesn't merely memorize the input. Moreso, care must be taken not to create too many deep layers, so the autoencoder does not overfit the input data and fail to generalize to out-of-sample examples. To optimize the training of a deep autoencoder, the weights of the symmetrical neural layers are shared in a technique called *tying*.

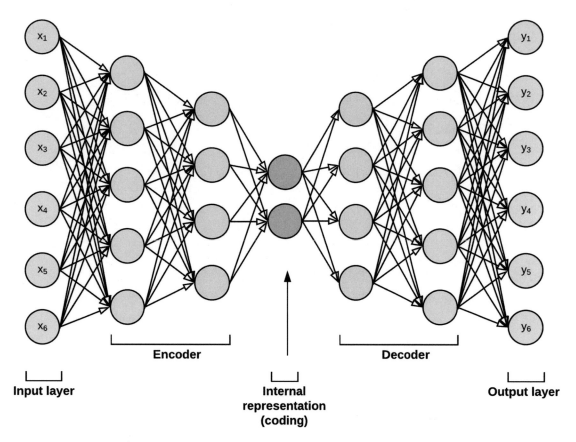

Figure 37-2. *Stacked or deep autoencoder. The hidden layers are added symmetrically at both the Encoder and Decoder*

Stacked Autoencoders with TensorFlow 2.0

The code example in this section shows how to implement an autoencoder network using TensorFlow 2.0. For simplicity, the MNIST handwriting dataset is used to create reconstructions of the original images. In this example, a stacked autoencoder is implemented with the original and reconstructed image shown in Figure 37-3. The code listing is presented in the following, and corresponding notes on the code are shown thereafter.

```
# import TensorFlow 2.0 with GPU
!pip install -q tf-nightly-gpu-2.0-preview

# import packages
import tensorflow as tf
```

```python
import numpy as np
import matplotlib.pyplot as plt

# import dataset
(x_train, _), (x_test, _) = tf.keras.datasets.mnist.load_data()

# change datatype to float
x_train = x_train.astype('float32')
x_test = x_test.astype('float32')

# scale the dataset from 0 -> 255 to 0 -> 1
x_train /= 255
x_test /= 255

# flatten the 28x28 images into vectors of size 784
x_train = x_train.reshape((len(x_train), np.prod(x_train.shape[1:])))
x_test = x_test.reshape((len(x_test), np.prod(x_test.shape[1:])))

# create the autoencoder model
def model_fn():
  model_input = tf.keras.layers.Input(shape=(784,))
  encoded = tf.keras.layers.Dense(units=512, activation='relu')(model_input)
  encoded = tf.keras.layers.Dense(units=128, activation='relu')(encoded)
  encoded = tf.keras.layers.Dense(units=64, activation='relu')(encoded)
  coding_layer = tf.keras.layers.Dense(units=32)(encoded)
  decoded = tf.keras.layers.Dense(units=64, activation='relu')(coding_layer)
  decoded = tf.keras.layers.Dense(units=128, activation='relu')(decoded)
  decoded = tf.keras.layers.Dense(units=512, activation='relu')(decoded)
  decoded_output = tf.keras.layers.Dense(units=784)(decoded)

  # the autoencoder model
  autoencoder_model = tf.keras.Model(inputs=model_input, outputs=decoded_output)

  # compile the model
  autoencoder_model.compile(optimizer='adam',
                loss='binary_crossentropy',
                metrics=['accuracy'])

  return autoencoder_model
```

```python
# build the model
autoencoder_model = model_fn()

# print autoencoder model summary
autoencoder_model.summary()

# train the model
autoencoder_model.fit(x_train, x_train, epochs=1000, batch_size=256,
                      shuffle=True, validation_data=(x_test, x_test))

# visualize reconstruction
sample_size = 6
test_image = x_test[:sample_size]
# reconstruct test samples
test_reconstruction = autoencoder_model.predict(test_image)

plt.figure(figsize = (8,25))
plt.suptitle('Stacked Autoencoder Reconstruction', fontsize=16)
for i in range(sample_size):
    plt.subplot(sample_size, 2, i*2+1)
    plt.title('Original image')
    plt.imshow(test_image[i].reshape((28, 28)), cmap="Greys",
    interpolation="nearest", aspect='auto')
    plt.subplot(sample_size, 2, i*2+2)
    plt.title('Reconstructed image')
    plt.imshow(test_reconstruction[i].reshape((28, 28)), cmap="Greys",
    interpolation="nearest", aspect='auto')
plt.show()
```

From the preceding code listing, take note of the following:

- Observe the arrangement of the encoder layers and the decoder layers of the stacked autoencoder. Specifically note how the corresponding layer arrangement of the encoder and the decoder has the same number of neurons.

- The loss error measures the squared difference between the inputs into the autoencoder network and the decoder output.

The image in Figure 37-3 contrasts the reconstructed images from the autoencoder network with the original images in the dataset.

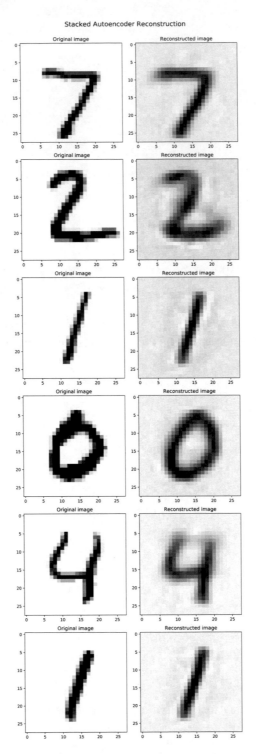

Figure 37-3. *Stacked autoencoder reconstruction. Left: Original image. Right: Reconstructed image.*

Denoising Autoencoders

Denoising autoencoders add a different type of constraint to the network by imputing some Gaussian noise into the inputs. This noise injection forces the autoencoder to learn the uncorrupted form of the input features; by doing so, the autoencoder learns the internal representation of the dataset without memorizing the inputs.

Another way a denoising autoencoder constrains the input is by deactivating some input neurons in a similar fashion to the Dropout technique. Denoising autoencoders use an overcomplete network architecture. This means that the dimensions of the hidden Encoder and Decoder layers are not restricted; hence, they are overcomplete. An illustration of a denoising autoencoder architecture is shown in Figure 37-4.

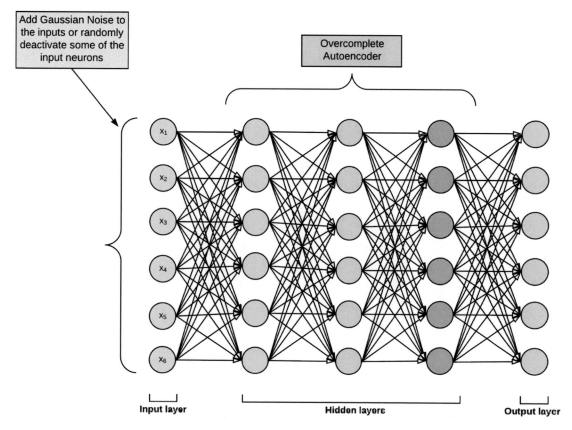

Figure 37-4. *Denoising autoencoder. Constraint is applied by either adding Gaussian noise or by switching off some a random selection of the input neurons.*

This chapter discussed how deep neural networks can be employed in an unsupervised fashion to reconstruct the inputs to the network as the network's output. This is the final chapter in Part 6 that provides a general theoretical background to deep neural networks and how they are implemented in TensorFlow 2.0. In Part 7, we will discuss doing advanced analytics and machine learning on Google Cloud Platform.

PART VII

Advanced Analytics/ Machine Learning on Google Cloud Platform

CHAPTER 38

Google BigQuery

BigQuery is a Google-managed data warehouse product that is highly scalable, fast, and optimized for data analytics with rudimentary in-built machine learning capabilities as part of the product offering. It is also one of Google's many serverless products. This means that you do not physically manage the infrastructure assets and the overhead responsibilities/costs. It is only used to solve the business use case, and it just works in a highly performant manner.

BigQuery is suited for storing and analyzing structured data. The idea of structured data is that it must have a schema that describes the columns or fields of the dataset. CSV or JSON files are examples of structured data formats. BigQuery differentiates itself from other relational databases in that it can store a collection of other fields (or columns) as a record type, and a particular field in a row can have more than one value. These features make BigQuery more expressive for storing datasets without the flat row constraint of relational databases.

Similar to relational databases, BigQuery organizes rows into *tables*, and are accessed using the familiar Structured Query Language (SQL) for databases. However, individual rows in a table cannot be updated by running a SQL Update statement. Tables can only be appended to or entirely re-written. Meanwhile, a group of tables in BigQuery is organized into *datasets*.

When a query is executed in BigQuery, it runs in parallel on thousands of cores. This feature greatly accelerates the performance of query execution and consequently the speed of gaining insights from your data. This ability for massive parallel execution is one of the major reasons individuals, companies, and institutions are migrating to BigQuery as their data warehouse of choice.

Also BigQueryML is a powerful platform for building machine learning models inside of BigQuery. The models take advantage of automated feature engineering and hyper-parameter optimization and are automatically updated based on changes to the underlying dataset. This feature is extremely powerful and lowers the threshold

© Ekaba Bisong 2019
E. Bisong, *Building Machine Learning and Deep Learning Models on Google Cloud Platform*,
https://doi.org/10.1007/978-1-4842-4470-8_38

of business intelligence and analytics personnel to more easily harness the predictive power of using machine learning for business forecasting and decision-making.

What BigQuery Is Not

As powerful and widely purposed as BigQuery is, it may not be properly suited for some use cases:

- BigQuery is not a replacement for a relational database. Some business use cases may involve a large number of table row updates; in such an instance, BigQuery is most likely not the data storage solution of choice, as relational databases are well suited for such highly transactional tasks. GCP offers the Cloud SQL and Cloud Spanner as parts of its managed relational products.

- BigQuery is not a NoSQL database. Data stored in BigQuery must have a schema. NoSQL is a schema-less data storage solution. GCP also has Cloud BigTable and Cloud Datastore, which are highly scalable and performant managed NoSQL products.

Getting Started with BigQuery

BigQuery can be accessed and used via a variety of ways; they include

- The BigQuery web UI

- The command-line tool, **'bq'**

- The client API libraries for programmatic access

In this section, we will introduce BigQuery by working with the web UI, because it gives a graphical view of the datasets and tables within BigQuery and is good for quick execution of queries on the query engine.

To open BigQuery from the GCP dashboard, click the triple dash on the top-left corner and select **BigQuery** from the product section labeled **Big Data** as shown in Figure 38-1.

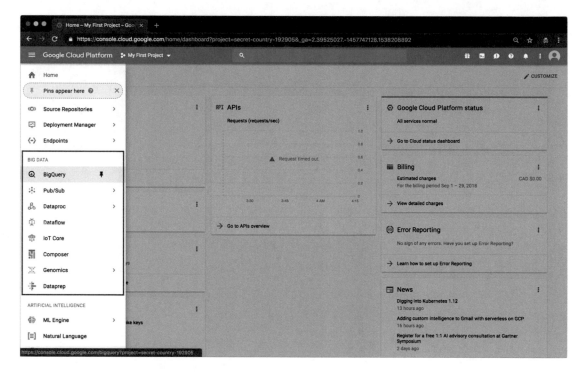

Figure 38-1. *Open BigQuery*

The BigQuery web UI dashboard is as shown in Figure 38-2.

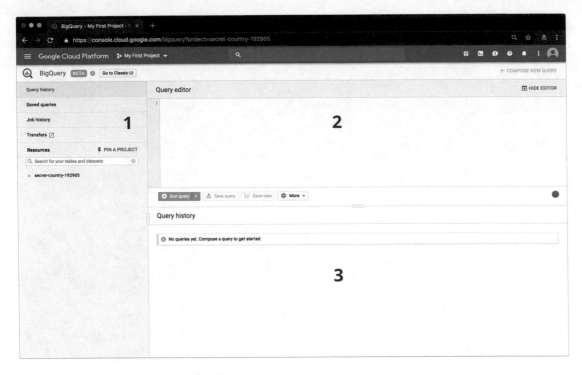

Figure 38-2. *BigQuery web UI*

In Figure 38-2, there are three labeled sections of the BigQuery web UI that we'll briefly explain:

1. The navigation panel: This panel contains a set of BigQuery resources such as

    -   Query history: For viewing previous queries

    -   Saved queries: For storing frequently used queries

    -   Job history: For viewing BigQuery jobs such as loading, copying, and exporting of data

    -   Transfers: Link to the BigQuery Data Transfer Service UI

    -   Resources: Shows a list of pinned projects and their containing Datasets

2. The Query editor: This is where queries are composed using the familiar SQL database language.

3. The Details panel: This panel shows the details of projects, datasets, and table when clicked in the **Resources** tab. Also, this panel shows the results of executed queries.

Public Datasets

BigQuery comes with access to some public datasets; we will use these datasets to explore working with BigQuery. To view the public datasets, go to

```
https://console.cloud.google.com/bigquery?p=bigquery-public-
data&page=project.
```

The public datasets will now show in the Resources section of the navigation panel (see Figure 38-3).

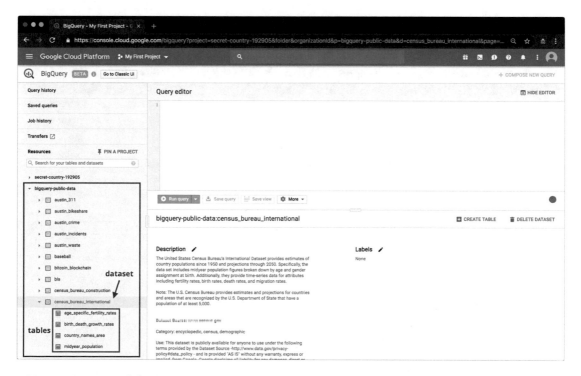

Figure 38-3. *Public Datasets*

Running Your First Query

For our first query, we will work with the **'census_bureau_international'** dataset which "provides estimates of country populations since 1950 and projections through 2050." In this query, we select a country and their life expectancy (for both sexes) in the year 2018.

```
SELECT
  country_name,
  life_expectancy
FROM
  `bigquery-public-data.census_bureau_international.mortality_life_
  expectancy`
WHERE
  year = 2018
ORDER BY
  life_expectancy DESC
```

A sample of the query result is shown in Figure 38-4 under **Query results**.

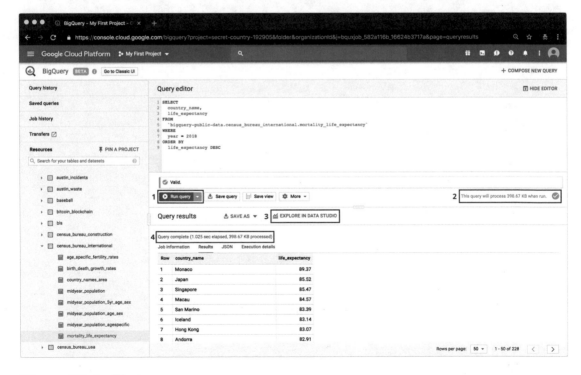

Figure 38-4. *First query*

After typing the query in the **Query editor**, the following should be noted, as numbered in Figure 38-4:

1. Click the **'Run query'** button to execute the query.

2. The green **status indicator** shows that the query is a valid SQL statement and shows by the side an estimate of the query size estimation.

3. The query results can be easily analyzed and visualized using Data Studio.

4. We can see that the query completed in just over a second.

Loading Data into BigQuery

In this simple data ingestion example, we will load a CSV file stored on Google Cloud Storage (GCS) into BigQuery. In GCP, Google Cloud Storage is a general-purpose storage location for all variety of file types and is preferred as a staging area or an archival repository for data. Let's walk through the following steps.

Staging the Data in GCS

Let's go through the steps to stage the data in Google Cloud Storage:

1. Activate Cloud Shell as shown in Figure 38-5.

Figure 38-5. *Activate Google Cloud Shell*

2. Create a bucket on GCS (remember to give the bucket a unique name).

    ```
    gsutil mb gs://my-test-data
    ```

3. Transfer data into bucket. The CSV data used in this example is a crypto-currency dataset stored in the code repository. Use the 'gsutil cp' command to move the dataset to GCS bucket.

    ```
    gsutil cp crypto-markets.csv gs://my-test-data
    ```

4. Show the transferred data in the bucket.

    ```
    gsutil ls gs://my-test-data/
    ```

Loading Data Using the BigQuery Web UI

Let's go through the following steps to load data into BigQuery using the web UI:

1. In the navigation panel, click the project name, and then click **CREATE DATASET** in the Details panel (see Figure 38-6).

Figure 38-6. *Create Dataset*

2. Type 'crypto_data' as the **DatasetID,** and select 'United States (US)' as the data location (see Figure 38-7).

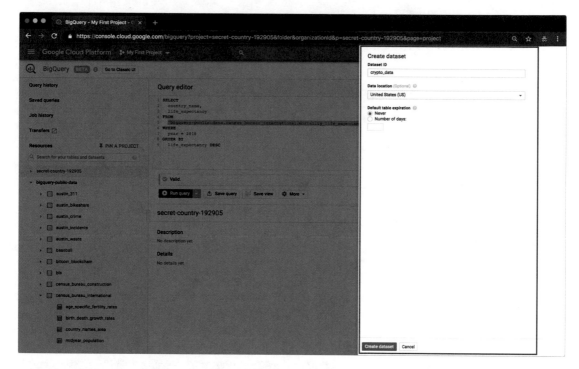

Figure 38-7. *Create Dataset parameters*

3. Next, click the newly created Dataset in the navigation panel, and
    then click **CREATE TABLE** in the Details panel (see Figure 38-8).

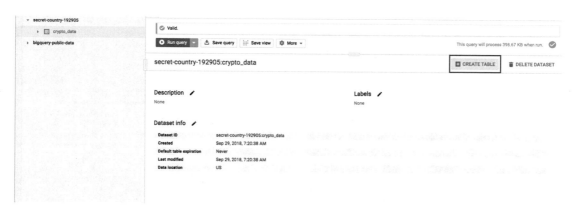

Figure 38-8. *Create Table*

4. We'll create a table from a CSV file stored on Google Cloud Storage. On the Create Table page, select the following parameters as shown in Figure 38-9:

    a. Select **'Google Cloud Storage'** for Source Data.

    b. Select the file **'crypto-markets.csv'** from the bucket **'my-test-data'**.

    c. Choose **CSV** as the file format.

    d. Type **'markets'** as the Destination table.

    e. Toggle 'Edit as Text' and enter the following as the schema:

        slug,symbol,name,date,ranknow,open,high,low,close,volume,market, close_ratio,spread

    f. Expand 'Advanced options' and set 'Header rows to skip' to 1.

    g. Click **Create table**.

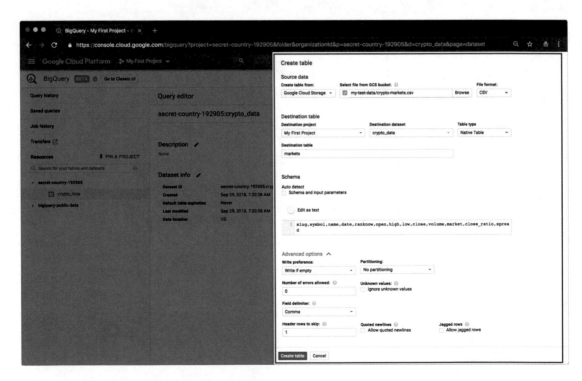

Figure 38-9. *Create table options*

Click **Job history** in the navigation panel to view the status of the loading job (see Figure 38-10).

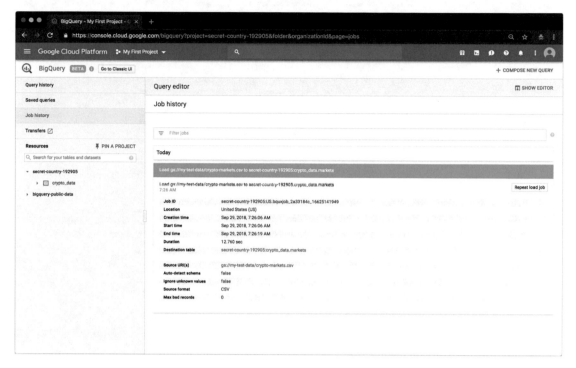

Figure 38-10. *BigQuery loading job*

A preview of the created table is as shown in Figure 38-11.

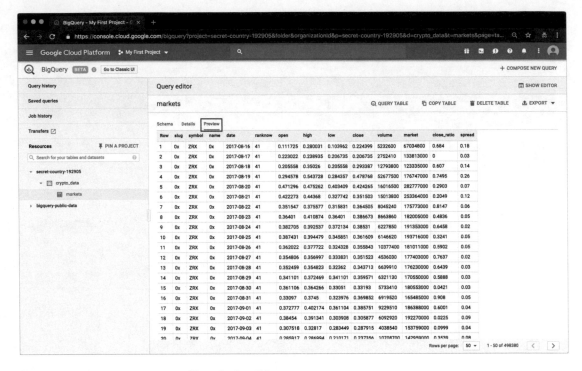

Figure 38-11. *Preview of loaded table*

The bq Command-Line Utility

Let's go through some useful commands on the Cloud Shell terminal with the 'bq' utility:

- List the projects that can be accessed.

```
bq ls -p

        projectId            friendlyName
----------------------  ------------------
secret-country-192905   My First Project
```

- List datasets in the default project.

```
bq ls

   datasetId
-------------
crypto_data
```

- List tables in a Dataset.

```
bq ls crypto_data

  tableId    Type    Labels   Time Partitioning
 --------- ------- -------- -------------------
  markets    TABLE
```

- List the recent executed jobs. This includes both load jobs and queries executed.

```
bq ls -j

jobId                          Job Type  State     Start Time        Duration
--------------------------    --------  --------  ---------------   --------
bquxjob_767fb332_16625172a52   load      SUCCESS   29 Sep 07:29:27   0:00:10
bquxjob_2a33184c_16625141949   load      SUCCESS   29 Sep 07:26:06   0:00:13
bquxjob_582a116b_16624b3717a   query     SUCCESS   29 Sep 05:41:20   0:00:01
bquxjob_7b18cd73_16624a0f378   query     SUCCESS   29 Sep 05:40:32   0:00:01
```

Loading Data Using the Command-Line bq Utility

The following commands walk through loading a dataset into BigQuery using the bq utility via the terminal:

- Create a new Dataset.

```
bq mk crypto_data_terminal

Dataset 'secret-country-192905:crypto_data_terminal' successfully
created.
```

- List the datasets to confirm creation of new Dataset.

```
bq ls

        datasetId
 ---------------------
  crypto_data
  crypto_data_terminal
```

- Load data as a Table into the newly created Dataset. We load the file using the 'bq load' command. This command loads data in a new or existing table. In our example, we load the data from the GCS bucket 'gs://my-test-data/crypto-markets.csv' into a newly created table named 'markets_terminal' with the schema "slug,symbol,name,date, ranknow,open,high,low,close,volume,market,close_ratio,spread"

```
bq load crypto_data_terminal.markets_terminal gs://my-test-data/
crypto-markets.csv slug,symbol,name,date,ranknow,open,high,low,
close,volume,market,close_ratio,spread
```

- List the tables in the dataset.

```
bq ls crypto_data_terminal

        tableId           Type    Labels   Time Partitioning
 ------------------- ------- -------- --------------------
   markets_terminal   TABLE
```

- Examine the table schema.

```
bq show crypto_data_terminal.markets_terminal

Table secret-country-192905:crypto_data_terminal.markets_terminal

    Last modified              Schema           Total Rows    Total
 Bytes    Expiration  Time Partitioning   Labels
   ---------------- ----------------------- ------------ ---------
 ---- ------------ ------------------- --------
    29 Sep 09:12:24   |- slug: string            498381        52777964
                      |- symbol: string
                      |- name: string
                      |- date: string
                      |- ranknow: string
                      |- open: string
                      |- high: string
                      |- low: string
                      |- close: string
                      |- volume: string
```

```
|- market: string
|- close_ratio: string
|- spread: string
```

- Delete a table.

```
bq rm crypto_data_terminal.markets_terminal
```

- Delete a Dataset. This command will delete a Dataset with all its containing tables.

```
bq rm -r crypto_data_terminal
```

BigQuery SQL

In this section, we'll have an overview of SQL by executing some examples that gives a broad perspective of what can be achieved with SQL. New users who have not used SQL before will benefit from this section. Also, SQL is amazingly easy and intuitive to use that non-technical people like personnel in marketing and sales are experts at this even sometimes more than programmers. It is an expressive declarative language.

BigQuery works with both the standard SQL which supports SQL 2011 standard and the legacy SQL syntax which is a non-standard variant of SQL. However, standard SQL is the preferred query syntax for BigQuery. In experimenting with SQL, we will work with the **census_bureau_international** public dataset. The following queries are available in the chapter notebook of the book repository.

Filtering

The following query selects the fertility rate for each country in the year 2018 from the 'age_specific_fertility_rates' table in the 'census_bureau_international' dataset. The resulting table is arranged in descending order.

```
bq query --use_legacy sql=false 'SELECT
  country_name AS country,
  total_fertility_rate AS fertility_rate
FROM
  `bigquery-public-data.census_bureau_international.age_specific_fertility_
  rates`
```

```
WHERE
  year = 2018
ORDER BY
  fertility_rate DESC
LIMIT
  10'
```

```
Waiting on bqjob_r142a3f484f713c4a_0000016626f7f063_1 ... (0s) Current
status: DONE
+-------------+----------------+
|   country   | fertility_rate |
+-------------+----------------+
| Niger       |         6.3504 |
| Angola      |         6.0945 |
| Burundi     |          5.934 |
| Mali        |            5.9 |
| Chad        |            5.9 |
| Somalia     |          5.702 |
| Uganda      |           5.62 |
| Zambia      |          5.582 |
| Malawi      |         5.4286 |
| South Sudan |           5.34 |
+-------------+----------------+
```

In the preceding query, the SQL command SELECT is used to select fields or columns from the table. What follows after the SELECT keyboard is the list of the column names separated by a comma. The keyword AS is used to give an alternative name to the column that will be displayed in the resulting table when the query is executed. The keyword FROM is used to point to the table from which the data is being retrieved. In BigQuery, using the standard SQL, the table name is prefixed by the database name and the project ID is surrounded by a pair of backticks (i.e., 'project_id.database_name. table_name').

The keyword WHERE is used to filter the rows returned from the query. The keyword ORDER BY is used to arrange the retrieved data in either ascending or descending order by a specified column or set of columns. The keyword LIMIT truncates the results retrieved from the query.

Aggregation

The following query selects the average population for each country between the years 2000 and 2018 from the 'midyear_population' table in the 'census_bureau_international' dataset. The resulting table is arranged in descending order.

```
bq query --use_legacy_sql=false 'SELECT
  country_name AS country,
  AVG(midyear_population) AS average_population
FROM
  `bigquery-public-data.census_bureau_international.midyear_population`
WHERE
  year >= 2000 AND year <= 2018
GROUP BY
  country
ORDER BY
  average_population DESC
LIMIT
  20'
```

```
Waiting on bqjob_r95be3d17e726415_000001662890a68f_1 ... (1s) Current
status: DONE
+-----------------+---------------------+
|     country     | average_population  |
+-----------------+---------------------+
| China           | 1.3285399873157892E9 |
| India           |  1.154912377105263E9 |
| United States   | 3.0594302226315784E8 |
| Indonesia       | 2.3984691394736844E8 |
| Brazil          |  1.930978929473684E8 |
| Pakistan        | 1.8112083526315784E8 |
| Nigeria         | 1.6255564478947365E8 |
| Bangladesh      |  1.447749475789474E8 |
| Russia          | 1.4330035963157892E8 |
| Japan           | 1.2727527184210527E8 |
| Mexico          | 1.1269223210526317E8 |
| Philippines     |          9.1357295E7 |
```

```
| Vietnam          |   8.83786184736842E7 |
| Ethiopia         |  8.460339989473683E7 |
| Germany          |  8.168817173684208E7 |
| Egypt            |  8.064017099999999E7 |
| Iran             |  7.427240431578948E7 |
| Turkey           |  7.389499394736844E7 |
| Congo (Kinshasa) |   6.82958565263158E7 |
| Thailand         |  6.619103463157895E7 |
+------------------+----------------------+
```

In the preceding query, the fields retrieved using the SELECT command are passed through an aggregation function to give the average of the mid-year population for the years between 2000 and 2018 inclusive. In order to mix aggregated field and non-aggregated fields, we need the GROUP BY command to group the result by one or more columns, or else only a single result will be returned because of the aggregated function.

Joins

The following query selects the average population for each country and their life expectancy for the year 2018. The data is joined from the 'midyear_population' table and the 'mortality_life_expectancy' table in the 'census_bureau_international' dataset. The resulting table is grouped by country name and year and arranged in descending order.

```
bq query --use_legacy_sql=false 'SELECT
  midyearpop.country_name AS country,
  midyearpop.year AS year,
  AVG(midyearpop.midyear_population) AS population,
  AVG(mortality.life_expectancy) AS life_expectancy
FROM
  `bigquery-public-data.census_bureau_international.midyear_population` AS
  midyearpop
JOIN
  `bigquery-public-data.census_bureau_international.mortality_life_
  expectancy` AS mortality
ON
  midyearpop.country_name = mortality.country_name
```

```
WHERE
  midyearpop.year = 2018
GROUP BY
  country, year
ORDER BY
  population DESC
LIMIT
  20'
```

Waiting on bqjob_r4ecdb3f115b3f5d3_0000016628b526ea_1 ... (0s) Current
status: DONE

```
+-----------------+------+--------------+--------------------+
|     country     | year |  population  |   life_expectancy  |
+-----------------+------+--------------+--------------------+
| China           | 2018 | 1.384688986E9 |  75.58754098360653 |
| India           | 2018 | 1.296834042E9 |  69.15033333333334 |
| United States   | 2018 |   3.29256465E8 |  82.25324324324323 |
| Indonesia       | 2018 |   2.62787403E8 |  70.89647887323946 |
| Brazil          | 2018 |   2.08846892E8 |  71.26444444444446 |
| Pakistan        | 2018 |   2.07862518E8 |  66.57942857142856 |
| Nigeria         | 2018 |   2.03452505E8 | 53.483061224489774 |
| Bangladesh      | 2018 |   1.59453001E8 |  69.93685714285715 |
| Russia          | 2018 |   1.42122776E8 |  71.61112903225805 |
| Japan           | 2018 |   1.26168156E8 |   85.6562295081967 |
| Mexico          | 2018 |   1.25959205E8 |              75.22 |
| Ethiopia        | 2018 |   1.08386391E8 | 59.355633802816925 |
| Philippines     | 2018 |   1.05893381E8 |  69.13042253521127 |
| Egypt           | 2018 |    9.9413317E7 |   73.8963636363636 |
| Vietnam         | 2018 |    9.7040334E7 |   74.0014516129032 |
| Congo (Kinshasa)| 2018 |    8.5281024E7 | 56.483376623376614 |
| Iran            | 2018 |    8.3024745E7 |  72.58799999999997 |
| Turkey          | 2018 |    8.1257239E7 |  73.33577464788735 |
| Germany         | 2018 |    8.0457737E7 |  80.61900000000001 |
| Thailand        | 2018 |    6.8615858E7 |  75.35032786885246 |
+-----------------+------+--------------+--------------------+
```

The JOIN command is used to bring together or concatenate data from two or more tables by matching their respective rows. The command uses the ON clause to determine what column will be used for the matching.

Subselect

The following query selects the average population for each country and their life expectancy for the year 2018. The data is joined from the 'midyear_population' table and the 'mortality_life_expectancy' table in the 'census_bureau_international' dataset. The query uses a subselect statement in the first FROM clause to filter by year and specific countries. The resulting table is grouped by country name and year and arranged in descending order. The general idea of a subselect statement is to be able to create more complex queries without using intermediate tables.

```
bq query --use_legacy_sql=false 'SELECT
  midyearpop.country_name AS country,
  midyearpop.year AS year,
  AVG(midyearpop.midyear_population) AS population,
  AVG(mortality.life_expectancy) AS life_expectancy
FROM (
  SELECT
    country_name,
    year,
    midyear_population
  FROM
    `bigquery-public-data.census_bureau_international.midyear_population`
  WHERE
    year = 2018
    AND (country_name LIKE "Nigeria"
    OR country_name LIKE "Egypt")) AS midyearpop
JOIN
  `bigquery-public-data.census_bureau_international.mortality_life_
  expectancy` AS mortality
```

```
ON
  midyearpop.country_name = mortality.country_name
GROUP BY
  country,
  year
ORDER BY
  population DESC
LIMIT
  20'

Waiting on bqjob_r5d381c26fcb6480e_0000016628e220c3_1 ... (0s) Current
status: DONE
+---------+------+-------------+--------------------+
| country | year |  population |   life_expectancy  |
+---------+------+-------------+--------------------+
| Nigeria | 2018 | 2.03452505E8 | 53.483061224489774 |
| Egypt   | 2018 | 9.9413317E7  |  73.8963636363636  |
+---------+------+-------------+--------------------+
```

The Case Against Running Select *

In BigQuery, it is ill-advised to run the SELECT * command, which is used in SQL to retrieve all the columns from the table. This command is rather expensive in BigQuery especially if your table contains terabytes of data. If instead you want to have a feel for the columns and their entries in your dataset, you can execute the command 'bq head [table_name]' to retrieve the first few rows of the table. As an example, we used the command in the following example listing to retrieve the first few rows of the 'market' table we earlier loaded from GCS in the 'crypto_data' dataset.

```
bq head crypto_data.markets
```

slug	symbol	name	date	ranknow	open	high	low	close	volume
market		close_ratio	spread						
0x	ZRX	0x	2017-08-16	41	0.111725	0.280031	0.103962	0.224399	5232600
67034800		0.684	0.18						
0x	ZRX	0x	2017-08-17	41	0.223022	0.238935	0.206735	0.206735	2752410
133813000		0	0.03						
0x	ZRX	0x	2017-08-18	41	0.205558	0.35026	0.205558	0.293387	12793800
123335000		0.607	0.14						
......									
......									
0x	ZRX	0x	2017-08-28	41	0.352459	0.354823	0.32362	0.343713	6639910
176230000		0.6439	0.03						

Using BigQuery with Notebooks on AI Cloud Instance and Google Colab

BigQuery integrates well with Notebooks on Google Notebook AI Instance and Google Colab. In this section, we'll go through executing on BigQuery datasets and tables from Notebooks. There are a couple of ways to interact with BigQuery from Notebooks, but one quick and easy method is the use of the **'%bigquery'** magic command from the BigQuery client library, **'google-cloud-bigquery'**, to run queries with minimal syntax.

The **%%bigquery** magic runs a SQL query and returns the results as a pandas DataFrame. Here, we use the '%%bigquery' magic command to interact with BigQuery. To begin, open a Notebook on GCP AI Notebook Instance or from Colab:

1. If running on Google Colab, authenticate the notebook by running the code

    from google.colab import auth
    auth.authenticate_user()
    print('Authenticated')

2. Import Pandas and Matplotlib.

    ```
    import pandas as pd
    import matplotlib.pyplot as plt
    ```

3. Store the following query output as a Pandas DataFrame named **'litcoin_crypto'**. Place your project id after the **'--project'** attribute. Be sure to update the FROM field with your dataset and table IDs.

    ```
    %%bigquery --project ekabasandbox litcoin_crypto
    SELECT
      symbol,
      date,
      close,
      open,
      high,
      low,
      spread
    ```

```
FROM
  `crypto_data.markets`
WHERE
  symbol = 'LTC'
LIMIT 10
```

	symbol	date	close	open	high	low	spread
0	LTC	2013-04-28	4.35	4.3	4.4	4.18	0.22
1	LTC	2013-05-07	3.33	3.37	3.41	2.94	0.47
2	LTC	2013-05-03	3.04	3.39	3.45	2.4	1.05
3	LTC	2013-05-04	3.48	3.03	3.64	2.9	0.74
4	LTC	2013-05-05	3.59	3.49	3.69	3.35	0.34
5	LTC	2013-05-06	3.37	3.59	3.78	3.12	0.66
6	LTC	2013-05-02	3.37	3.78	4.04	3.01	1.03
7	LTC	2013-05-01	3.8	4.29	4.36	3.52	0.84
8	LTC	2013-04-29	4.38	4.37	4.57	4.23	0.34
9	LTC	2013-04-30	4.3	4.4	4.57	4.17	0.4

4. The variable 'litcoin_crypto' is a Pandas DataFrame. Now, let's
    modify the data attributes and plot a bar chart.

```
# convert columns to numeric
litcoin_crypto = litcoin_crypto.apply(pd.to_numeric,
errors='ignore')
```

```
# check the datatypes
litcoin_crypto.dtypes
```

```
symbol      object
date        object
close       float64
open        float64
high        float64
low         float64
spread      float64
dtype: object
```

5. Plot the bar chart with the variable 'date' on the x axis and closing
    price on the y axis (see Figure 38-12).

```
# plot the bar chart
litcoin_crypto.plot(kind='bar', x='date', y='close')
plt.show()
```

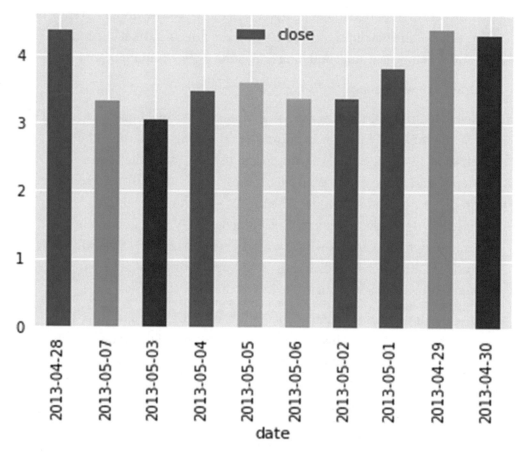

Figure 38-12. *Litcoin crypto-currency bar chart plot*

BigQueryML

BigQuery machine learning makes it quick and easy to harness the power of machine
learning on your datasets in BigQuery by using simple standard SQL commands. This
functionality includes the capability to train and test models on the datasets by using
subsets of the data, as well as the capability for automatic hyper-parameter tuning of the
learning models.

At this time of writing, the following learning models are available in BigQuery:

- Linear regression

- Binary and multi-class logistic regression

In this section, we'll work with BigQuery ML using the Notebook instance on Colab on Google AI VMs to build a predictive model using the 'market' table in the 'crypto_data' dataset that we earlier imported into BigQuery. This model will attempt to predict the next day's closing price of the Bitcoin crypto-currency given a set of market attributes. The data processing and machine learning modeling is all done using standard SQL:

1. Open a new notebook.

2. Select features for training the ML model. In the SQL code, we use the 'LEAD()' function to return the value of the next row. The offset of 1 indicates that we want to get the next value that is one step ahead in the query. With this, it is easy to adjust the query to predict a 2- to n-day window. The LEAD() function is a window function that moves over a rowset. Hence, the OVER() function is used to define a window within a query, while the PARTITION BY and ORDER BY clauses divide the query results into partitions and define the arrangement of the rows within each partition.

   We use the 'params' variable to sample half of the data and store it in the 'TRAIN' set. This makes sure that the rest of the dataset is not used in model training and can be used to check that the model generalizes well during the model evaluation phase.

   Be sure to update the FROM field with your dataset and table IDs.

   ```
   %%bigquery --project ekabasandbox btc_market
   WITH
       params AS (
       SELECT
         1 AS TRAIN,
         2 AS EVAL ),

       btc_market AS (
       SELECT
         symbol,
   ```

```
    date,
    open,
    high,
    low,
    close,
    spread,
    cast(LEAD(close, 1) OVER (PARTITION BY symbol ORDER BY symbol
    DESC) AS NUMERIC) AS next_day_close
  FROM
    `crypto_data.markets`,
    params
  WHERE
    symbol = 'BTC'
    AND MOD(ABS(FARM_FINGERPRINT(CAST(date AS STRING))),4) =
    params.TRAIN )

SELECT
  *
FROM
  btc_market
WHERE
  next_day_close IS NOT NULL
```

3. Display the first ten rows of the query.

```
btc_market.head(10)
```

	symbol	date	open	high	low	close	spread	next_day_close
0	BTC	2013-05-05	112.9	118.8	107.14	115.91	11.66	112.3
1	BTC	2013-05-06	115.98	124.66	106.64	112.3	18.02	112.67
2	BTC	2013-05-09	113.2	113.46	109.26	112.67	4.2	115.24
3	BTC	2013-05-11	117.7	118.68	113.01	115.24	5.67	111.5
4	BTC	2013-05-14	117.98	119.8	110.25	111.5	9.55	114.22
5	BTC	2013-05-15	111.4	115.81	103.5	114.22	12.31	121.99
6	BTC	2013-05-19	123.21	124.5	119.57	121.99	4.93	123.89
7	BTC	2013-05-22	122.89	124	122	123.89	2	133.2
8	BTC	2013-05-24	126.3	133.85	125.72	133.2	8.13	131.98
9	BTC	2013-05-25	133.1	133.22	128.9	131.98	4.32	133.48

4. The trained model is stored in a BigQuery dataset. In this case,
    we'll create a BigQuery dataset to store the model.

```
from google.cloud import bigquery
client = bigquery.Client(project='ekabasandbox')
# create a BigQuery dataset to store your ML model
dataset = client.create_dataset('btc_crypto')
print('Dataset: `{}` created.'.format(dataset.dataset_id))
```

```
Dataset: `btc_crypto` created.
```

5. After preparing our training dataset, now it is time to train the
    model. Be sure to update the FROM field with your dataset and
    table IDs.

```
%%bigquery --project ekabasandbox model
CREATE OR REPLACE MODEL `btc_crypto.market_closing_model`
OPTIONS
  (model_type='linear_reg',
    labels=['next_day_close']) AS
WITH
  params AS (
  SELECT
    1 AS TRAIN,
    2 AS EVAL ),
  btc_market AS (
  SELECT
    CAST(open AS NUMERIC) AS open,
    CAST(high AS NUMERIC) AS high,
    CAST(low AS NUMERIC) AS low,
    CAST(close AS NUMERIC) AS close,
    CAST(spread AS NUMERIC) AS spread,
    CAST(LEAD(close, 1) OVER (PARTITION BY symbol ORDER BY symbol
    DESC) AS NUMERIC) AS next_day_close
  FROM
    `crypto_data.markets`,
    params
```

```
WHERE
  symbol = 'BTC'
  AND MOD(ABS(FARM_FINGERPRINT(CAST(date AS STRING))),4) =
  params.TRAIN )
SELECT
  *
FROM
  btc_market
WHERE
  next_day_close IS NOT NULL
```

6. Check that the created model exists in the Dataset 'btc_crypto'. We prefix the exclamation sign ('!') in a Notebook cell to execute bash commands.

```
!bq ls btc_crypto

        tableId            Type    Labels   Time Partitioning
--------------------- ------- -------- -------------------
  market_closing_model   MODEL
```

7. Evaluate the model to estimate the performance of the model. The RMSE metric is evaluated in BigQuery calling the 'mean_squared_error' field of the trained model and passing it through the 'SQRT()' function. To evaluate the model, pass the model through the function 'ML.EVALUATE()'. This time we select the remaining subset of the dataset and store it in 'params.EVAL'.

8. Be sure to update the FROM field with your dataset and table IDs.

```
%%bigquery --project ekabasandbox rmse
SELECT
  SQRT(mean_squared_error) AS rmse
FROM
  ML.EVALUATE(MODEL `btc_crypto.market_closing_model`,
    (
    WITH
      params AS (
```

```
        SELECT
          1 AS TRAIN,
          2 AS EVAL ),
        btc_market AS (
        SELECT
          CAST(open AS NUMERIC) AS open,
          CAST(high AS NUMERIC) AS high,
          CAST(low AS NUMERIC) AS low,
          CAST(close AS NUMERIC) AS close,
          CAST(spread AS NUMERIC) AS spread,
          CAST(LEAD(close, 1) OVER (PARTITION BY symbol ORDER BY
          symbol DESC) AS NUMERIC) AS next_day_close
        FROM
          `crypto_data.markets`,
          params
        WHERE
          symbol = 'BTC'
          AND MOD(ABS(FARM_FINGERPRINT(CAST(date AS STRING))),4) =
          params.EVAL )
      SELECT
        *
      FROM
        btc_market
      WHERE
        next_day_close IS NOT NULL ))

        rmse
0     393.265715
```

9. Predict the next day's closing prices for the Bitcoin crypto-
   currency using the trained model. Be sure to update the FROM
   field with your dataset and table IDs.

```
%%bigquery --project ekabasandbox predict
SELECT
    *
FROM
```

```
ml.PREDICT(MODEL `btc_crypto.market_closing_model`,
  (
  WITH
    params AS (
    SELECT
      1 AS TRAIN,
      2 AS EVAL ),
    btc_market AS (
    SELECT
      CAST(close AS NUMERIC) AS close,
      date,
      CAST(open AS NUMERIC) AS open,
      CAST(high AS NUMERIC) AS high,
      CAST(low AS NUMERIC) AS low,
      CAST(spread AS NUMERIC) AS spread,
      CAST(LEAD(close, 1) OVER (PARTITION BY symbol ORDER BY
      symbol DESC) AS NUMERIC) AS next_day_close
    FROM
      `crypto_data.markets`,
      params
    WHERE
      symbol = 'BTC'
      AND MOD(ABS(FARM_FINGERPRINT(CAST(date AS STRING))),4) =
      params.EVAL )
  SELECT
    *
  FROM
    btc_market
  WHERE
    next_day_close IS NOT NULL ))
```

predict	predicted_next_day_close	close	date	open	high	low	spread	next_day_close
0	193.523361	116.99	2013-05-01	139	139.89	107.72	32.17	112.5
1	162.505189	112.5	2013-05-04	98.1	115	92.5	22.5	111.5
2	158.389055	111.5	2013-05-07	112.25	113.44	97.7	15.74	117.2
3	158.700481	117.2	2013-05-10	112.8	122	111.55	10.45	115
...	...	...	...	...	...	...	...	...
388	4491.052680	4703.39	2017-08-31	4555.59	4736.05	4549.4	186.65	4597.12
389	4422.931411	4597.12	2017-09-06	4376.59	4617.25	4376.59	240.66	4122.94
390	4163.348876	4122.94	2017-09-10	4229.34	4245.44	3951.04	294.4	4161.27
391	4029.355833	4161.27	2017-09-11	4122.47	4261.67	4099.4	162.27	4130.81
...	...	...	...	...	...	...	...	...
416	14723.798445	15201	2018-01-03	14978.2	15572.8	14844.5	728.3	15599.2
417	15421.170791	15599.2	2018-01-04	15270.7	15739.7	14522.2	1217.5	14595.4

This chapter provided an overview of working with Google BigQuery as a data warehouse and analytics platform on GCP. It covered working with BigQuery from Notebooks hosted on Google Colab or on GCP AI Instances and included how to work with BigQuery ML to build machine learning predictive models using SQL commands.

The next chapter will introduce Cloud Dataprep for visually exploring and transforming large datasets on GCP.

## CHAPTER 39

Google Cloud Dataprep

Google Cloud Dataprep is a managed cloud service for quick data exploration and transformation. Dataprep makes it easy to clean and transform large datasets for analysis. It is auto-scalable as it takes advantage of the distributed processing capabilities of Google Cloud Dataflow.

Typically Cloud Dataprep is aimed at easing the data preparation process. Datasets from real-world use cases are often messy and untidy. In this form, it cannot be used for downstream analytics or machine learning modeling. Hence, a large portion of the modeling process involves preparing and cleaning the data. Programming libraries earlier discussed like Pandas are centrally used for carrying out data preparation. However, Google Cloud Dataprep provides a simple visual interface for performing data cleaning. The ability to re-organize the dataset for modeling quickly without coding provides an instant appeal for Dataprep, as this can greatly speed up the time spent in data preparation as part of the overall modeling pipeline. The other good part is that Dataprep can work with petabyte scale data as it is built on a serverless infrastructure. Dataprep can be used for processing structured and unstructured datasets.

In this section, we'll go through a brief tour of Google Dataprep by using it to prepare our 'crypto_markets.csv' dataset already stored on Google Cloud Storage.

Getting Started with Cloud Dataprep

From the GCP dashboard, click the triple dash at the top-left corner and scroll down to 'Dataprep' under the **BIG DATA** section as seen in Figure 39-1.

© Ekaba Bisong 2019
E. Bisong, *Building Machine Learning and Deep Learning Models on Google Cloud Platform*,
https://doi.org/10.1007/978-1-4842-4470-8_39

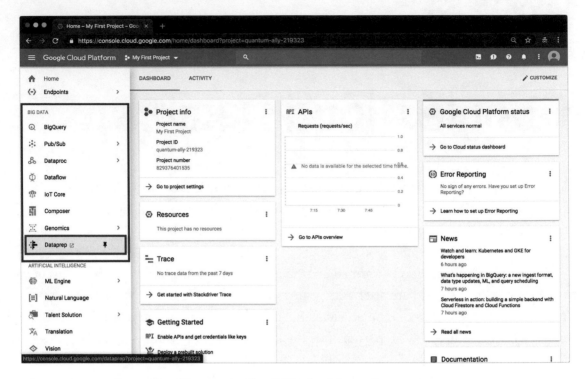

Figure 39-1. *Open Dataprep via the GCP dashboard*

Dataprep is a service offered on GCP in alliance with the company Trifacta. To begin using Dataprep, agree and accept all the license agreements (see Figure 39-2). Dataprep creates a bucket on GCS to store the files that are uploaded to Dataprep and the outputs of its transformation (see Figure 39-3). The Dataprep dashboard is shown in Figure 39-4.

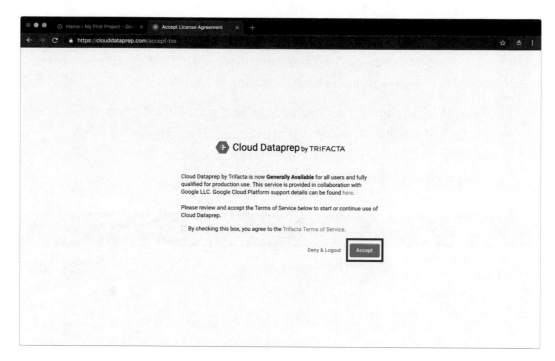

Figure 39-2. *Trifacta license agreement*

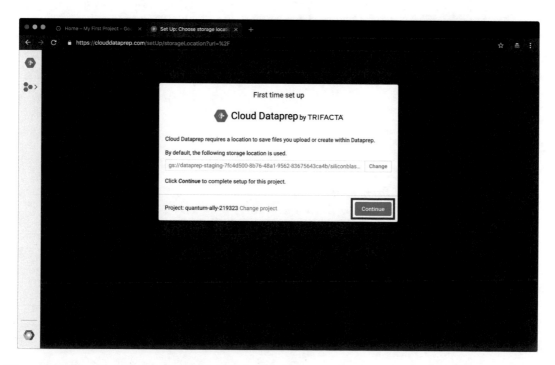

Figure 39-3. *Dataprep GCS location setup*

Using Flows to Transform Data

A Dataprep flow is an object created to organize and manage the datasets and operations that are involved in data cleaning and transformation process:

1. We begin by creating a flow by clicking the 'Create Flow' button in the top-right corner of the Dataprep dashboard (see Figure 39-4). Enter the user-defined flow name and click 'Create' as shown in Figure 39-5. The Flow page is shown in Figure 39-6.

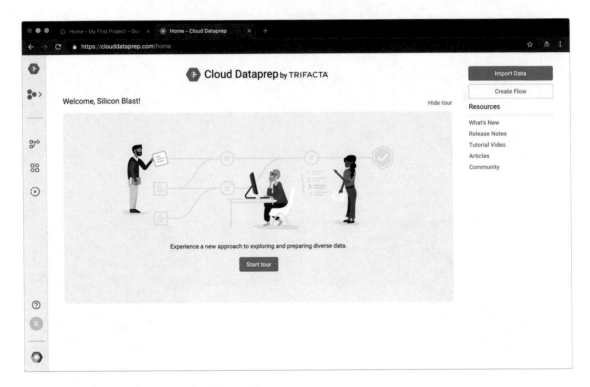

Figure 39-4. *Dataprep dashboard*

Figure 39-5. *Create Flow*

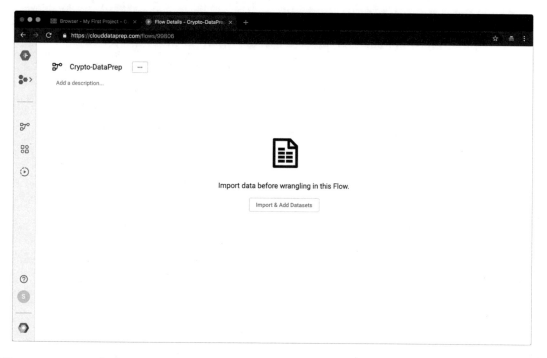

Figure 39-6. *Flow page*

2. Let's start by placing our dataset in a GCS bucket. We'll do so by running the following commands on the terminal.

    Create a new bucket.

    ```
    gsutil mb gs://my-dataprep-data
    ```

3. Transfer data from GitHub to the bucket.

    ```
    gsutil cp crypto-markets.csv gs://my-dataprep-data
    ```

4. Next, we'll transfer our 'crypto-market' dataset from the 'my-dataprep-data' bucket to the Dataprep staging bucket. We can quickly do this by executing the following code on the terminal.

    ```
    gsutil cp -r gs://my-dataprep-data gs://dataprep-staging-7fc4d500-8b76-48a1-9562-83675643ca4b
    ```

    ```
    Copying gs://my-dataprep-data/crypto-markets.csv [Content-
    Type=application/octet-stream]...
    / [1 files][ 47.0 MiB/ 47.0 MiB]
    Operation completed over 1 objects/47.0 MiB.
    ```

5. Next, we'll import and add Datasets to the Flow. Datasets can be uploaded directly to Dataprep which will then be stored to the bucket Dataprep generated on start-up. Also, Dataprep can import datasets already stored in BigQuery or GCS. In this case, we will import the 'crypto-market' dataset that we earlier transferred to the Dataprep staging bucket which is in the folder 'my-dataprep-data' (see Figure 39-7). Figure 39-8 shows the dataset loading into Dataprep.

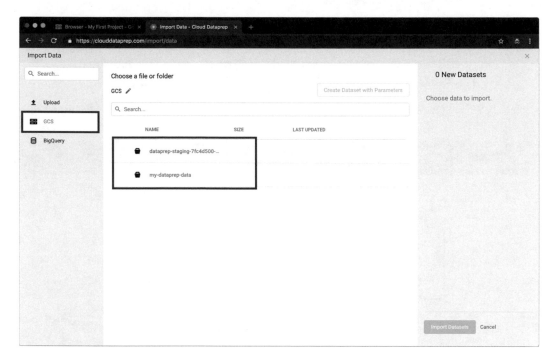

Figure 39-7. *Import Dataset from GCS to Dataprep*

Figure 39-8. *Loading Dataset to Dataprep*

6. Next, we'll create a recipe. A Dataprep recipe contains the transformation steps taken to clean and process a Dataset. This recipe is later executed as a Dataflow job to operate on the Dataset and come up with results. Click the 'Add New Recipe' button to create a recipe. The recipe is in the bounded red box in Figure 39-9.

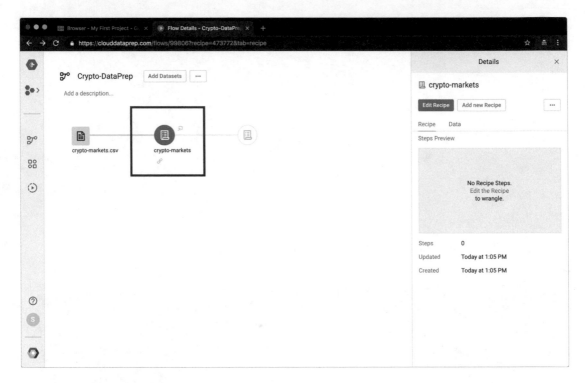

Figure 39-9. *Dataset recipe*

7. Then click the 'Edit Recipe' button to open the 'Transformation Grid' where we carry out various cleaning and processing steps on the Dataset.

8. For the example in this section, we'll carry out a simple transformation process by dropping some unused columns and then removing all rows in the dataset except those for Bitcoin crypto-currency:

a. Remove the 'slug' column. Click 'Add' within the red box to drop the column (see Figure 39-10).

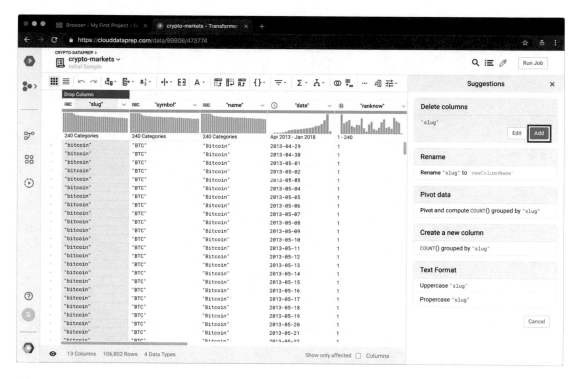

Figure 39-10. *Remove 'slug' column*

b. Remove the 'name' column (see Figure 39-11).

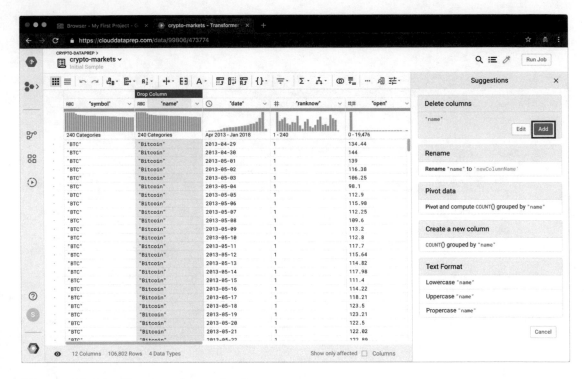

Figure 39-11. *Remove 'name' column*

    c. Next, we'll filter the rows in the dataset to retain only the Bitcoin records (see Figures 39-12 and 39-13).

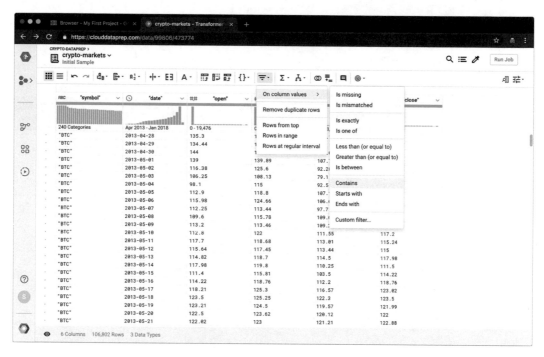

Figure 39-12. *Filter rows using Dataprep*

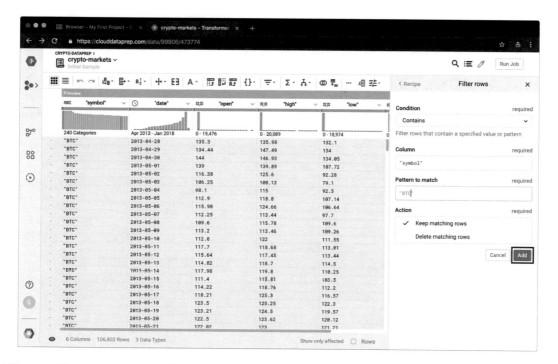

Figure 39-13. *Remove all rows except the Bitcoin records*

9.   Figure 39-14 shows the dataset transformation recipes. Click 'Run
     Job' in Figure 39-14 and also in Figure 39-15 to run the job on
     Cloud Dataflow.

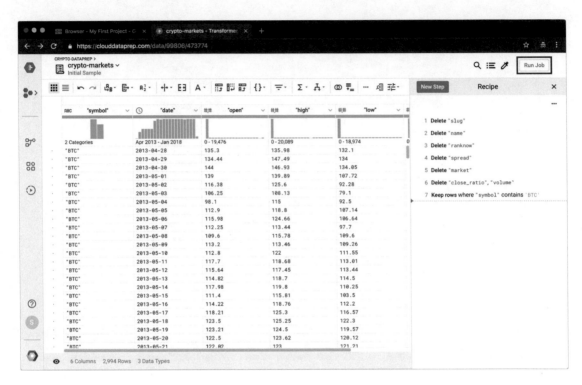

Figure 39-14. *View transformation recipes*

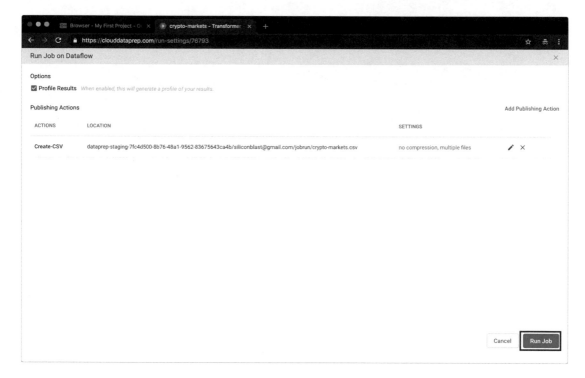

Figure 39-15. *Run Job on Dataflow*

10.   Figure 39-16 shows the running job, and Figure 39-17 shows the
completed job after some minutes.

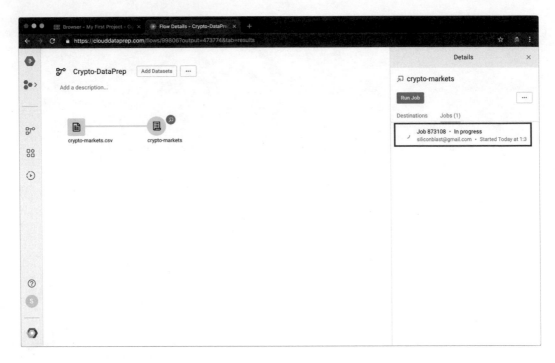

Figure 39-16. *Job running on Dataflow*

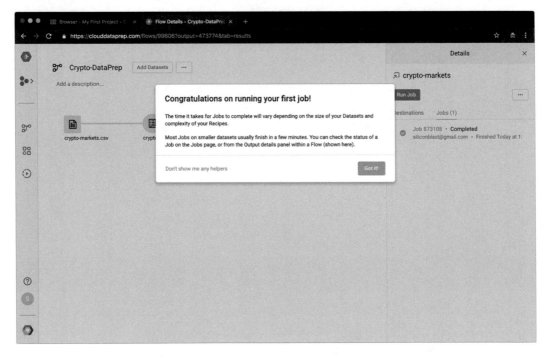

Figure 39-17. *Completed job*

11. View the results of the job (see Figure 39-18).

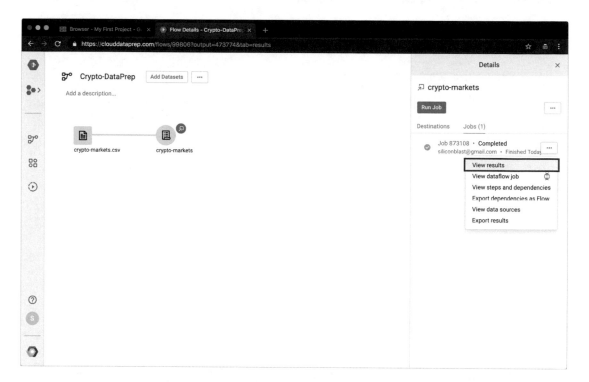

Figure 39-18. *View job result*

12. From the Results page shown in Figure 39-19, we can export the
results back to GCS (see Figure 39-20).

Figure 39-19. *Job Results page*

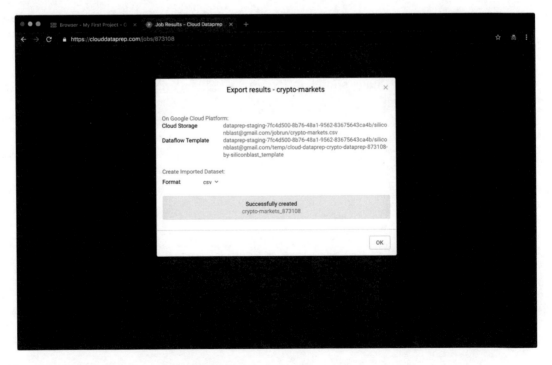

Figure 39-20. *Export completed jobs*

This chapter provides an example overview of working with Dataprep to visually explore and transform large datasets on GCP by using the Google Cloud Dataflow infrastructure for distributed processing. In the next chapter, we will introduce working with Cloud Dataflow for building custom data transformation pipelines.

## CHAPTER 40

Google Cloud Dataflow

Google Cloud Dataflow provides a serverless, parallel, and distributed infrastructure for running jobs for batch and stream data processing. One of the core strengths of Dataflow is its ability to almost seamlessly handle the switch from processing of batch historical data to streaming datasets while elegantly taking into consideration the perks of streaming processing such as windowing. Dataflow is a major component of the data/ML pipeline on GCP. Typically, Dataflow is used to transform humongous datasets from a variety of sources such as Cloud Pub/Sub or Apache Kafka to a sink such as BigQuery or Google Cloud Storage.

Critical to Dataflow is the use of the Apache Beam programming model for building the parallel data processing pipelines for batch and stream operations. The data processing pipelines built with the Beam SDKs can be executed on various processing backends such as Apache Apex, Apache Spark, Apache Flink, and of course Google Cloud Dataflow. In this section, we will build data transformation pipelines using the Beam Python SDK. As of this time of writing, Beam also supports building data pipelines using Java, Go, and Scala languages.

Beam Programming

Apache Beam provides a set of broad concepts to simplify the process of building a transformation pipeline for distributed batch and stream jobs. We'll go through these concepts providing simple code samples:

- A Pipeline: A Pipeline object wraps the entire operation and prescribes the transformation process by defining the input data source to the pipeline, how that data will be transformed, and where the data will be written. Also, the Pipeline object indicates the distributed processing backend to execute on. Indeed, a Pipeline

© Ekaba Bisong 2019
E. Bisong, *Building Machine Learning and Deep Learning Models on Google Cloud Platform*,
https://doi.org/10.1007/978-1-4842-4470-8_40

is the central component of a Beam execution. Code for creating a pipeline is as shown in the following:

```
import apache_beam as beam
from apache_beam.options.pipeline_options import PipelineOptions

p = beam.Pipeline(options=PipelineOptions())
```

In the preceding code snippet, the Pipeline object is configured using 'PipelineOptions' to set the required fields. This can be done both programmatically and from the command line.

- A PCollection: A PCollection is used to define a data source. The data source can either be *bounded* or *unbounded.* A bounded data source refers to batch or historical data, whereas an unbounded data source refers to streaming data. Beam uses a technique called *windowing* to partition unbounded PCollections into finite logical segments using some attribute of the data such as a timestamp. PCollections can also be created from in-memory data where PCollections are both the inputs and outputs for a particular step in the pipeline. Let's see an example of reading a csv data from an external source:

```
lines = p | 'ReadMyFile' >> beam.io.ReadFromText('gs://gcs_bucket/
my_data.csv')
```

The pipe operator '|' in the preceding code is also called the apply method and is used to apply the PCollection to the pipeline instantiated as 'p'.

- A PTransform: A PTransform refers to a particular transformation task carried out on one or more PCollections in the pipeline. PTransforms can be applied to PCollections as follows.

```
[Output PCollection] = [Input PCollection] | [Transform]
```

Note that while a PTransform creates a new PCollection, it does not modify or alter the input collection. A number of core Beam transforms include

- ParDo: For parallel processing

- GroupByKey: For processing collections of key/value pairs

- CoGroupByKey: For a relational join of two or more key/value PCollections with the same key type

- Combine: For combining collections of elements or values in your data

- Flatten: For merging multiple PCollection objects

- Partition: Splits a single PCollection into smaller collections

- I/O transforms: These are PTransforms that read or write data to different external storage systems. Some of the currently available I/O transforms working with Beam Python SDK include

  - avroio: For reading from and writing to an Avro file

  - textio: For reading from and writing to text files

For a simple linear pipeline with sequential transformation, the processing graph looks like what is shown in Figure 40-1.

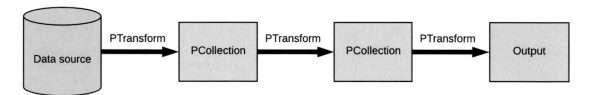

Figure 40-1. *A simple linear Pipeline with sequential transforms*

Building a Simple Data Processing Pipeline

In this simple Beam application, we will build a Dataflow pipeline to preprocess a CSV file from a GCS bucket and write the output back to GCS. This example selects certain features and rows that are of interest to the downstream modeling task. Here, we considered the 'crypto-markets.csv' dataset. In the data preprocessing pipeline, we removed data attributes that may not be relevant for analytics/model building and we also filtered records pertaining to 'bitcoin'. The steps that follow create a simple Beam pipeline and execute in on Google Dataflow:

1. Enable the GCP Cloud Dataflow API and Cloud Resource Manager API from the APIs & Services dashboard.

2. Open a new Notebook.

3. Note that at this time of writing, Apache Beam only works with Python version 2.7, so be sure to switch the kernel for your Python interpreter. Add the following code blocks in the Notebook cell.

4. If running on Google Colab, first authenticate the notebook with GCP.

```
from google.colab import auth
auth.authenticate_user()
print('Authenticated')

# configure GCP project. Change to your project ID
project_id = 'ekabasandbox'
!gcloud config set project {project_id}
```

5. Install the Apache beam library and other important setup packages.

```
%%bash
pip install apache-beam[gcp]
```

6. After installing, change the notebook runtime type to Python 2.

7. Next, reset the notebook kernel before running the code to import the relevant libraries.

```
import apache_beam as beam
from apache_beam.io import ReadFromText
from apache_beam.io import WriteToText
```

8. Assign the parameters for the pipeline. Replace the relevant parameters with your entries.

```
# parameters
staging_location = 'gs://enter_bucket_name/staging' # change this
temp_location = 'gs://enter_bucket_name/temp' # change this
```

```python
job_name = 'dataflow-crypto'
project_id = enter_project_id' # change this
source_bucket = 'enter_bucket_name' # change this
target_bucket = 'enter_bucket_name' # change this
```

9. Method to build and run the pipeline.

```python
def run(project, source_bucket, target_bucket):
    import csv

    options = {
        'staging_location': staging_location,
        'temp_location': temp_location,
        'job_name': job_name,
        'project': project,
        'max_num_workers': 24,
        'teardown_policy': 'TEARDOWN_ALWAYS',
        'no_save_main_session': True,
        'runner': 'DataflowRunner'
    }
    options = beam.pipeline.PipelineOptions(flags=[], **options)

    crypto_dataset = 'gs://{}/crypto-markets.csv'.format(source_
    bucket)
    processed_ds = 'gs://{}/transformed-crypto-bitcoin'.
    format(target_bucket)

    pipeline = beam.Pipeline(options=options)

    # 0:slug, 3:date, 5:open, 6:high, 7:low, 8:close
    rows = (
        pipeline |
            'Read from bucket' >> ReadFromText(crypto_dataset) |
            'Tokenize as csv columns' >> beam.Map(lambda line:
            next(csv.reader([line]))) |
            'Select columns' >> beam.Map(lambda fields:
            (fields[0], fields[3], fields[5], fields[6],
            fields[7], fields[8])) |
```

```
                    'Filter bitcoin rows' >> beam.Filter(lambda row: row[0] ==
                    'bitcoin')
                )

          combined = (
              rows |
                    'Write to bucket' >> beam.Map(lambda (slug, date,
                    open, high, low, close): '{},{},{},{},{},{}'.format(
                        slug, date, open, high, low, close)) |
                    WriteToText(
                        file_path_prefix=processed_ds,
                        file_name_suffix=".csv", num_shards=2,
                        shard_name_template="-SS-of-NN",
                        header='slug, date, open, high, low, close')
                )

          pipeline.run()
```

10. Run the pipeline.

```
    if __name__ == '__main__':
        print 'Run pipeline on the cloud'
        run(project=project_id, source_bucket=source_bucket,
        target_bucket=target_bucket)
```

The image in Figure 40-2 shows the Dataflow pipeline created as a result of this job.

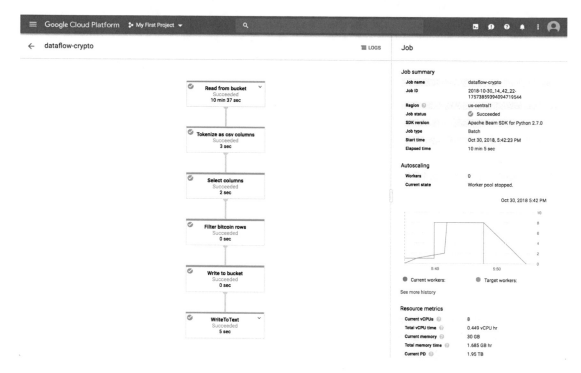

Figure 40-2. *Preprocessing Pipeline on Google Cloud Dataflow*

More complex and advanced uses of Google Cloud Dataflow are beyond the scope of this book as they are more in the area of building big data pipelines for large-scale data transformation. However, this section is included because big data transformation is an important component for the design and productionalization of machine learning models when solving a particular business use case at scale. It is important for readers to get a feel of working with these sort of technologies.

This chapter provides an introduction to building large-scale big data transformation pipelines using Python Apache Beam programming model that runs on Google Dataflow computing infrastructure. The next chapter will cover using Google Cloud Machine Learning Engine to train and deploy large-scale models.

CHAPTER 41

Google Cloud Machine Learning Engine (Cloud MLE)

The Google Cloud Machine Learning Engine, simply known as Cloud MLE, is a managed Google infrastructure for training and serving "large-scale" machine learning models. Cloud ML Engine is a part of GCP AI Platform. This managed infrastructure can train large-scale machine learning models built with TensorFlow, Keras, Scikit-learn, or XGBoost. It also provides modes of serving or consuming the trained models either as an online or batch prediction service. Using online prediction, the infrastructure scales in response to request throughout, while with the batch mode, Cloud MLE can provide inference for TBs of data.

Two important features of Cloud MLE is the ability to perform distribution training and automatic hyper-parameter tuning of your models while training. The big advantage of automatic hyper-parameter tuning is the ability to find the best set of parameters that minimize the model cost or loss function. This saves time of development hours in iterative experiments.

The Cloud MLE Train/Deploy Process

The high-level overview of the train/deploy process on Cloud MLE is depicted in Figure 41-1:

1.   The data for training/inference is kept on GCS.

2.   The execution script uses the application logic to train the model on Cloud MLE using the training data.

© Ekaba Bisong 2019

E. Bisong, *Building Machine Learning and Deep Learning Models on Google Cloud Platform*, https://doi.org/10.1007/978-1-4842-4470-8_41

3. The trained model is stored on GCS.

4. A prediction service is created on Cloud MLE using the trained model.

5. The external application sends data to the deployed model for inference.

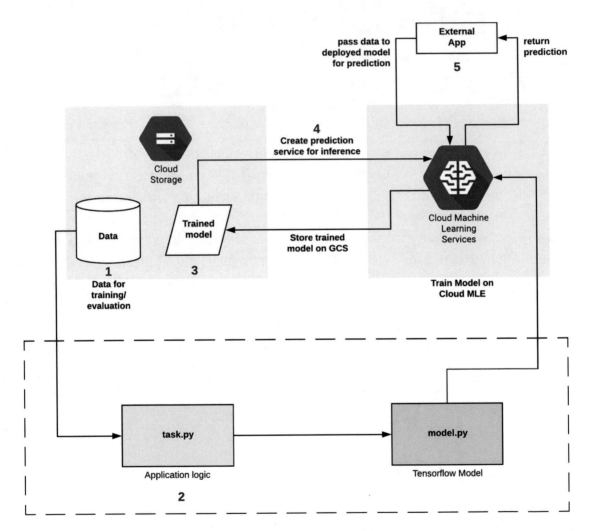

Figure 41-1. *The train/deploy process on Cloud MLE*

Preparing for Training and Serving on Cloud MLE

In this contrived example, we'll use the famous Iris dataset to train and serve a TensorFlow model using the Estimator API on Cloud MLE. To begin, let's walk through the following steps:

1. Create a bucket on GCS by running the gsutil mb command on the cloud terminal. Replace it with unique bucket name.

   export bucket_name=iris-dataset'
   gsutil mb gs://$bucket_name

2. Transfer training and test data from the code repository to the GCP bucket.

3. Move the train data.

   gsutil cp *train_data.csv gs://$bucket_name*

4. Move the train data.

   gsutil cp *test_data.csv gs://$bucket_name*

5. Move the hold-out data for batch predictions.

   gsutil cp *hold_out_test.csv gs://$bucket_name*

6. Enable the Cloud Machine Learning API to be able to create and use machine learning models on GCP Cloud MLE:

   a. Go to APIs & Services.

   b. Click "Enable APIs & Services".

   c. Search for "Cloud Machine Learning Engine".

   d. Click ENABLE API as shown in Figure 41-2.

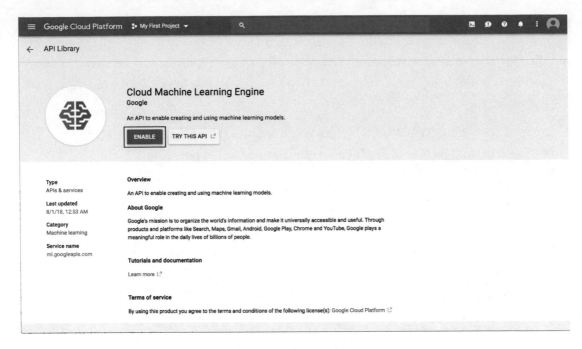

Figure 41-2. *Enable Cloud Machine Learning APIs*

Packaging the Code for Training on Cloud MLE

The code for training on Cloud MLE must be prepared as a python package. The
recommended project structure is explained as follows:

IrisCloudML: [project name as parent folder]

- Trainer: [folder containing the model and execution code]

    - __init__.py: [an empty special python file indicating that the
      containing folder is a Python package]

    - model.py: [script contains the logic of the model written in
      TensorFlow, Keras, etc.]

    - task.py: [script contains the application that orchestrates or
      manages the training job]

- scripts: [folder containing scripts to execute jobs on Cloud MLE]

    - distributed-training.sh: [script to run a distributed training job on
      Cloud MLE]

- hyper-tune.sh: [script to run a training job with hyper-parameter tuning on Cloud MLE]

- single-instance-training.sh: [script to run a single instance training job on Cloud MLE]

- online-prediction.sh: [script to execute an online prediction job on Cloud MLE]

- create-prediction-service.sh: [script to create a prediction service on Cloud MLE]

- hptuning_config: [configuration file for hyper-parameter tuning on Cloud MLE]

- gpu_hptuning_config.yaml: [configuration file for hyper-parameter tuning with GPU training on Cloud MLE]

NOTE: FOLLOW THESE INSTRUCTIONS TO RUN THE EXAMPLES FOR TRAINING ON CLOUD MACHINE LEARNING ENGINE

---

1. Launch a Notebook Instance on GCP AI Platform.

2. Pull the code repository.

3. Navigate to the book folder. Run the scripts in the sub-folder `tensorflow`.

4. Should you choose to work with Google Colab, authenticate the user by running the code

```
from google.colab import auth
   auth.authenticate_user()
```

The TensorFlow Model

Now let's briefly examine the TF model code in the file '**model.py**'.

```
import six
```

```python
import tensorflow as tf
from tensorflow.python.estimator.model_fn import ModeKeys as Modes

# Define the format of your input data including unused columns.
CSV_COLUMNS = [
    'sepal_length', 'sepal_width', 'petal_length',
    'petal_width', 'class'
]
CSV_COLUMN_DEFAULTS = [[0.0], [0.0], [0.0], [0.0], ["‟"]]
LABEL_COLUMN = 'class'
LABELS = ['setosa', 'versicolor', 'virginica']

# Define the initial ingestion of each feature used by your model.
# Additionally, provide metadata about the feature.
INPUT_COLUMNS = [
    # Continuous base columns.
    tf.feature_column.numeric_column('sepal_length'),
    tf.feature_column.numeric_column('sepal_width'),
    tf.feature_column.numeric_column('petal_length'),
    tf.feature_column.numeric_column('petal_width')
]

UNUSED_COLUMNS = set(CSV_COLUMNS) - {col.name for col in INPUT_COLUMNS} - \
    {LABEL_COLUMN}

def build_estimator(config, hidden_units=None, learning_rate=None):
    """Deep NN Classification model for predicting flower class.
    Args:
        config: (tf.contrib.learn.RunConfig) defining the runtime
        environment for
          the estimator (including model_dir).
        hidden_units: [int], the layer sizes of the DNN (input layer first)
        learning_rate: (int), the learning rate for the optimizer.
    Returns:
        A DNNClassifier
    """

    (sepal_length, sepal_width, petal_length, petal_width) = INPUT_COLUMNS
```

```
    columns = [
        sepal_length,
        sepal_width,
        petal_length,
        petal_width,
    ]

    return tf.estimator.DNNClassifier(
      config=config,
      feature_columns=columns,
      hidden_units=hidden_units or [256, 128, 64],
      n_classes = 3,
      optimizer=tf.train.AdamOptimizer(learning_rate)
    )

def parse_label_column(label_string_tensor):
  """Parses a string tensor into the label tensor.
  Args:
    label_string_tensor: Tensor of dtype string. Result of parsing the CSV
      column specified by LABEL_COLUMN.
  Returns:
    A Tensor of the same shape as label_string_tensor, should return
    an int64 Tensor representing the label index for classification tasks,
    and a float32 Tensor representing the value for a regression task.
  """
  # Build a Hash Table inside the graph
  table = tf.contrib.lookup.index_table_from_tensor(tf.constant(LABELS))

  # Use the hash table to convert string labels to ints and one-hot encode
  return table.lookup(label_string_tensor)

# [START serving-function]

def csv_serving input_fn():
    """Build the serving inputs."""
    csv_row = tf.placeholder(shape=[None], dtype=tf.string)
    features = _decode_csv(csv_row)
```

```python
        # Ignore label column
        features.pop(LABEL_COLUMN)
        return tf.estimator.export.ServingInputReceiver(features,
                                              {'csv_row': csv_row})

def json_serving_input_fn():
    """Build the serving inputs."""
    inputs = {}
    for feat in INPUT_COLUMNS:
        inputs[feat.name] = tf.placeholder(shape=[None], dtype=feat.dtype)

    return tf.estimator.export.ServingInputReceiver(inputs, inputs)

# [END serving-function]

SERVING_FUNCTIONS = {
  'JSON': json_serving_input_fn,
  'CSV': csv_serving_input_fn
}

def _decode_csv(line):
    """Takes the string input tensor and returns a dict of rank-2 tensors."""

    # Takes a rank-1 tensor and converts it into rank-2 tensor
    row_columns = tf.expand_dims(line, -1)
    columns = tf.decode_csv(row_columns, record_defaults=CSV_COLUMN_DEFAULTS)
    features = dict(zip(CSV_COLUMNS, columns))

    # Remove unused columns
    for col in UNUSED_COLUMNS:
      features.pop(col)
    return features

def input_fn(filenames,
          num_epochs=None,
          shuffle=True,
          skip_header_lines=1,
          batch_size=200):
```

```
"""Generates features and labels for training or evaluation.
This uses the input pipeline based approach using file name queue
to read data so that entire data is not loaded in memory.
"""
dataset = tf.data.TextLineDataset(filenames).skip(skip_header_lines).map(
  _decode_csv)

if shuffle:
    dataset = dataset.shuffle(buffer_size=batch_size * 10)
iterator = dataset.repeat(num_epochs).batch(
    batch_size).make_one_shot_iterator()
features = iterator.get_next()
return features, parse_label_column(features.pop(LABEL_COLUMN))
```

The code for the most part is self-explanatory; however, the reader should take note of the following points:

- The function 'build_estimator' uses the canned Estimator API to train a 'DNNClassifier' model on Cloud MLE. The learning rate and hidden units of the model can be adjusted and tuned as a hyper-parameter during training.

- The methods 'csv_serving_input_fn' and 'json_serving_input_fn' define the serving inputs for CSV and JSON serving input formats.

- The method 'input_fn' uses the TensorFlow Dataset API to build the input pipelines for training and evaluation on Cloud MLE. This method calls the private method _decode_csv() to convert the CSV columns to Tensors.

The Application Logic

Let's see the application logic in the file '**task.py**'.

```
import argparse
import json
import os
```

```python
import tensorflow as tf
from tensorflow.contrib.training.python.training import hparam

import trainer.model as model

def _get_session_config_from_env_var():
    """Returns a tf.ConfigProto instance that has appropriate device_
    filters set.
    """

    tf_config = json.loads(os.environ.get('TF_CONFIG', '{}'))

    if (tf_config and 'task' in tf_config and 'type' in tf_config['task'] and
        'index' in tf_config['task']):
        # Master should only communicate with itself and ps
        if tf_config['task']['type'] == 'master':
            return tf.ConfigProto(device_filters=['/job:ps', '/job:master'])
        # Worker should only communicate with itself and ps
        elif tf_config['task']['type'] == 'worker':
            return tf.ConfigProto(device_filters=[
                '/job:ps',
                '/job:worker/task:%d' % tf_config['task']['index']
            ])
    return None

def train_and_evaluate(hparams):
    """Run the training and evaluate using the high level API."""

    train_input = lambda: model.input_fn(
        hparams.train_files,
        num_epochs=hparams.num_epochs,
        batch_size=hparams.train_batch_size
    )

    # Don't shuffle evaluation data
    eval_input = lambda: model.input_fn(
        hparams.eval_files,
        batch_size=hparams.eval_batch_size,
        shuffle=False
    )
```

```python
    train_spec = tf.estimator.TrainSpec(
        train_input, max_steps=hparams.train_steps)

    exporter = tf.estimator.FinalExporter(
        'iris', model.SERVING_FUNCTIONS[hparams.export_format])
    eval_spec = tf.estimator.EvalSpec(
        eval_input,
        steps=hparams.eval_steps,
        exporters=[exporter],
        name='iris-eval')

    run_config = tf.estimator.RunConfig(
        session_config=_get_session_config_from_env_var())
    run_config = run_config.replace(model_dir=hparams.job_dir)
    print('Model dir %s' % run_config.model_dir)
    estimator = model.build_estimator(
        learning_rate=hparams.learning_rate,
        # Construct layers sizes with exponential decay
        hidden_units=[
            max(2, int(hparams.first_layer_size * hparams.scale_factor**i))
            for i in range(hparams.num_layers)
        ],
        config=run_config)

    tf.estimator.train_and_evaluate(estimator, train_spec, eval_spec)

if __name__ == '__main__':
    parser = argparse.ArgumentParser()
    # Input Arguments
    parser.add_argument(
        '--train-files',
        help='GCS file or local paths to training data',
        nargs='+',
        default='gs://iris-dataset/train_data.csv')
    parser.add_argument(
        '--eval-files',
        help='GCS file or local paths to evaluation data',
```

```
        nargs='+',
        default='gs://iris-dataset/test_data.csv')
    parser.add_argument(
        '--job-dir',
        help='GCS location to write checkpoints and export models',
        default='/tmp/iris-estimator')
    parser.add_argument(
        '--num-epochs',
        help="""\
        Maximum number of training data epochs on which to train.
        If both --max-steps and --num-epochs are specified,
        the training job will run for --max-steps or --num-epochs,
        whichever occurs first. If unspecified will run for --max-steps.\
        """,
        type=int)
    parser.add_argument(
        '--train-batch-size',
        help='Batch size for training steps',
        type=int,
        default=20)
    parser.add_argument(
        '--eval-batch-size',
        help='Batch size for evaluation steps',
        type=int,
        default=20)
    parser.add_argument(
        '--learning_rate',
        help='The training learning rate',
        default=1e-4,
        type=int)
    parser.add_argument(
        '--first-layer-size',
        help='Number of nodes in the first layer of the DNN',
        default=256,
        type=int)
```

```
parser.add_argument(
    '--num-layers', help='Number of layers in the DNN', default=3,
    type=int)
parser.add_argument(
    '--scale-factor',
    help='How quickly should the size of the layers in the DNN decay',
    default=0.7,
    type=float)
parser.add_argument(
    '--train-steps',
    help="""\
    Steps to run the training job for. If --num-epochs is not specified,
    this must be. Otherwise the training job will run indefinitely.\
    """,
    default=100,
    type=int)
parser.add_argument(
    '--eval-steps',
    help='Number of steps to run evalution for at each checkpoint',
    default=100,
    type=int)
parser.add_argument(
    '--export-format',
    help='The input format of the exported SavedModel binary',
    choices=['JSON', 'CSV'],
    default='CSV')
parser.add_argument(
    '--verbosity',
    choices=['DEBUG', 'ERROR', 'FATAL', 'INFO', 'WARN'],
    default='INFO')

args, _ = parser.parse_known_args()

# Set python level verbosity
tf.logging.set_verbosity(args.verbosity)
# Set C++ Graph Execution level verbosity
```

```
os.environ['TF_CPP_MIN_LOG_LEVEL'] = str(
    tf.logging.__dict__[args.verbosity] / 10)

# Run the training job
hparams = hparam.HParams(**args.__dict__)
train_and_evaluate(hparams)
```

Note the following in the preceding code:

- The method '_get_session_config_from_env_var()' defines the configuration for the runtime environment on Cloud MLE for the Estimator.

- The method 'train_and_evaluate()' does a number of orchestration events including

    - Routing training and evaluation datasets to the model function in 'model.py'

    - Setting up the runtime environment of the Estimator

    - Passing hyper-parameters to the Estimator model

- The line of code "if __name__ == '__main__':" defines the entry point of the Python script via the terminal session. In this script, the code will receive inputs from the terminal through the 'argparse. ArgumentParser()' method.

Training on Cloud MLE

The training execution codes are bash commands stored in a shell script. Shell scripts end with the suffix '.sh'.

Running a Single Instance Training Job

The bash codes for executing training on a single instance on Cloud MLE is shown in the following. Change the bucket names accordingly.

```
DATE=`date '+%Y%m%d_%H%M%S'`
export JOB_NAME=iris_$DATE
```

```
export GCS_JOB_DIR=gs://iris-dataset/jobs/$JOB_NAME
export TRAIN_FILE=gs://iris-dataset/train_data.csv
export EVAL_FILE=gs://iris-dataset/test_data.csv

echo $GCS_JOB_DIR

gcloud ai-platform jobs submit training $JOB_NAME \
                              --stream-logs \
                              --runtime-version 1.8 \
                              --job-dir $GCS_JOB_DIR \
                              --module-name trainer.task \
                              --package-path trainer/ \
                              --region us-central1 \
                              -- \
                              --train-files $TRAIN_FILE \
                              --eval-files $EVAL_FILE \
                              --train-steps 5000 \
                              --eval-steps 100
```

This code is stored in the file 'single-instance-training.sh' and executed by running the command on the terminal.

```
source ./scripts/single-instance-training.sh

'Output:'
gs://iris-dataset/jobs/iris_20181112_010123
Job [iris_20181112_010123] submitted successfully.
INFO    2018-11-12 01:01:25 -0500    service    Validating job
                                                requirements...
INFO    2018-11-12 01:01:26 -0500    service    Job creation request
                                                has been successfully
                                                validated.
INFO    2018-11-12 01:01:26 -0500    service    Job iris_20181112_010123 is
                                                queued.
INFO    2018-11-12 01:01:26 -0500    service    Waiting for job to be
                                                provisioned.
```

```
INFO     2018-11-12 01:05:32 -0500     service       Waiting for training
                                                      program to start.
...
INFO     2018-11-12 01:09:05 -0500     ps-replica-2  Module completed;
                                                      cleaning up.
INFO     2018-11-12 01:09:05 -0500     ps-replica-2  Clean up finished.
INFO     2018-11-12 01:09:55 -0500     service       Finished tearing
                                                      down training
                                                      program.
INFO     2018-11-12 01:10:53 -0500     service       Job completed
                                                      successfully.
endTime: '2018-11-12T01:08:35'
jobId: iris_20181112_010123
startTime: '2018-11-12T01:07:34'
state: SUCCEEDED
```

Running a Distributed Training Job

The code for initiating distributed training on Cloud MLE is shown in the following, and the code is stored in the file 'distributed-training.sh'. For a distributed job, the attribute '- -scale-tier' is set to a tier above the basic machine type. Change the bucket names accordingly.

```
export SCALE_TIER=STANDARD_1 # BASIC | BASIC_GPU | STANDARD_1 | PREMIUM_1 |
BASIC_TPU
DATE=`date '+%Y%m%d_%H%M%S'`
export JOB_NAME=iris_$DATE
export GCS_JOB_DIR=gs://iris-dataset/jobs/$JOB_NAME
export TRAIN_FILE=gs://iris-dataset/train_data.csv
export EVAL_FILE=gs://iris-dataset/test_data.csv

echo $GCS_JOB_DIR

gcloud ai-platform jobs submit training $JOB_NAME \
                            --stream-logs \
                            --scale-tier $SCALE_TIER \
                            --runtime-version 1.8 \
```

```
                              --job-dir $GCS_JOB_DIR \
                              --module-name trainer.task \
                              --package-path trainer/ \
                              --region us-central1 \
                              -- \
                              --train-files $TRAIN_FILE \
                              --eval-files $EVAL_FILE \
                              --train-steps 5000 \
                              --eval-steps 100
```

The following executes a distributed training job.

```
source ./scripts/distributed-training.sh
```

Running a Distributed Training Job with Hyper-parameter Tuning

To run a training job with hyper-parameter tuning, add the '- -config' attribute and link to the '.yaml' hyper-parameter configuration file. The code for running the job is the same, but with the attribute '- -config' added. Change the bucket names accordingly.

```
export SCALE_TIER=STANDARD_1 # BASIC | BASIC_GPU | STANDARD_1 | PREMIUM_1 |
BASIC_TPU
DATE=`date '+%Y%m%d_%H%M%S'`
export JOB_NAME=iris_$DATE
export HPTUNING_CONFIG=hptuning_config.yaml
export GCS_JOB_DIR=gs://iris-dataset/jobs/$JOB_NAME
export TRAIN_FILE=gs://iris-dataset/train_data.csv
export EVAL_FILE=gs://iris-dataset/test_data.csv

echo $GCS_JOB_DIR

gcloud ai-platform jobs submit training $JOB_NAME \
                              --stream-logs \
                              --scale-tier $SCALE_TIER \
                              --runtime-version 1.8 \
                              --config $HPTUNING_CONFIG \
```

```
                              --job-dir $GCS_JOB_DIR \
                              --module-name trainer.task \
                              --package-path trainer/ \
                              --region us-central1 \
                              -- \
                              --train-files $TRAIN_FILE \
                              --eval-files $EVAL_FILE \
                              --train-steps 5000 \
                              --eval-steps 100
```

hptuning_config.yaml File

This file contains the hyper-parameter and the ranges we wish to explore in tuning our training job on Cloud MLE. The goal of the tuning job is to 'MAXIMIZE' the 'accuracy' metric.

```
trainingInput:
  hyperparameters:
    goal: MAXIMIZE
    hyperparameterMetricTag: accuracy
    maxTrials: 4
    maxParallelTrials: 2
    params:
      - parameterName: learning-rate
        type: DOUBLE
        minValue: 0.00001
        maxValue: 0.005
        scaleType: UNIT_LOG_SCALE
      - parameterName: first-layer-size
        type: INTEGER
        minValue: 50
        maxValue: 500
        scaleType: UNIT_LINEAR_SCALE
      - parameterName: num-layers
        type: INTEGER
        minValue: 1
```

```
      maxValue: 15
      scaleType: UNIT_LINEAR_SCALE
    - parameterName: scale-factor
      type: DOUBLE
      minValue: 0.1
      maxValue: 1.0
      scaleType: UNIT_REVERSE_LOG_SCALE
```

Execute Training Job with Hyper-parameter Tuning

Run the following code on the terminal to launch a distributed training job.

```
source ./scripts/hyper-tune.sh
```

```
gs://iris-dataset/jobs/iris_20181114_190121
Job [iris_20181114_190121] submitted successfully.
INFO    2018-11-14 12:41:07 -0500    service    Validating job
                                                requirements...
INFO    2018-11-14 12:41:07 -0500    service    Job creation request
                                                has been successfully
                                                validated.
INFO    2018-11-14 12:41:08 -0500    service    Job iris_20181114_190121 is
                                                queued.
INFO    2018-11-14 12:41:18 -0500    service    Waiting for job to be
                                                provisioned.
INFO    2018-11-14 12:41:18 -0500    service    Waiting for job to be
                                                provisioned.
...
INFO    2018-11-14 12:56:38 -0500    service    Finished tearing down
                                                training program.
INFO    2018-11-14 12:56:45 -0500    service    Finished tearing down
                                                training program.
INFO    2018-11-14 12:57:37 -0500    service    Job completed successfully.
INFO    2018-11-14 12:57:43 -0500    service    Job completed successfully.
endTime: '2018-11-14T13:04:34'
jobId: iris_20181114_190121
```

```
startTime: '2018-11-14T12:41:12'
state: SUCCEEDED
```

The job details of the hyper-parameter training job is shown in Figure 41-3.

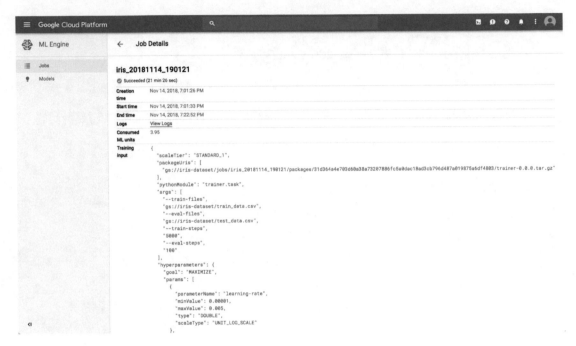

Figure 41-3. *Job details: Hyper-parameter distributed training job on Cloud MLE*

Under **'Training output'**, the first **'trialID'** contains the hyper-parameter set that minimizes the cost function and performs best on the evaluation metric. Observe that the trial run within the red box has the highest accuracy value in the **'objectiveValue'** attribute. This is illustrated in Figure 41-4.

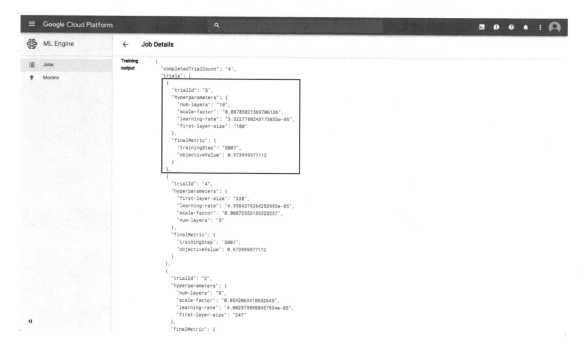

Figure 41-4. *Choosing the best hyper-parameter set*

Making Predictions on Cloud MLE

To make predictions on Cloud MLE, we first create a prediction instance. To do this, run the code in 'create-prediction-service.sh' as shown in the following. The variable 'MODEL_BINARIES' points to the folder location on GCS that stores the trained model for the hyper-parameter setting with '**trialID** = 2'.

```
export MODEL_VERSION=v1
export MODEL_NAME=iris
export MODEL_BINARIES=$GCS_JOB_DIR/3/export/iris/1542241126

# Create a Cloud ML Engine model
gcloud ai-platform models create $MODEL_NAME

# Create a model version
gcloud ai-platform versions create $MODEL_VERSION \
    --model $MODEL_NAME \
    --origin $MODEL_BINARIES \
    --runtime-version 1.8
```

Run the following code to create the prediction service.

```
source ./scripts/create-prediction-service.sh
```

```
Creating model...
Created ml engine model [projects/quantum-ally-219323/models/iris].
Creating model version...
Creating version (this might take a few minutes)......done.
```

The version details of the created model is as seen in Figure 41-5.

Figure 41-5. *Created model for serving on Cloud MLE*

Run Batch Prediction

Now let's run a batch prediction job on Cloud MLE. The code to execute a batch prediction call on Cloud MLE is provided in the following and stored in 'run-batch-predictions.sh'.

```
export JOB_NAME=iris_prediction
export MODEL_NAME=iris
```

```
export MODEL_VERSION=v1
export TEST_FILE=gs://iris-dataset/hold_out_test.csv

# submit a batched job
gcloud ai-platform jobs submit prediction $JOB_NAME \
        --model $MODEL_NAME \
        --version $MODEL_VERSION \
        --data-format TEXT \
        --region $REGION \
        --input-paths $TEST_FILE \
        --output-path $GCS_JOB_DIR/predictions

# stream job logs
echo "Job logs..."
gcloud ai-platform jobs stream-logs $JOB_NAME

# read output summary
echo "Job output summary:"
gsutil cat $GCS_JOB_DIR/predictions/prediction.results-00000-of-00001
```

Execute the code with the command

```
source ./scripts/run-batch-prediction.sh
```

```
Job [iris_prediction] submitted successfully.

jobId: iris_prediction
state: QUEUED
Job logs...
INFO    2018-11-12 14:48:18 -0500    service    Validating job
                                                requirements...
INFO    2018-11-12 14:48:18 -0500    service    Job creation request
                                                has been successfully
                                                validated.
INFO    2018-11-12 14:48:19 -0500    service    Job iris_prediction is
                                                queued.
Job output summary:
Job output summary:
```

{"classes": ["0", "1", "2"], "scores": [8.242315743700601e-06, 0.9921771883964539, 0.007814492098987103]}
{"classes": ["0", "1", "2"], "scores": [2.7296309657032225e-09, 0.015436310321092606, 0.9845637083053589]}
{"classes": ["0", "1", "2"], "scores": [5.207379217608832e-06, 0.9999237060546875, 7.100913353497162e-05]}
....
{"classes": ["0", "1", "2"], "scores": [0.999919056892395, 8.089694165391847e-05, 9.295699552171275e-16]}
{"classes": ["0", "1", "2"], "scores": [0.9999765157699585, 2.3535780201200396e-05, 1.2826575252518792e-17]}
{"classes": ["0", "1", "2"], "scores": [1.8082465658153524e-06, 0.7016969919204712, 0.29830116033554077]}

The prediction job details on Cloud MLE is as shown in Figure 41-6.

Figure 41-6. *Batch prediction job details*

Training with GPUs on Cloud MLE

Training models on GPUs can greatly reduce the processing time. In order to use GPUs on Cloud MLE, we make the following changes to our code example:

1. Change the scale tier to **'CUSTOM'.** The CUSTOM tier makes a number of GPU accelerators available, namely:

   a. standard_gpu: A single NVIDIA Tesla K80 GPU

   b. complex_model_m_gpu: Four NVIDIA Tesla K80 GPUs

   c. complex_model_l_gpu: Eight NVIDIA Tesla K80 GPUs

   d. standard_p100: A single NVIDIA Tesla P100 GPU

   e. complex_model_m_p100: Four NVIDIA Tesla P100 GPUs

   f. standard_v100: A single NVIDIA Tesla V100 GPU

   g. large_model_v100: A single NVIDIA Tesla V100 GPU

   h. complex_model_m_v100: Four NVIDIA Tesla V100 GPUs

   i. complex_model_l_v100: Eight NVIDIA Tesla V100 GPUs

2. Add the following parameters to the '.yaml' file to configure the GPU instance.

```
trainingInput:
  scaleTier: CUSTOM
  masterType: complex_model_m_gpu
  workerType: complex_model_m_gpu
  parameterServerType: large_model
  workerCount: 2
  parameterServerCount: 3
```

3. The full configuration file in 'gpu_hptuning_config.yaml' now looks like this:

```
trainingInput:
  scaleTier: CUSTOM
  masterType: complex_model_m_gpu
  workerType: complex_model_m_gpu
```

```
        parameterServerType: large_model
        workerCount: 2
        parameterServerCount: 3
        hyperparameters:
          goal: MAXIMIZE
          hyperparameterMetricTag: accuracy
          maxTrials: 4
          maxParallelTrials: 2
          params:
            - parameterName: learning-rate
              type: DOUBLE
              minValue: 0.00001
              maxValue: 0.005
              scaleType: UNIT_LOG_SCALE
            - parameterName: first-layer-size
              type: INTEGER
              minValue: 50
              maxValue: 500
              scaleType: UNIT_LINEAR_SCALE
            - parameterName: num-layers
              type: INTEGER
              minValue: 1
              maxValue: 15
              scaleType: UNIT_LINEAR_SCALE
            - parameterName: scale-factor
              type: DOUBLE
              minValue: 0.1
              maxValue: 1.0
              scaleType: UNIT_REVERSE_LOG_SCALE
```

Note that running GPUs on Cloud MLE is only available in the following regions:

- us-east1

- us-central1

- us-west1

- asia-east1

- europe-west1

- europe-west4

The updated execution code for training with GPUs on Cloud MLE is saved as 'gpu-hyper-tune.sh' (code shown in the following).

```
export SCALE_TIER=CUSTOM
DATE=`date '+%Y%m%d_%H%M%S'`
export JOB_NAME=iris_$DATE
export HPTUNING_CONFIG=gpu_hptuning_config.yaml
export GCS_JOB_DIR=gs://iris-dataset/jobs/$JOB_NAME
export TRAIN_FILE=gs://iris-dataset/train_data.csv
export EVAL_FILE=gs://iris-dataset/test_data.csv

echo $GCS_JOB_DIR

gcloud ai-platform jobs submit training $JOB_NAME \
                                --stream-logs \
                                --scale-tier $SCALE_TIER \
                                --runtime-version 1.8 \
                                --config $HPTUNING_CONFIG \
                                --job-dir $GCS_JOB_DIR \
                                --module-name trainer.task \
                                --package-path trainer/ \
                                --region us-central1 \
                                -- \
                                --train-files $TRAIN_FILE \
                                --eval-files $EVAL_FILE \
                                --train-steps 5000 \
                                --eval-steps 100
```

To execute the code, run

```
source ./scripts/gpu-hyper-tune.sh

gs://iris-dataset/jobs/iris_20181112_211040
Job [iris_20181112_211040] submitted successfully.
...
INFO    2018-11-12 21:35:36 -0500    ps-replica-2    4    Module completed;
                                                          cleaning up.
INFO    2018-11-12 21:35:36 -0500    ps-replica-2    4    Clean up finished.
INFO    2018-11-12 21:36:18 -0500    service     Finished tearing down
                                                 training program.
INFO    2018-11-12 21:36:25 -0500    service     Finished tearing down
                                                 training program.
INFO    2018-11-12 21:37:11 -0500    service     Job completed successfully.
INFO    2018-11-12 21:37:11 -0500    service     Job completed successfully.
endTime: '2018-11-12T21:38:26'
jobId: iris_20181112_211040
startTime: '2018-11-12T21:10:47'
state: SUCCEEDED
```

Scikit-learn on Cloud MLE

This section will provide a walk-through of training a Scikit-learn model on Google Cloud MLE using the same Iris dataset example. We'll begin by moving the appropriate data files from the GitHub repository of this book to GCS.

Move the Data Files to GCS

Walk through the following steps to move the data files to GCS:

1. Create bucket to hold the datasets.

   ```
   gsutil mb gs://iris-sklearn
   ```

2. Run the following commands on the terminal to move the training and testing datasets to the buckets:

Train set features.

```
gsutil cp X_train.csv gs://iris-sklearn
```

Train set targets.

```
gsutil cp y_train.csv gs://iris-sklearn
```

Test sample for online prediction.

```
gsutil cp test-sample.json gs://iris-sklearn
```

Prepare the Training Scripts

The code for training a Scikit-learn model on Cloud MLE is also prepared as a python package. The project structure is as follows:

Iris_SklearnCloudML: [project name as parent folder]

- Trainer: [folder containing the model and execution code]

  - _init_.py: [an empty special python file indicating that the containing folder is a Python package]

  - model.py: [file contains the logic of the model written in Scikit-learn]

- scripts: [folder containing scripts to execute jobs on Cloud MLE]

  - single-instance-training.sh: [script to run a single instance training job on Cloud MLE]

  - online-prediction.sh: [script to execute an online prediction job on Cloud MLE]

  - create-prediction-service.sh: [script to create a prediction service on Cloud MLE]

- config.yaml: [configuration file for specifying model version]

The model code for training on Cloud MLE with Scikit-learn (shown in the following) is stored in the file 'model.py'. The machine learning algorithm used in this model is the Random forest Classifier.

```python
# [START setup]
import datetime
import os
import subprocess
import sys
import pandas as pd

from sklearn.ensemble import RandomForestClassifier
from sklearn.externals import joblib
from tensorflow.python.lib.io import file_io

# Fill in your Cloud Storage bucket name
BUCKET_ID = 'iris-sklearn'
# [END setup]

# [START download-and-load-into-pandas]
iris_data_filename = 'gs://iris-sklearn/X_train.csv'
iris_target_filename = 'gs://iris-sklearn/y_train.csv'

# Load data into pandas
with file_io.FileIO(iris_data_filename, 'r') as iris_data_f:
    iris_data = pd.read_csv(filepath_or_buffer=iris_data_f,
                      header=None, sep=',').values

with file_io.FileIO(iris_target_filename, 'r') as iris_target_f:
    iris_target = pd.read_csv(filepath_or_buffer=iris_target_f,
                      header=None, sep=',').values

iris_target = iris_target.reshape((iris_target.size,))
# [END download-and-load-into-pandas]

# [START train-and-save-model]
# Train the model
classifier = RandomForestClassifier()
classifier.fit(iris_data, iris_target)
```

```
# Export the classifier to a file
model = 'model.joblib'
joblib.dump(classifier, model)
# [END train-and-save-model]

# [START upload-model]
# Upload the saved model file to Cloud Storage
model_path = os.path.join('gs://', BUCKET_ID, 'model', datetime.datetime.
now().strftime(
    'iris_%Y%m%d_%H%M%S'), model)
subprocess.check_call(['gsutil', 'cp', model, model_path], stderr=sys.
stdout)
# [END upload-model]
```

Take note of the following points in the preceding code block:

- The code uses the 'file.io' module from the package 'tensorflow. python.lib.io' to stream a file stored on Cloud Storage.

- The rest of the code runs the classifier to build the model and exports the model to a bucket location on GCS. Cloud MLE will read from this bucket when building a prediction service for online predictions.

Execute a Scikit-learn Training Job on Cloud MLE

The bash code for executing a training job for the Scikit-learn model is presented in the following and is saved in the file 'single-instance-training.sh'.

```
export SCALE_TIER=BASIC # BASIC | BASIC_GPU | STANDARD_1 | PREMIUM_1 |
BASIC_TPU
DATE=`date '+%Y%m%d_%H%M%S'`
export JOB_NAME=iris_sklearn_$DATE
export GCS_JOB_DIR=gs://iris-sklearn/jobs/$JOB_NAME

echo $GCS_JOB_DIR

gcloud ml-engine jobs submit training $JOB_NAME \
                                --stream-logs \
                                --scale-tier $SCALE_TIER \
```

```
                                  --runtime-version 1.8 \
                                  --job-dir $GCS_JOB_DIR \
                                  --module-name trainer.model \
                                  --package-path trainer/ \
                                  --region us-central1 \
                                  --python-version 3.5
```

The following code runs a training job to build a Scikit-learn Random forest model.

```
source ./scripts/single-instance-training.sh
```

```
gs://iris-sklearn/jobs/iris_sklearn_20181119_000349
Job [iris_sklearn_20181119_000349] submitted successfully.
INFO    2018-11-19 00:03:51 -0500    service    Validating job
                                                requirements...
INFO    2018-11-19 00:03:52 -0500    service    Job creation request
                                                has been successfully
                                                validated.
INFO    2018-11-19 00:03:52 -0500    service    Job iris_sklearn_20181119_
                                                000349 is queued.
INFO    2018-11-19 00:03:52 -0500    service    Waiting for job to be
                                                provisioned.
INFO    2018-11-19 00:03:54 -0500    service    Waiting for training
                                                program to start.
...
INFO    2018-11-19 00:05:19 -0500    master-replica-0    Module
                                                           completed;
                                                           cleaning up.
INFO    2018-11-19 00:05:19 -0500    master-replica-0    Clean up
                                                           finished.
INFO    2018-11-19 00:05:19 -0500    master-replica-0    Task completed
                                                           successfully.
endTime: '2018-11-19T00:09:38'
jobId: iris_sklearn_20181119_000349
startTime: '2018-11-19T00:04:29'
state: SUCCEEDED
```

Create a Scikit-learn Prediction Service on Cloud MLE

The code for creating a prediction service is shown in the following, and is saved in the file 'create-prediction-service.sh'.

```
export MODEL_VERSION=v1
export MODEL_NAME=iris_sklearn
export REGION=us-central1

# Create a Cloud ML Engine model
echo "Creating model..."
gcloud ml-engine models create $MODEL_NAME --regions=$REGION

# Create a model version
echo "Creating model version..."
gcloud ml-engine versions create $MODEL_VERSION \
    --model $MODEL_NAME \
    --config config.yaml
```

The preceding code references a configuration file 'config.yaml'. This file (as shown in the following) holds the configuration for the Scikit-learn model. Let's briefly go through the attributes listed:

- deploymentUri: This points to the bucket location of the Scikit-learn model.

- runtime version: This attribute specifies the Cloud MLE runtime version.

- framework: This attribute is of particular importance as it specifies the model framework in use; this can be SCIKIT_LEARN, XGBOOST, or TENSORFLOW. For this example, it is set to SCIKIT_LEARN.

- pythonVersion: This attribute specifies the Python version in use.

The 'config.yaml' is as defined in the following:

```
deploymentUri: "gs://iris-sklearn/iris_20181119_050517"
runtimeVersion: '1.8'
framework: "SCIKIT_LEARN"
pythonVersion: "3.5"
```

Run the following command to create a prediction service.

```
source ./scripts/create-prediction-service.sh

Creating model...
Created ml engine model [projects/quantum-ally-219323/models/iris_sklearn].
Creating model version...
Creating version (this might take a few minutes)......done.
```

Make Online Predictions from the Scikit-learn Model

The code to make an online prediction from the Scikit-learn model is shown in the following and is stored in the file 'online-prediction.sh'. In online predictions, the input data is passed directly as a JSON string.

```
export JOB_NAME=iris_sklearn_prediction
export MODEL_NAME=iris_sklearn
export MODEL_VERSION=v1
export TEST_FILE_GCS=gs://iris-sklearn/test-sample.json
export TEST_FILE=./test-sample.json

# download file
gsutil cp $TEST_FILE_GCS .

# submit an online job
gcloud ml-engine predict --model $MODEL_NAME \
        --version $MODEL_VERSION \
        --json-instances $TEST_FILE

echo "0 -> setosa, 1 -> versicolor, 2 -> virginica"
```

The input data stored as a JSON string is shown in the following.

```
[5.1, 3.5, 1.4, 0.2]
```

Run the following command to execute an online prediction request to the hosted model on Cloud MLE.

```
source ./scripts/online-prediction.sh
```

```
Copying gs://iris-sklearn/test-sample.json...
/ [1 files][   20.0 B/   20.0 B]
Operation completed over 1 objects/20.0 B.
[0]
0 -> setosa, 1 -> versicolor, 2 -> virginica
```

In this chapter, we discuss training large-scale models using Google Cloud Machine Learning Engine, which is a part of the Google AI Platform. In the examples in this chapter, we trained the models using the Estimator High-level API and Scikit-learn. It is important to mention that the Keras high-level API can also be used to train large-scale models on Cloud MLE.

In the next chapter, we will cover training custom image recognition models with Google Cloud AutoML.

## CHAPTER 42

Google AutoML: Cloud Vision

Google Cloud AutoML Vision facilitates the creation of custom vision models for image recognition use cases. This managed service works with the concepts of transfer learning and neural architecture search under the hood to find the best network architecture and the optimal hyper-parameter configuration of that architecture that minimizes the loss function of the model. This chapter will go through a sample project of building a custom image recognition model using Google Cloud AutoML Vision. In this chapter, we will build an image model to recognize select cereal boxes.

Enable AutoML Cloud Vision on GCP

Step through the following steps to enable AutoML Cloud Vision on GCP:

1. Open Cloud Vision by clicking the triple dash at the top-left corner of the GCP dashboard. Select **Vision** under the product section **ARTIFICIAL INTELLIGENCE** as shown in Figure 42-1.

© Ekaba Bisong 2019

E. Bisong, *Building Machine Learning and Deep Learning Models on Google Cloud Platform*, https://doi.org/10.1007/978-1-4842-4470-8_42

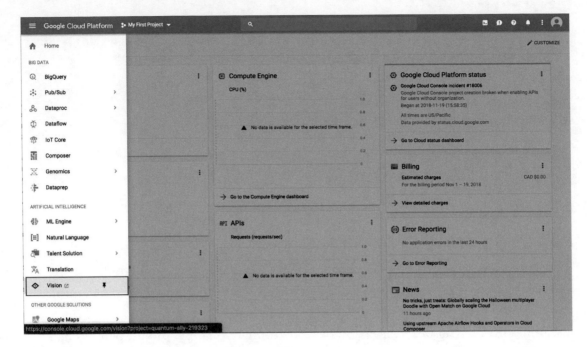

Figure 42-1. *Open Google AutoML: Cloud Vision*

2. Select the Google user account on which to activate AutoML as shown in Figures 42-2 and 42-3.

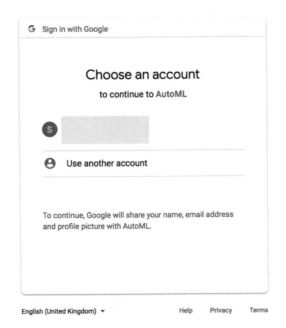

Figure 42-2. *Select account to authenticate AutoML*

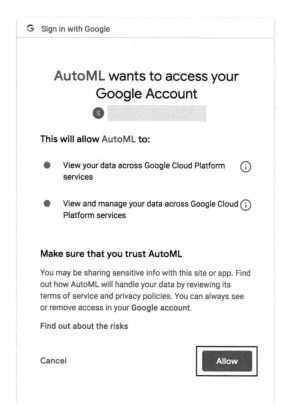

Figure 42-3. *Authenticate AutoML*

3. After authentication, the Google Cloud Vision Welcome page opens up (see Figure 42-4).

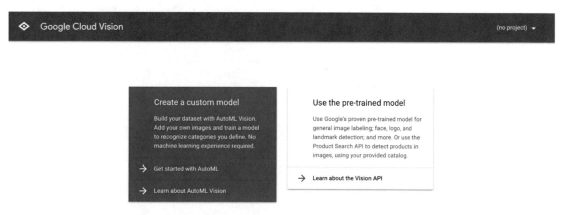

Figure 42-4. *Cloud Vision Welcome page*

4.   From the drop-down menu, select the **Project ID** (with billing enabled) that will be used to set up AutoML (see Figure 42-5).

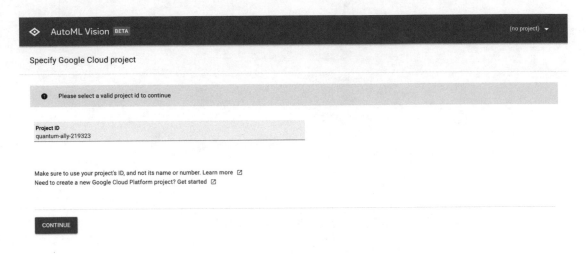

Figure 42-5. Select Project ID for configuring AutoML

5.   The final configuration step is to enable the AutoML API on the GCP project and to create a GCS bucket for storing the output models. Click **'SET UP NOW'** to automatically complete the configuration as shown in Figure 42-6.

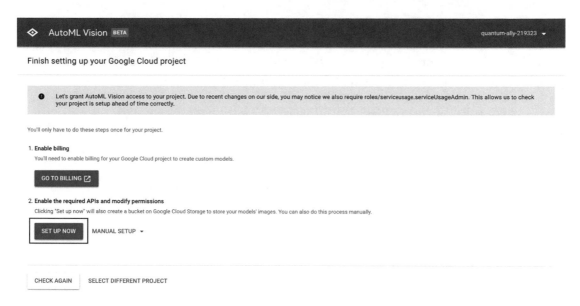

Figure 42-6. *Automatically complete AutoML configuration*

6. When the configuration is complete, the AutoML Vision
    Dashboard is activated (see Figure 42-7).

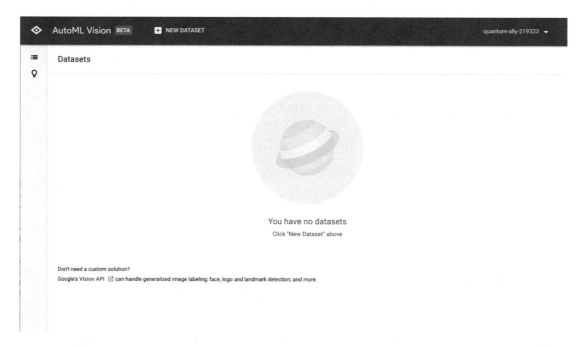

Figure 42-7. *Automatically complete AutoML configuration*

Preparing the Training Dataset

Before building a custom image recognition model with AutoML Cloud Vision, the dataset must be prepared in a particular format; they include

1. For training, JPEG, PNG, WEBP, GIF, BMP, TIFF, and ICO image formats are supported with a maximum size of 30mb per image.

2. For inference, the image formats JPEG, PNG, and GIF are supported with each image being of maximum size 1.5mb.

3. It is best to place each image category into containing sub-folder within an image folder For example:

    - [image-directory]

        - [image-class-1-dir]

        - [image-class-2-dir]

        - ...

        - [image-class-n-dir]

4. Next, a CSV must be created that points to the paths of the images and their corresponding label. AutoML uses the CSV file to point to the location of the training images and their labels. The CSV file is placed in the same GCS bucket containing the image files. Use the bucket automatically created when AutoML Vision was configured. In our case, this bucket is named 'gs://quantum-ally-219323-vcm'. We use the following code segment to create the CSV file used in the cereal classifier example.

```
import os
import numpy as np
import pandas as pd

directory = 'cereal_photos/

data = []

# go through sub-directories in the image directory and get the
image paths
```

```
for subdir, dirs, files in os.walk(directory):
    for file in files:
        filepath = subdir + os.sep + file

        if filepath.endswith(".jpg"):
            entry = ['{}/{}'.format('gs://quantum-ally-219323-
            vcm',filepath), os.path.basename(subdir)]
            data.append(entry)

# convert to Pandas DataFrame
data_pd = pd.DataFrame(np.array(data))

# export CSV
data_pd.to_csv("data.csv", header=None, index=None)
```

5. The preceding code will result in a CSV looking like the following
   sample:

```
gs://quantum-ally-219323-vcm/cereal_photos/apple_cinnamon_
cheerios/001.jpg,apple_cinnamon_cheerios
gs://quantum-ally-219323-vcm/cereal_photos/apple_cinnamon_
cheerios/002.jpg,apple_cinnamon_cheerios
gs://quantum-ally-219323-vcm/cereal_photos/apple_cinnamon_
cheerios/003.jpg,apple_cinnamon_cheerios
...
gs://quantum-ally-219323-vcm/cereal_photos/none_of_the_above/
images_(97).jpg,none_of_the_above
gs://quantum-ally-219323-vcm/cereal_photos/none_of_the_above/
images_(98).jpg,none_of_the_above
gs://quantum-ally-219323-vcm/cereal_photos/none_of_the_above/
images_(99).jpg,none_of_the_above
...
gs://quantum-ally-219323-vcm/cereal_photos/sugar_crisp/001.
jpg,sugar_crisp
gs://quantum-ally-219323-vcm/cereal_photos/sugar_crisp/002.
jpg,sugar_crisp
gs://quantum-ally-219323-vcm/cereal_photos/sugar_crisp/003.
jpg,sugar_crisp
```

The first part is the image path or URI, while the other is the image label.

6. When preparing the image dataset, it is useful to have a '**None_ of_the_above'** image class. This class will contain random images that do not belong to any of the predicted classes. Adding this class can have an overall effect on the model accuracy.

7. Clone the GitHub book repository to the Notebook instance.

8. Navigate to the folder chapter and copy the image files to the GCS bucket.

```
gsutil cp -r cereal_photos gs://quantum-ally-219323-vcm
```

9. Copy the CSV data file containing the image paths and their labels to the GCS bucket.

```
gsutil cp data.csv gs://quantum-ally-219323-vcm/cereal_photos/
```

Building Custom Image Models on Cloud AutoML Vision

In AutoML for Cloud Vision, a dataset contains the images that will be used in building the classifier and their corresponding labels. This section will walk through creating a dataset and building a custom image model on AutoML Vision.

1. From the Cloud AutoML Vision Dashboard, click **NEW DATASET** as shown in Figure 42-8.

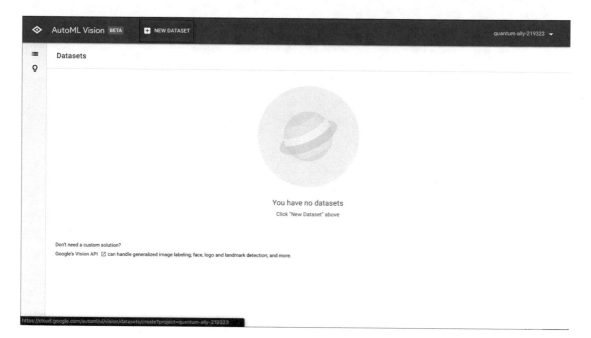

Figure 42-8. *New Dataset on AutoML Vision*

2. To create a Dataset on Cloud AutoML Vision, set the following parameters as shown in Figure 42-9:

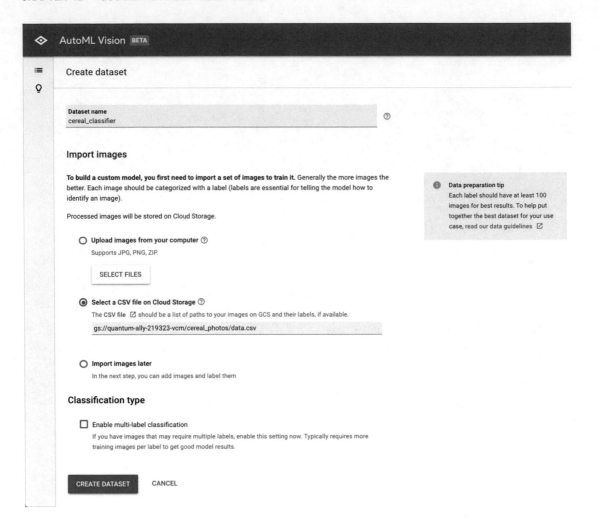

Figure 42-9. *Create a Dataset on Cloud AutoML Vision*

    a. Dataset name: cereal_classifier.

    b. Select a CSV file on Cloud Storage (this is the CSV file placed on the bucket created when Cloud AutoML was configured that contains the path to the images): gs://quantum-ally-219323-vcm/cereal_photos/data.csv.

    c. Click **CREATE DATASET** to begin importing images (see Figure 42-10).

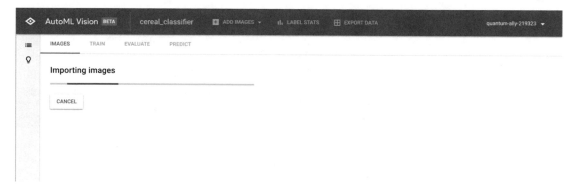

Figure 42-10. *Cloud AutoML Vision: Importing images*

3. After importing the Dataset, click **TRAIN** (see Figure 42-11) to
    initiate the process of building a custom image recognition model.

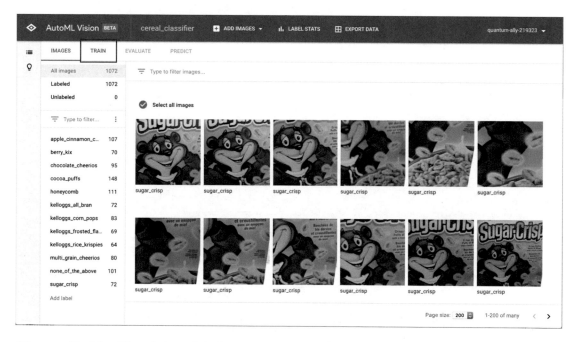

Figure 42-11. *Cloud AutoML Vision: Imported images and their labels*

4. In machine learning, more labeled training examples boost the
    performance of the model. Likewise, when using AutoML, there
    should be at least 100 training examples for each image class. In
    the example used in this section, some classes do not have up to

100 examples, so AutoML gives a warning as seen in Figure 42-12. However, for the purposes of this exercise, we will continue with training. Click **START TRAINING**.

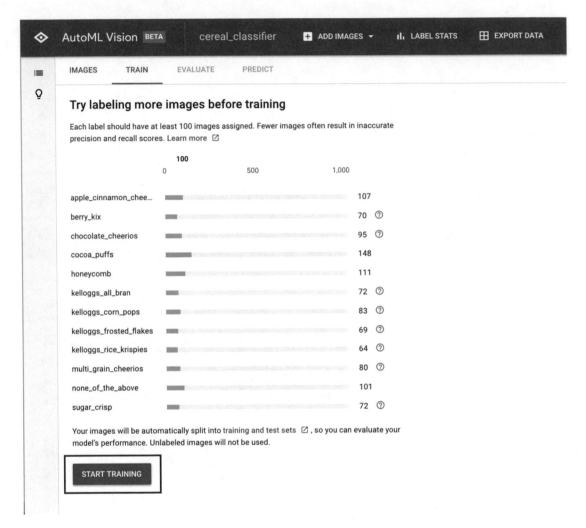

Figure 42-12. *Cloud AutoML Vision requesting for more training examples per image class*

5. Choose how long the model will be trained. More training time might have an effect on the model accuracy, but this may cost more for running on Cloud AutoML's machines (see Figure 42-13). Again, click **START TRAINING** to begin building the model (see Figure 42-14).

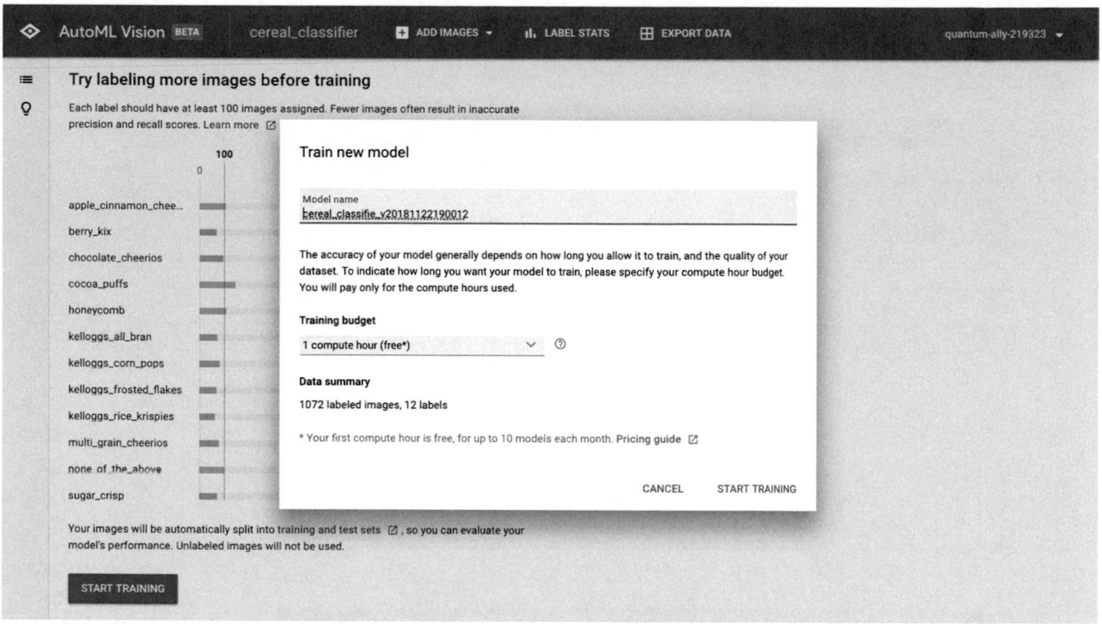

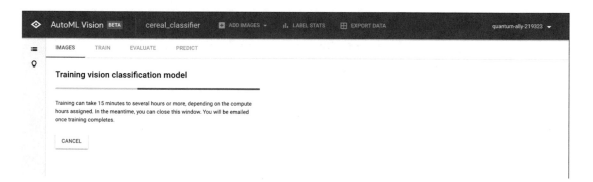

Figure 42-13. *Select training budget*

Figure 42-14. *Training vision model on Cloud AutoML Vision*

6.   The training summary is shown in Figure 42-15.

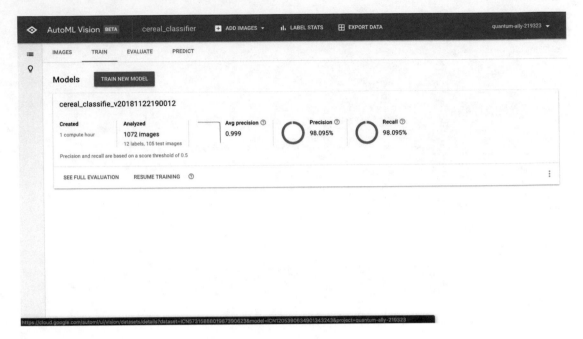

Figure 42-15. *Cloud AutoML Vision: Training summary*

7. AutoML Vision uses the set-aside test images to evaluate the
    quality of the model after training as seen in Figure 42-16. The F1
    plot showing the trade-off between precision and recall is shown
    in Figure 42-17. Also, a visual confusion matrix is provided to
    further evaluate the model quality (see Figure 42-18).

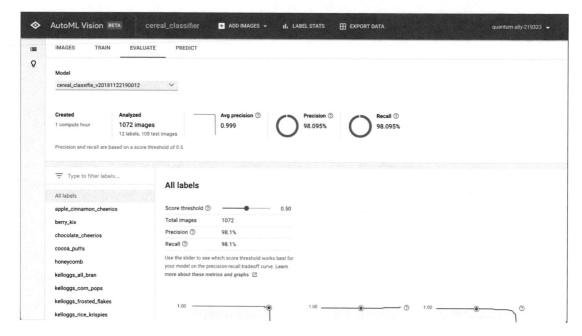

Figure 42-16. *Cloud AutoML Vision: Model evaluation*

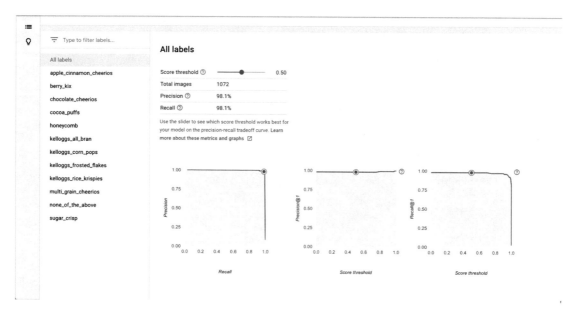

Figure 42-17. *F1 evaluation matrix on Cloud AutoML Vision*

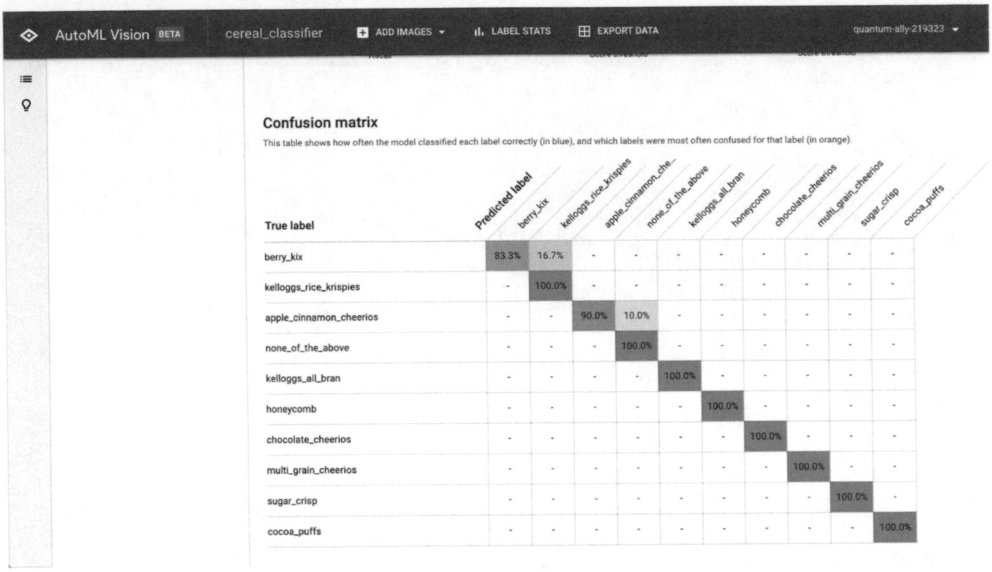

Figure 42-18. *Confusion matrix for model evaluation on Cloud AutoML Vision*

8. The custom image recognition model is exposed as a REST or
    Python API for integration into software applications as a prediction
    service (see Figure 42-19). We can test our model by uploading a
    sample image for classification as shown in Figure 42-20.

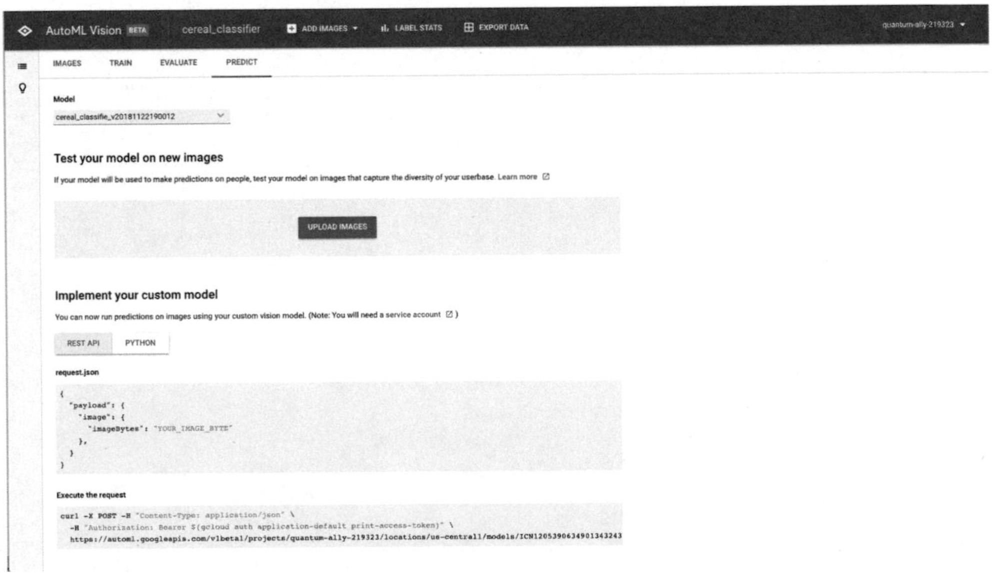

Figure 42-19. *Cloud AutoML Vision: Model as a prediction service*

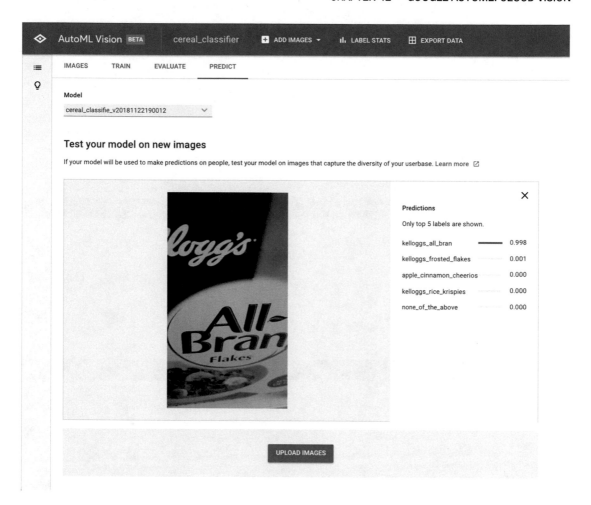

Figure 42-20. *Test prediction service on Cloud AutoML Vision*

9. To delete a model, click the triple dash and select Models to navigate to the Models Dashboard (see Figure 42-21). At the side of the model, click the triple dot and select Delete model (see Figure 42-22). Confirm deletion as shown in Figure 42-23. Note, however, that API calls affiliated with a deleted model will cease to be operational.

Figure 42-21. *Return to Models dashboard*

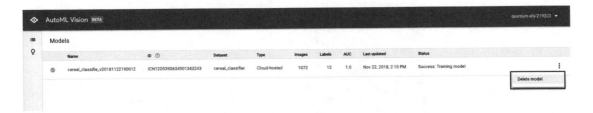

Figure 42-22. *Select model to delete*

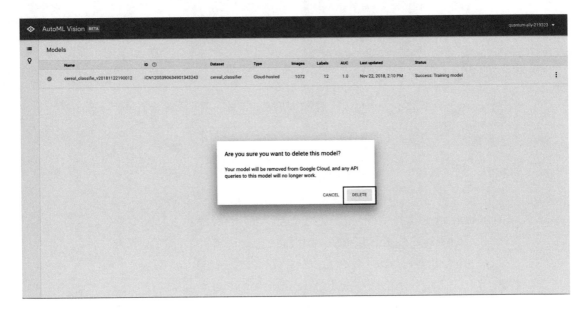

Figure 42-23. *Delete a model on Cloud AutoML Vision*

This chapter covered building and deploying custom image classification models using Google AutoML Cloud Vision. In the next chapter, we will discover how to build and deploy custom text classification models with Google Cloud AutoML for natural language processing.

CHAPTER 43

Google AutoML: Cloud Natural Language Processing

This chapter will build a language toxicity classification model to classify and recognize toxic and non-toxic or clean phrases using Google Cloud AutoML for natural language processing (NLP). The data used in this project is from the Toxic Comment Classification Challenge on Kaggle by Jigsaw and Google. The data is modified to have a sample of 16,000 toxic and 16,000 non-toxic words as inputs to build the model on AutoML NLP.

Enable AutoML NLP on GCP

The following steps will enable AutoML NLP on GCP:

1. Click the triple dash in the top-left corner of the interface and select **Natural Language** under the category ARTIFICIAL INTELLIGENCE as shown in Figure 43-1.

© Ekaba Bisong 2019
E. Bisong, *Building Machine Learning and Deep Learning Models on Google Cloud Platform*,
https://doi.org/10.1007/978-1-4842-4470-8_43

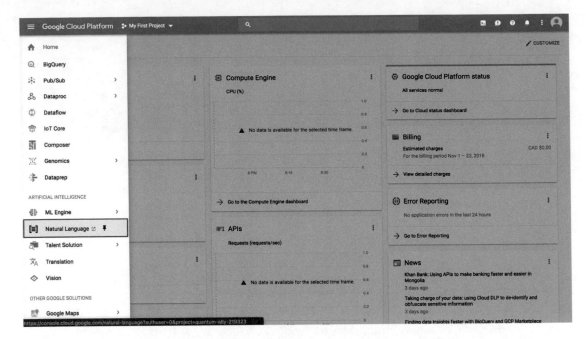

Figure 43-1. *Open Cloud AutoML for Natural Language*

2. From the screen that follows, click **Get started with AutoML** (see
    Figure 43-2).

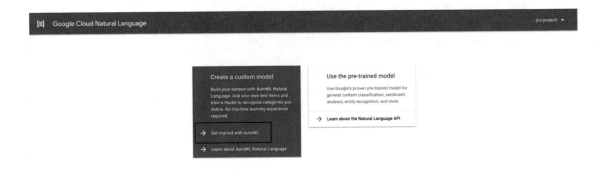

Figure 43-2. *Click Get started with Cloud AutoML NLP*

3. Click **SET UP NOW** to automatically setup the GCP project for working with Cloud AutoML NLP (see Figure 43-3). This process involves activating the API for AutoML and creating a bucket on GCP for storing the data input and output models. We will use this bucket in the next section.

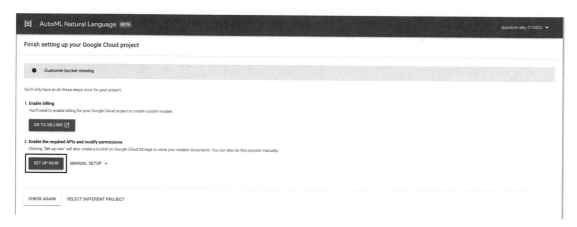

Figure 43-3. *Auto-configure Cloud AutoML NLP*

4. After configuration, the Cloud AutoML NLP Dashboard is activated (see Figure 43-4).

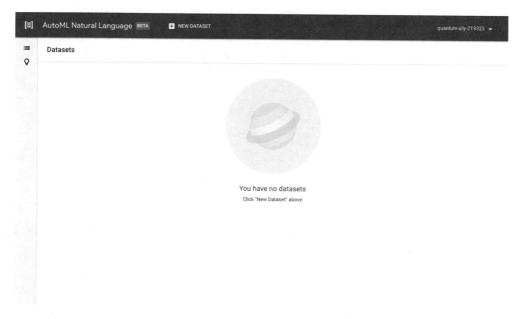

Figure 43-4. *AutoML NLP dashboard*

Preparing the Training Dataset

Let's step through preparing the dataset for building a custom language classification model with Cloud AutoML NLP:

1. The training input can either be a document in (.txt) format or as an in-line text in a (.csv) file. Multiple texts can be grouped as a compressed (.zip) file.

2. For this project, text files are placed in sub-folders with their grouped output labels as the folder names. This is later used to create a CSV file containing the data file path and their labels. For example:

    -   [files]

        -   [toxic]

        -   [clean]

3. Next, a CSV must be generated that points to the paths of the images and their corresponding label. Just like Cloud Vision, Cloud NLP uses the CSV file to point to the location of the training documents or words and their corresponding labels. The CSV file is placed in the same GCS bucket created AutoML NLP was configured. In our case, this bucket is named 'gs://quantum-ally-219323-lcm'. The following code segment prepares the data and produces a CSV file.

```
import numpy as np
import pandas as pd
import re
import pathlib
import os

# read the Toxic Comment Classification training dataset
data = pd.read_csv('./data/train.csv')

# add clean column label
data['clean'] = (1 - data.iloc[:, 2:].sum(axis=1) >= 1).
astype(int)
```

```
# merge all other non-clean comments to toxic
data.loc[data['clean'] == 0, ['toxic']] = 1

# select dataframe of clean examples
data_clean = data[data['clean'] == 1].sample(n=20000)
# select dataframe of toxic examples
data_toxic = data[data['toxic'] == 1].sample(n=16000)

# join into one dataframe
data = pd.concat([data_clean, data_toxic])

# remove unused columns
data.drop(['severe_toxic', 'obscene', 'threat', 'insult',
'identity_hate'], axis=1, inplace=True)

# create text documents and place them in their folder classes.
for index, row in data.iterrows():
    comment_text = re.sub(r'[^\w\s]','',row['comment_text']).
    rstrip().lstrip().strip()
    classes = "
    if (row['toxic'] == 1):
        classes = 'toxic'
    else:
        classes = 'clean'

    pathlib.Path("./file/{}".format(classes)).mkdir(parents=True,
    exist_ok=True)
    with open("./file/{}/text_{}.txt".format(classes,index), "w")
    as text_file:
        text_file.write(comment_text)

data_path = []
directory = 'file/'

# create data csv
for subdir, dirs, files in os.walk(directory):
    for file in files:
        filepath = subdir + os.sep + file
```

```
        if filepath.endswith(".txt"):
            entry = ['{}/{}'.format('gs://quantum-ally-219323-
            lcm',filepath), os.path.basename(subdir)]
            data_path.append(entry)

    # convert to Pandas DataFrame
    data_pd = pd.DataFrame(np.array(data_path))

    # export data to csv
    data_pd.to_csv("data.csv", header=None, index=None)
```

4. The preceding code will result in a CSV looking like the following
    sample:

    ```
    gs://quantum-ally-219323-lcm/file/clean/text_100055.txt,clean
    gs://quantum-ally-219323-lcm/file/clean/text_100059.txt,clean
    gs://quantum-ally-219323-lcm/file/clean/text_100077.txt,clean
    ...
    gs://quantum-ally-219323-lcm/file/toxic/text_141122.txt,toxic
    gs://quantum-ally-219323-lcm/file/toxic/text_141138.txt,toxic
    gs://quantum-ally-219323-lcm/file/toxic/text_141143.txt,toxic
    ```

    The first part is the image path or URI, while the other is the
    document label.

5. When preparing the text dataset, it is useful to have a '**None_of_
    the_above'** class. This class will contain documents that do not
    belong to any of the predicted classes. Adding this class can have
    an overall effect on the model accuracy.

6. Navigate to the folder chapter and copy the image files to the GCS
    bucket. The flag **-m** initiates parallel uploads to speed up upload
    time of large document sizes to GCP.

    ```
    gsutil -m cp -r file gs://quantum-ally-219323-lcm
    ```

7. Copy the CSV data file containing the document paths and their
    labels to the GCS bucket.

    ```
    gsutil cp data.csv gs://quantum-ally-219323-lcm/file/
    ```

Building a Custom Language Classification Model on Cloud AutoML NLP

This section will walk through creating a document dataset and building a custom language classification model on AutoML Vision.

1. From the Cloud AutoML NLP dashboard, click **NEW DATASET** as shown in Figure 43-5.

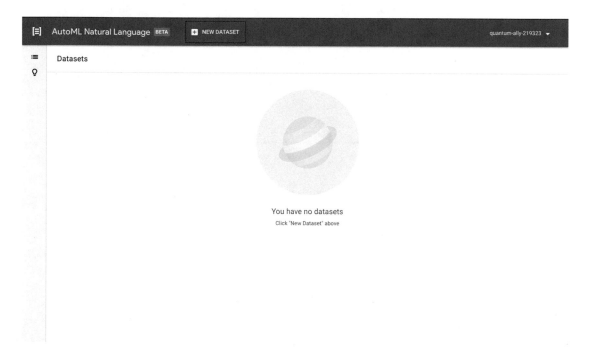

Figure 43-5. *New Dataset on AutoML NLP*

2. To create a Dataset on Cloud AutoML NLP, set the following parameters as shown in Figure 43-6:

Figure 43-6. *Create a Dataset on Cloud AutoML NLP*

    a. Dataset name: toxicity_dataset.

    b. Select a CSV file on Cloud Storage (this is the CSV file placed on the bucket created when Cloud AutoML was configured that contains the path to the text documents): gs://quantum-ally-219323-lcm/file/data.csv.

    c. Click **CREATE DATASET** to begin importing images (see Figure 43-7).

Figure 43-7. Cloud AutoML NLP: Importing text items

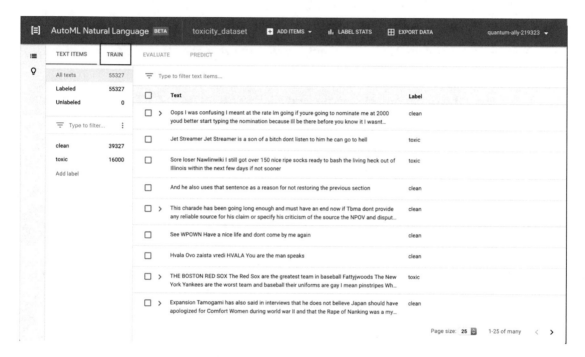

Figure 43-8. Cloud AutoML NLP: Imported text documents and their labels

3. After importing the Dataset, click **TRAIN** (see Figure 43-8) to initiate the process of building a custom language classification model.

4. In this example, we have a good enough number of training examples as seen in Figure 43-9, so hopefully, it makes sense to expect a good language classification model. Click **START TRAINING** to begin the training job.

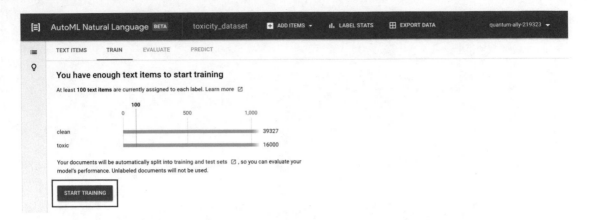

Figure 43-9. *Cloud AutoML NLP checking the adequacy of training examples*

5. Accept the default model name, and click **START TRAINING** (see Figure 43-10) to begin building the model as seen in Figure 43-11. Note that this training might take about an hour to complete. When done, the user will get an email of completion.

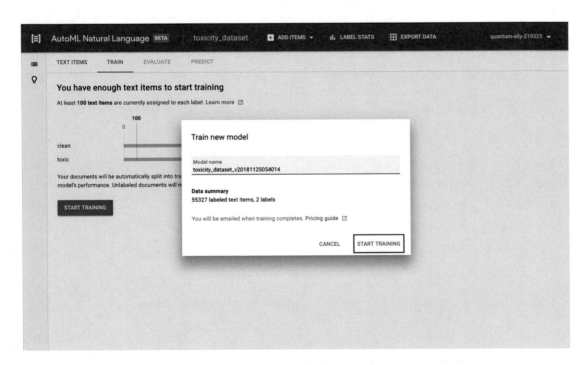

Figure 43-10. *Accept the Model name and click on "Start Training"*

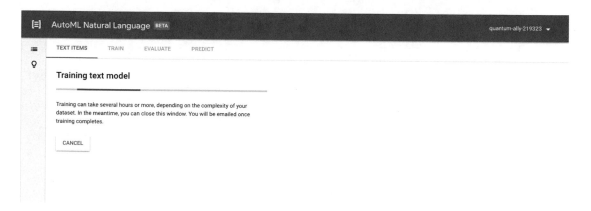

Figure 43-11. *Training the text classification model on Cloud AutoML NLP*

6. The training summary is shown in Figure 43-12. The training
   phase lasted for approximately 4 hours and 45 minutes.

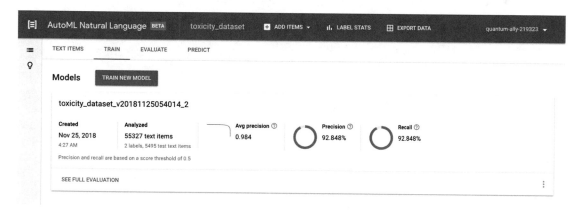

Figure 43-12. *Cloud AutoML NLP: Training summary*

7. AutoML NLP sets aside a portion of the documents as a test set
   in order to evaluate the quality of the model after training (see
   Figure 43-13). The F1 plot shows the trade-off between precision
   and recall. Also, a confusion matrix provides further insight into
   the model quality (see Figure 43-14).

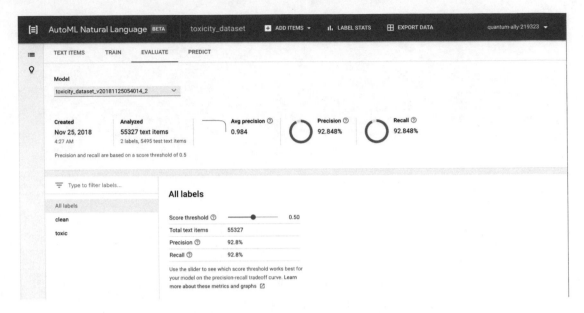

Figure 43-13. *Cloud AutoML NLP: Model evaluation*

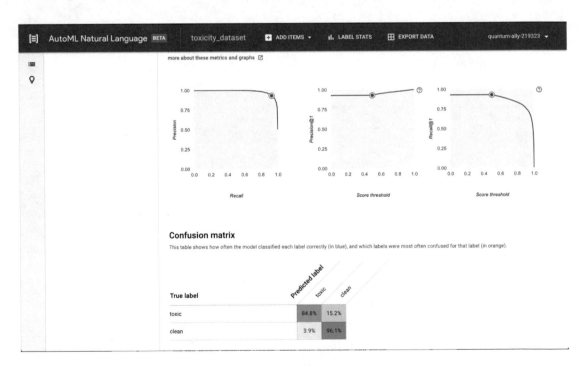

Figure 43-14. *F1 evaluation plot and confusion matrix on Cloud AutoML NLP*

8. The custom text classification model is exposed as a REST or Python API for integration into software applications as a prediction service (see Figure 43-15). We can test our model by uploading a sample image for classification. Figure 43-16 passes a clean text example to the model and it predicts correctly with a probability of 98%, while Figure 43-17 passes a toxic text example to the model. This example is also correctly classified with a probability score of 99%.

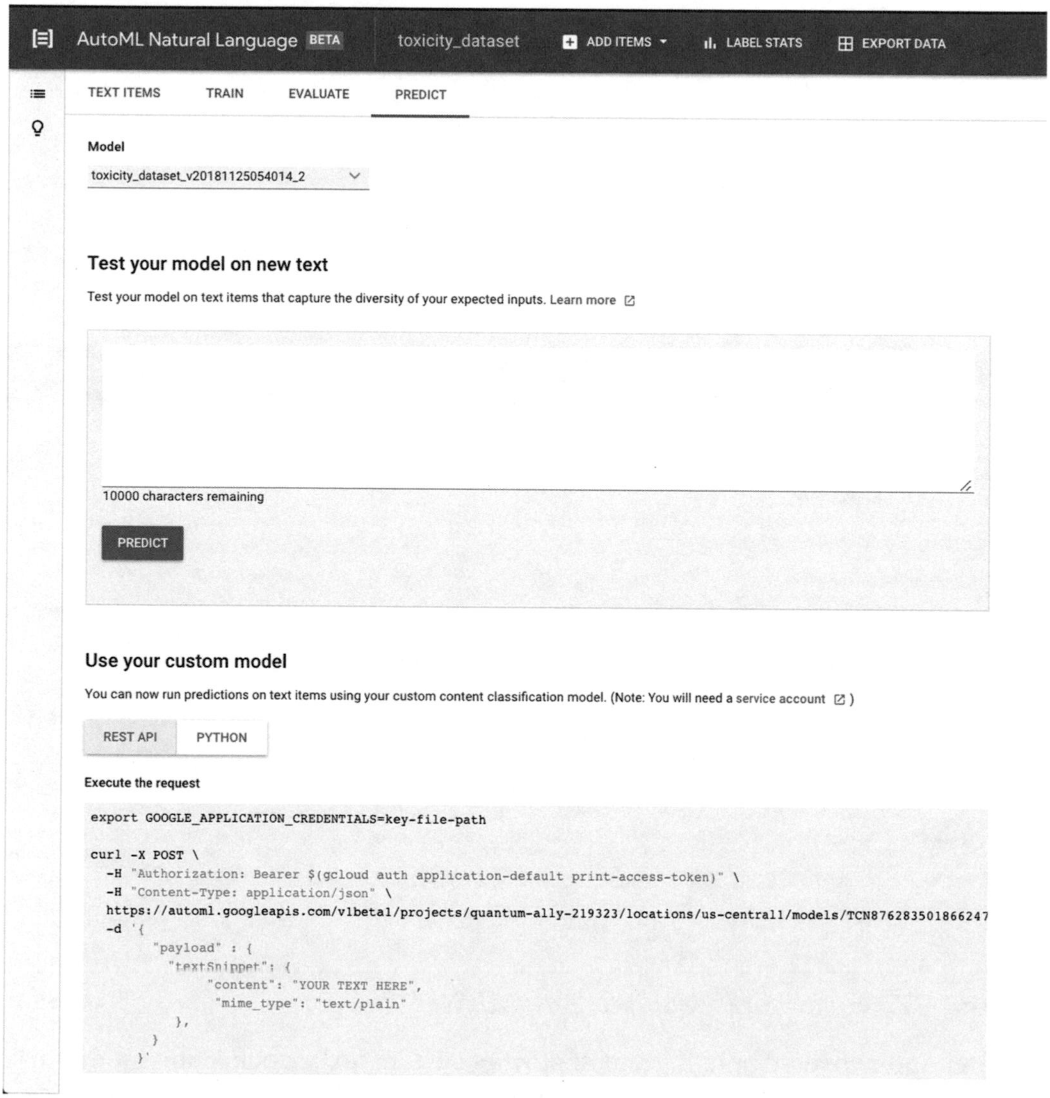

Figure 43-15. *Cloud AutoML NLP model as a prediction service*

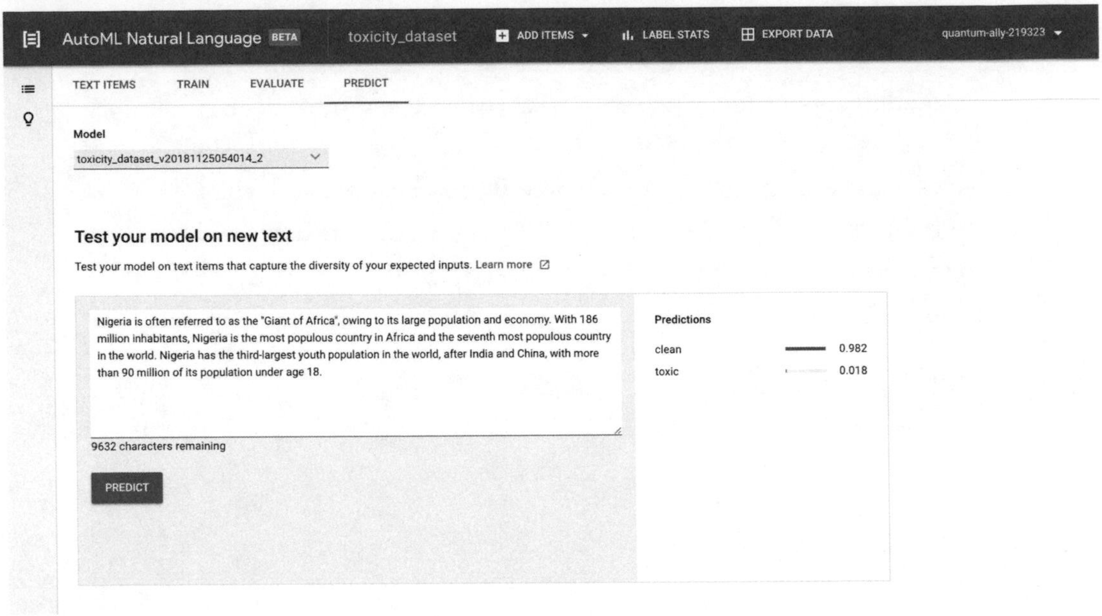

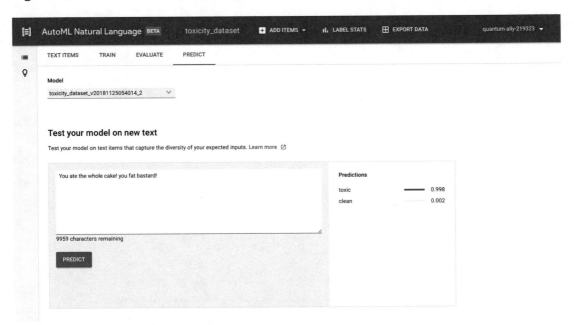

Figure 43-16. *Clean words example: AutoML NLP*

Figure 43-17. *Toxic words example: AutoML NLP*

This chapter covered building and deploying custom text classification models using Google AutoML Cloud Vision. In the next chapter, we will build an end-to-end data science product on GCP.

CHAPTER 44

Model to Predict the Critical Temperature of Superconductors

This chapter builds a regression machine learning model to predict the critical temperature of superconductors. The features for this dataset were derived based on the following superconductor properties:

- Atomic mass

- First ionization energy

- Atomic radius

- Density

- Electron affinity

- Fusion heat

- Thermal conductivity

- Valence

And for each property, the mean, weighted mean, geometric mean, weighted geometric mean, entropy, weighted entropy, range, weighted range, standard deviation, and weighted standard deviation are extracted. Thus, this results in a total number of $8 \times 10 - 80$ features. In addition to this, a feature that contains the number of elements in the superconductor is added to the design matrix. The predictor variable is the critical temperature of the superconductor. Hence, the dataset has a total of 81 features and 21,263 rows.

© Ekaba Bisong 2019
E. Bisong, *Building Machine Learning and Deep Learning Models on Google Cloud Platform*,
https://doi.org/10.1007/978-1-4842-4470-8_44

This dataset is made available by Kam Hamidieh of the University of Pennsylvania and submitted to the UCI Machine Learning Repository. The goal of this section is to demonstrate delivering an end-to-end machine learning modeling pipeline on Google Cloud Platform.

The Modeling Architecture on GCP

The goal of this end-to-end project is to demonstrate building a large-scale learning model on GCP using the components already discussed in this book. The modeling architecture is illustrated in Figure 44-1. Let's briefly explain the connections:

1. Stage the raw data on GCS.

2. Load data into BigQuery for analytics.

3. Exploratory data analysis.

4. Large-scale data processing with Dataflow.

5. Place transformed training and evaluation data on GCS.

6. Train the model on Cloud MLE.

7. Place the trained model output on GCS.

8. Deploy the model for inference on Cloud MLE.

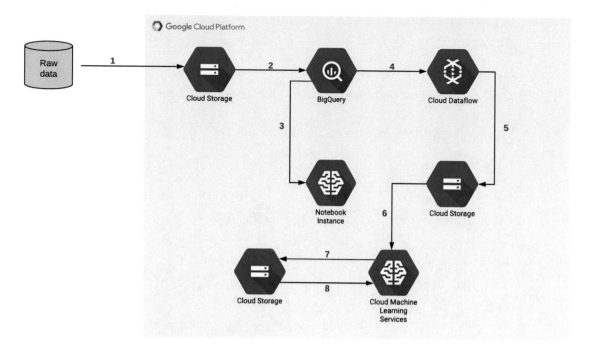

Figure 44-1. *Modeling architecture on GCP*

Stage Raw Data in GCS

Retrieve the raw data from the book code repository for modeling:

- Create a GCS bucket.

```
gsutil mb gs://superconductor
```

- Navigate to the chapter folder and transfer the raw data to GCS.

```
gsutil cp train.csv gs://superconductor/raw-data/
```

Load Data into BigQuery for Analytics

Move the dataset from Google Cloud Storage to BigQuery:

- Create a Dataset in BigQuery.

```
bq mk superconductor
```

- Load raw data from GCS as a Table into the newly created BigQuery
  Dataset.

  ```
  bq --location=US load --autodetect --source_format=CSV super
  conductor.superconductor gs://superconductor/raw-data/train.csv
  ```

- View created Table schema on BigQuery.

  ```
  bq show superconductor.superconductor

  Last modified          Schema               Total Rows    Total Bytes
  Expiration    Time Partitioning    Labels
  -------------  ---------------------  ----------  --------------
  ----------  -------------------  --------
    08 Dec 01:16:51   |- number_of_elements: string
  21264         25582000
                          |- mean_atomic_mass: string
                          |- wtd_mean_atomic_mass: string
                          |- wtd_mean_atomic_radius: string
                          |- gmean_atomic_radius: string
                          |- wtd_gmean_atomic_radius: string
                          |- entropy_atomic_radius: string
                          |- wtd_entropy_atomic_radius: string
                          ...
                          |- range_ThermalConductivity: string
                          |- wtd_range_ThermalConductivity: string
                          |- std_ThermalConductivity: string
                          |- wtd_std_ThermalConductivity: string
                          |- mean_Valence: string
                          |- wtd_std_Valence: string
                          |- critical_temp: string
  ```

Exploratory Data Analysis

The Table in BigQuery contains 21,264 rows. In the interest of speed and rapid iteration, we will not operate on all the rows of this dataset, but rather, we will select a thousand rows for data exploration, transformation, and machine learning spot checking.

```
import pandas as pd
%%bigquery --project ekabasandbox super_cond_df
WITH super_df AS (
SELECT
  number_of_elements, mean_atomic_mass, wtd_mean_atomic_mass,
  gmean_atomic_mass, wtd_gmean_atomic_mass, entropy_atomic_mass,
  wtd_entropy_atomic_mass, range_atomic_mass, wtd_range_atomic_mass,
  std_atomic_mass, wtd_std_atomic_mass, mean_fie, wtd_mean_fie,
  gmean_fie, wtd_gmean_fie, entropy_fie, wtd_entropy_fie, range_fie,
  wtd_range_fie, std_fie, wtd_std_fie, mean_atomic_radius, wtd_mean_atomic_
  radius,
  gmean_atomic_radius, wtd_gmean_atomic_radius, entropy_atomic_radius,
  wtd_entropy_atomic_radius, range_atomic_radius, wtd_range_atomic_radius,
  std_atomic_radius, wtd_std_atomic_radius, mean_Density, wtd_mean_Density,
  gmean_Density, wtd_gmean_Density, entropy_Density, wtd_entropy_Density,
  range_Density, wtd_range_Density, std_Density, wtd_std_Density, mean_
  ElectronAffinity,
  wtd_mean_ElectronAffinity, gmean_ElectronAffinity, wtd_gmean_
  ElectronAffinity
  entropy_ElectronAffinity, wtd_entropy_ElectronAffinity, range_
  ElectronAffinity,
  wtd_range_ElectronAffinity, std_ElectronAffinity, wtd_std_
  ElectronAffinity,
  mean_FusionHeat, wtd_mean_FusionHeat, gmean_FusionHeat, wtd_gmean_
  FusionHeat,
  entropy_FusionHeat, wtd_entropy_FusionHeat, range_FusionHeat,
  wtd_range_FusionHeat, std_FusionHeat, wtd_std_FusionHeat, mean_
  ThermalConductivity,
  wtd_mean_ThermalConductivity, gmean_ThermalConductivity, wtd_gmean_
  ThermalConductivity,
```

```
entropy_ThermalConductivity, wtd_entropy_ThermalConductivity, range_
ThermalConductivity,
wtd_range_ThermalConductivity, std_ThermalConductivity, wtd_std_
ThermalConductivity,
mean_Valence, wtd_mean_Valence, gmean_Valence, wtd_gmean_Valence,
entropy_Valence, wtd_entropy_Valence, range_Valence, wtd_range_Valence,
std_Valence, wtd_std_Valence, critical_temp, ROW_NUMBER() OVER (PARTITION
BY number_of_elements) row_num
FROM
  `superconductor.superconductor` )

SELECT
  *
FROM
  super_df
LIMIT
  1000
# Dataframe shape
super_cond_df.shape
```

Next, we'll explore the dataset to gain more understanding of the features and their relationships. This process is called exploratory data analysis (EDA).

- Check the column datatypes.

```
# check column datatypes
super_cond_df.dtypes
```

```
number_of_elements                  int64
mean_atomic_mass                  float64
wtd_mean_atomic_mass              float64
gmean_atomic_mass                 float64
wtd_gmean_atomic_mass             float64
entropy_atomic_mass               float64
wtd_entropy_atomic_mass           float64
                                    ...
range_Valence                       int64
wtd_range_Valence                 float64
```

```
std_Valence                          float64
wtd_std_Valence                      float64
critical_temp                        float64
row_num                                int64
Length: 82, dtype: object
```

From the results, all the data attributes are of numeric type:

- Next, we will use a tool called **pandas profiling**. This package produces a full range of exploratory data analytics for a Pandas DataFrame object. The result includes summary statistics of the dataset such as the number of variables, number of data observations, and number of missing values (if any). It also includes a histogram visualization for each attribute, descriptive statistics (such as the mean, mode, standard deviation, sum, median absolute deviation, coefficient of variation, kurtosis, and skewness), and quantile statistics (such as minimum value, Q1, median, Q3, maximum, range, and interquartile range). Also, the profile produces multivariate correlation graphs and produces a list of variables that are highly correlated.

  Import the pandas profiling library.

  ```
  # pandas profiling
  import pandas_profiling
  ```

  Run the profile and save the output.

  ```
  # run report
  profile_result = pandas_profiling.ProfileReport(super_cond_df)
  ```

To view the complete report, run the saved output variable:

```
profile_result
```

- Retrieve the rejected variables (i.e, attributes with high correlation).

  ```
  # get rejected variables (i.e, attributes with high correlation)
  rejected_vars = profile_result.get_rejected_variables
  ```

- Filter the dataset columns by removing the variables with high correlation.

```
# filter from attributes set
super_cond_df.drop(rejected_vars(), axis=1, inplace=True)
```

- Next, standardize the dataset values so that they fall within the same scale range (we'll be using Scikit-learn minmax_scale function). Standardizing the values improves the predictive performance of the model because the optimization algorithm can better minimize the cost function.

```
# scale the dataframe values
from sklearn.preprocessing import minmax_scale

dataset = pd.DataFrame(minmax_scale(super_cond_df), columns=
super_cond_df.columns)
```

- Also, the attribute values are normalized so that the distribution more closely resembles a normal or Gaussian distribution. This technique is also noted to have a positive impact on the model performance.

```
# normalize the dataframe
from sklearn.preprocessing import Normalizer

dataset = pd.DataFrame(Normalizer().fit(dataset).
transform(dataset), columns=dataset.columns)
```

- Plot the histogram distribution of the variables (see Figure 44-2).

```
# plot the histogram distribution of the variables
import matplotlib.pyplot as plt

%matplotlib inline

dataset.hist(figsize=(18, 18))
plt.show()
```

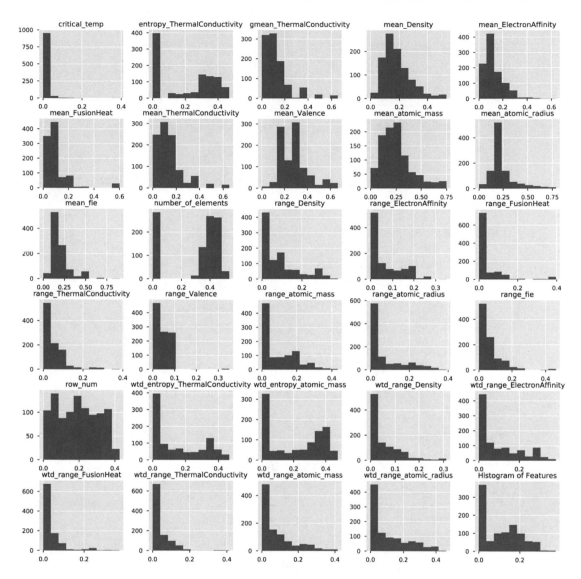

Figure 44-2. *Histogram showing variable distribution*

Spot Checking Machine Learning Algorithms

With our reduced dataset, let's sample a few candidate algorithms to have an idea on their performance and which is more likely to work best for this problem domain. Let's take the following steps:

- The dataset is split into a design matrix and their corresponding label vector.

```
# split features and labels
dataset_y = dataset['critical_temp']
dataset_X = dataset.drop(['critical_temp', 'row_num'], axis=1)
```

- Randomly split the dataset into a training set and a test set.

```
# train-test split
from sklearn.model_selection import train_test_split

X_train, X_test, y_train, y_test = train_test_split(dataset_X,
dataset_y, shuffle=True)
```

- Outline the candidate algorithms to create a model.

```
# spot-check ML algorithms
from sklearn.linear_model import LinearRegression
from sklearn.tree import DecisionTreeRegressor
from sklearn.ensemble import RandomForestRegressor
from sklearn.svm import SVR
from xgboost import XGBRegressor
from sklearn.neural_network import MLPRegressor
from sklearn.metrics import mean_squared_error
from math import sqrt
```

- Create a dictionary of the candidate algorithms.

```
ml_models = {
    'Linear Reg.': LinearRegression(),
    'Dec. Trees': DecisionTreeRegressor(),
    'Rand. Forest': RandomForestRegressor(),
    'SVM': SVR(),
    'XGBoost': XGBRegressor(),
    'NNets': MLPRegressor(warm_start=True, early_stopping=True,
            learning_rate='adaptive')
}
```

- For each candidate algorithm, train with the training set and evaluate on the hold-out test set.

```python
ml_results = {}
for name, model in ml_models.items():
    # fit model on training data
    model.fit(X_train, y_train)
    # make predictions for test data
    prediction = model.predict(X_test)
    # evaluate predictions
    rmse = sqrt(mean_squared_error(y_test, prediction))
    # append accuracy results to dictionary
    ml_results[name] = rmse
    print('RMSE: {} -> {}'.format(name, rmse))
```

```
'Output':
RMSE: SVM -> 0.0748587427887
RMSE: XGBoost -> 0.0222440358318
RMSE: Rand. Forest -> 0.0227742725953
RMSE: Linear Reg. -> 0.025615918858
RMSE: Dec. Trees -> 0.0269103025639
RMSE: NNets -> 0.0289585489638
```

- The plots of the model performances are shown in Figure 44-3.

```python
plt.plot(ml_results.keys(), ml_results.values(), 'o')
plt.title("RMSE estimates for ML algorithms")
plt.xlabel('Algorithms')
plt.ylabel('RMSE')
```

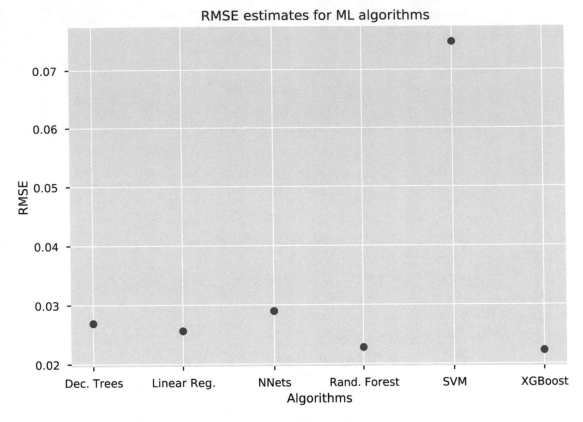

Figure 44-3. *RMSE estimates for ML algorithms*

Dataflow and TensorFlow Transform for Large-Scale Data Processing

In this section, we use Google Cloud Dataflow to carry out large-scale data processing on humongous datasets. Google Dataflow as earlier discussed is a serverless, parallel, and distributed infrastructure for running jobs for batch and stream data processing. Dataflow is a vital component in architecting a production pipeline for building and deploying large-scale machine learning products. In conjunction with Cloud Dataflow, we use TensorFlow Transform (TFT), a library built for preprocessing with Tensorflow. The goal of using TFT is to have a consistent set of transformation operations applied to the dataset when the model is trained and when it is served or deployed for consumption. In the following steps, each code block is executed in a Notebook cell:

- Import the relevant libraries. Remember that Apache Beam (as of now) only supports Python 2. Moreso, TFT only works with a specific combination of Tensorflow and Apache Beam packages. In this case, TFT 0.8.0 works with TF 1.8 and Apache Beam [GCP] 2.5.0. After importing the libraries, be sure to **restart the Notebook kernel**.

At this point, change the Notebook runtime type to Python 2.0.

```bash
%%bash
source activate py2env
pip install --upgrade tensorflow
pip install --upgrade apache-beam[gcp]
pip install --upgrade tensorflow_transform==0.8.0
apt-get install libsnappy-dev
pip install --upgrade python-snappy==0.5.1
```

Restart the kernel after you do a pip install.

- Connect to GCP.

```python
from google.colab import auth
auth.authenticate_user()
print('Authenticated')

# configure GCP project - update with your parameters
project_id = 'ekabasandbox'
bucket_name = 'superconductor'
region = 'us-central1'
tf_version = '1.8'

# configure gcloud
!gcloud config set project {project_id}
!gcloud config set compute/region {region}
```

- Create query method for retrieving training and testing datasets from BigQuery.

```python
def create_query(phase, EVERY_N=None):
    """
```

```
        EVERY_N: Integer. Sample one out of every N rows from the full
        dataset. Larger values will yield smaller sample
        phase: 1=train 2=valid
        """

        base_query = """
        WITH super_df AS (
          SELECT
            number_of_elements, mean_atomic_mass, wtd_mean_atomic_mass,
            gmean_atomic_mass, wtd_gmean_atomic_mass, entropy_atomic_
            mass,
            wtd_entropy_atomic_mass, range_atomic_mass, wtd_range_
            atomic_mass,
            std_atomic_mass, wtd_std_atomic_mass, mean_fie, wtd_mean_fie,
            gmean_fie, wtd_gmean_fie, entropy_fie, wtd_entropy_fie,
            range_fie,
            wtd_range_fie, std_fie, wtd_std_fie, mean_atomic_radius,
            wtd_mean_atomic_radius,
            gmean_atomic_radius, wtd_gmean_atomic_radius, entropy_
            atomic_radius,
            wtd_entropy_atomic_radius, range_atomic_radius, wtd_range_
            atomic_radius,
            std_atomic_radius, wtd_std_atomic_radius, mean_Density,
            wtd_mean_Density,
            gmean_Density, wtd_gmean_Density, entropy_Density, wtd_
            entropy_Density,
            range_Density, wtd_range_Density, std_Density, wtd_std_
            Density, mean_ElectronAffinity,
            wtd_mean_ElectronAffinity, gmean_ElectronAffinity, wtd_
            gmean_ElectronAffinity
            entropy_ElectronAffinity, wtd_entropy_ElectronAffinity,
            range_ElectronAffinity,
            wtd_range_ElectronAffinity, std_ElectronAffinity, wtd_std_
            ElectronAffinity,
            mean_FusionHeat, wtd_mean_FusionHeat, gmean_FusionHeat,
            wtd_gmean_FusionHeat,
```

```
        entropy_FusionHeat, wtd_entropy_FusionHeat, range_FusionHeat,
        wtd_range_FusionHeat, std_FusionHeat, wtd_std_FusionHeat,
        mean_ThermalConductivity,
        wtd_mean_ThermalConductivity, gmean_ThermalConductivity,
        wtd_gmean_ThermalConductivity,
        entropy_ThermalConductivity, wtd_entropy_
        ThermalConductivity, range_ThermalConductivity,
        wtd_range_ThermalConductivity, std_ThermalConductivity,
        wtd_std_ThermalConductivity,
        mean_Valence, wtd_mean_Valence, gmean_Valence, wtd_gmean_
        Valence,
        entropy_Valence, wtd_entropy_Valence, range_Valence, wtd_
        range_Valence,
        std_Valence, wtd_std_Valence, critical_temp, ROW_NUMBER()
        OVER (PARTITION BY number_of_elements) row_num
    FROM
        `superconductor.superconductor`)

    SELECT
        *
    FROM
        super_df
"""

if EVERY_N == None:
    if phase < 2:
        # training
        query = "{0} WHERE MOD(row_num,4) < 2".format(base_
        query)
    else:
        query = "{0} WHERE MOD(row_num,4) = {1}".format(base_
        query, phase)
else:
    query = "{0} WHERE MOD(row_num,{1}) = {2}".format(base_
    query, EVERY_N, phase)

return query
```

- Create requirements.txt file to install dependencies (in this case tensorflow_transform) on Dataflow worker machines.

```
%%writefile requirements.txt
tensorflow-transform==0.8.0
```

- The following code block uses Apache Beam to build a data preprocessing pipeline to transform the raw dataset into a form suitable for building a predictive model. The transformation is the same procedure as done earlier with the reduced dataset, which included removing columns that had a high correlation and scaling the dataset numeric values to be within the same range. The output of the preprocessing pipeline produces a training set and an evaluation set. The Beam pipeline also uses TensorFlow Transform to save the metadata (both raw and processed) of the data transformation, as well as the transformed graph which can later be used as part of the serving function of the deployed model. We made this example to include the use of TensorFlow Transform for reference purposes.

```
import datetime
import snappy
import tensorflow as tf
import apache_beam as beam
import tensorflow_transform as tft
from tensorflow_transform.beam import impl as beam_impl

def get_table_header(projection_fields):
    header = "
    for cnt, val in enumerate(projection_fields):
        if cnt > 0:
            header+=','+val
        else:
            header+=val
    return header

def preprocess_tft(inputs):
    result = {}
```

```
    for attr, value in inputs.items():
        result[attr] = tft.scale_to_0_1(value)

    return result

def cleanup(rowdict):
    # pull columns from BQ and create a line
    CSV_COLUMNS = 'number_of_elements,mean_atomic_mass,entropy_
    atomic_mass,wtd_entropy_atomic_mass,range_atomic_mass,wtd_
    range_atomic_mass,mean_fie,wtd_mean_fie,wtd_entropy_
    fie,range_fie,wtd_range_fie,mean_atomic_radius,wtd_mean_
    atomic_radius,range_atomic_radius,wtd_range_atomic_
    radius,mean_Density,entropy_Density,wtd_entropy_Density,range_
    Density,wtd_range_Density,mean_ElectronAffinity,wtd_
    entropy_ElectronAffinity,range_ElectronAffinity,wtd_range_
    ElectronAffinity,mean_FusionHeat,gmean_FusionHeat,entropy_
    FusionHeat,wtd_entropy_FusionHeat,range_FusionHeat,wtd_
    range_FusionHeat,mean_ThermalConductivity,wtd_mean_
    ThermalConductivity,gmean_ThermalConductivity,entropy_
    ThermalConductivity,wtd_entropy_ThermalConductivity,
    range_ThermalConductivity,wtd_range_ThermalConductivity,
    mean_Valence,wtd_mean_Valence,range_Valence,wtd_range_
    Valence,wtd_std_Valence,critical_temp'.split(',')

    def tofloat(value, ifnot):
        try:
            return float(value)
        except (ValueError, TypeError):
            return ifnot

    result = {
      k : tofloat(rowdict[k], -99) if k in rowdict else -99 for k
      in CSV_COLUMNS
    }

    row = ('{}'+',{}'*(len(result)-1)).format(result['number_of_
    elements'],result['mean_atomic_mass'],
```

```
            result['entropy_atomic_mass'], result['wtd_entropy_atomic_
            mass'],result['range_atomic_mass'],
            result['wtd_range_atomic_mass'],result['mean_fie'],
            result['wtd_mean_fie'],
            result['wtd_entropy_fie'],result['range_fie'],result['wtd_
            range_fie'],
            result['mean_atomic_radius'],result['wtd_mean_atomic_radius'],
            result['range_atomic_radius'],result['wtd_range_atomic_
            radius'],result['mean_Density'],
            result['entropy_Density'],result['wtd_entropy_Density'],
            result['range_Density'],
            result['wtd_range_Density'],result['mean_ElectronAffinity'],
            result['wtd_entropy_ElectronAffinity'],result['range_
            ElectronAffinity'],
            result['wtd_range_ElectronAffinity'],result['mean_
            FusionHeat'],result['gmean_FusionHeat'],
            result['entropy_FusionHeat'],result['wtd_entropy_
            FusionHeat'],result['range_FusionHeat'],
            result['wtd_range_FusionHeat'],result['mean_
            ThermalConductivity'],
            result['wtd_mean_ThermalConductivity'],result['gmean_
            ThermalConductivity'],
            result['entropy_ThermalConductivity'],result['wtd_entropy_
            ThermalConductivity'],
            result['range_ThermalConductivity'],result['wtd_range_
            ThermalConductivity'],
            result['mean_Valence'],result['wtd_mean_Valence'],
            result['range_Valence'],
            result['wtd_range_Valence'],result['wtd_std_Valence'],
            result['critical_temp'])
        yield row

def preprocess():
    import os
    import os.path
    import datetime
```

```
from apache_beam.io import WriteToText
from apache_beam.io import tfrecordio
from tensorflow_transform.coders import example_proto_coder
from tensorflow_transform.tf_metadata import dataset_metadata
from tensorflow_transform.tf_metadata import dataset_schema
from tensorflow_transform.beam import tft_beam_io
from tensorflow_transform.beam.tft_beam_io import transform_
fn_io

job_name = 'preprocess-features' + '-' + datetime.datetime.
now().strftime('%y%m%d-%H%M%S')

print 'Launching Dataflow job {} ... hang on'.format(job_name)
OUTPUT_DIR = 'gs://{0}/preproc_csv/'.format(bucket_name)
import subprocess
subprocess.call('gsutil rm -r {}'.format(OUTPUT_DIR).split())
EVERY_N = 3

options = {
    'staging_location': os.path.join(OUTPUT_DIR, 'tmp', 'staging'),
    'temp_location': os.path.join(OUTPUT_DIR, 'tmp'),
    'job_name': job_name,
    'project': project_id,
    'max_num_workers': 24,
    'teardown_policy': 'TEARDOWN_ALWAYS',
    'no_save_main_session': True,
    'requirements_file': 'requirements.txt'
}
opts = beam.pipeline.PipelineOptions(flags=[], **options)
RUNNER = 'DataflowRunner'

# set up metadata
raw_data_schema = {
    colname : dataset_schema.ColumnSchema(tf.float32, [],
    dataset_schema.FixedColumnRepresentation())
                    for colname in 'number_of_elements,mean_atomic_
                    mass,entropy_atomic_mass,wtd_entropy_atomic_
                    mass,range_atomic_mass,wtd_range_atomic_mass,
```

```
                      mean_fie,wtd_mean_fie,wtd_entropy_fie,range_
                      fie,wtd_range_fie,mean_atomic_radius,wtd_
                      mean_atomic_radius,range_atomic_radius,wtd_
                      range_atomic_radius,mean_Density,entropy_
                      Density,wtd_entropy_Density,range_Density,
                      wtd_range_Density,mean_ElectronAffinity,wtd_
                      entropy_ElectronAffinity,range_
                      ElectronAffinity,wtd_range_ElectronAffinity,
                      mean_FusionHeat,gmean_FusionHeat,entropy_
                      FusionHeat,wtd_entropy_FusionHeat,range_
                      FusionHeat,wtd_range_FusionHeat,
                      mean_ThermalConductivity,
                      wtd_mean_ThermalConductivity,
                      gmean_ThermalConductivity,
                      entropy_ThermalConductivity,
                      wtd_entropy_ThermalConductivity,
                      range_ThermalConductivity,wtd_range_
                      ThermalConductivity,mean_Valence,wtd_mean_
                      Valence,range_Valence,wtd_range_Valence,wtd_
                      std_Valence,critical_temp'.split(',')
}
raw_data_metadata = dataset_metadata.DatasetMetadata(dataset_
schema.Schema(raw_data_schema))

# run Beam
with beam.Pipeline(RUNNER, options=opts) as p:
    with beam_impl.Context(temp_dir=os.path.join
    (OUTPUT_DIR, 'tmp')):
        # save the raw data metadata
        _ = (raw_data_metadata
          | 'WriteInputMetadata' >> tft_beam_io.WriteMetadata(
              os.path.join(OUTPUT_DIR, 'metadata/rawdata_
              metadata'),
              pipeline=p))

        projection_fields = ['number_of_elements',
        'mean_atomic_mass', 'entropy_atomic_mass',
```

```
'wtd_entropy_atomic_mass',
'range_atomic_mass',
'wtd_range_atomic_mass', 'mean_
fie', 'wtd_mean_fie',
'wtd_entropy_fie', 'range_fie',
'wtd_range_fie',
'mean_atomic_radius', 'wtd_mean_
atomic_radius',
'range_atomic_radius', 'wtd_
range_atomic_radius', 'mean_
Density',
'entropy_Density', 'wtd_entropy_
Density', 'range_Density',
'wtd_range_Density', 'mean_
ElectronAffinity',
'wtd_entropy_ElectronAffinity',
'range_ElectronAffinity',
'wtd_range_ElectronAffinity',
'mean_FusionHeat', 'gmean_
FusionHeat',
'entropy_FusionHeat', 'wtd_
entropy_FusionHeat', 'range_
FusionHeat',
'wtd_range_FusionHeat', 'mean_
ThermalConductivity',
'wtd_mean_ThermalConductivity',
'gmean_ThermalConductivity',
'entropy_ThermalConductivity',
'wtd_entropy_
ThermalConductivity',
'range_ThermalConductivity',
'wtd_range_ThermalConductivity',
'mean_Valence', 'wtd_mean_
Valence', 'range_Valence',
'wtd_range_Valence', 'wtd_std_
Valence', 'critical_temp']
```

```
header = get_table_header(projection_fields)

# analyze and transform training
raw_data = (p
  | 'train_read' >> beam.io.Read(beam.
  io.BigQuerySource(query=create_query(1, EVERY_N),
  use_standard_sql=True)))

raw_dataset = (raw_data, raw_data_metadata)
transformed_dataset, transform_fn = (
    raw_dataset | beam_impl.AnalyzeAndTransformDataset
    (preprocess_tft))
transformed_data, transformed_metadata = transformed_
dataset

_ = (transformed_data
    | 'train_filter' >> beam.FlatMap(cleanup)
    | 'WriteTrainData' >> beam.io.Write(beam.
      io.WriteToText(
        file_path_prefix=os.path.join(OUTPUT_DIR,
        'data', 'train'),
        file_name_suffix=".csv",
        shard_name_template="-SS-of-NN",
        header=header,
        num_shards=1)))

# transform eval data
raw_test_data = (p
  | 'eval_read' >> beam.io.Read(beam.
    io.BigQuerySource(query=create_query(2, EVERY_N),
    use_standard_sql=True)))

raw_test_dataset = (raw_test_data, raw_data_metadata)
transformed_test_dataset = (
    (raw_test_dataset, transform_fn) | beam_impl.
    TransformDataset())
transformed_test_data, _ = transformed_test_dataset
```

```
_ = (transformed_test_data
        | 'eval_filter' >> beam.FlatMap(cleanup)
        | 'WriteTestData' >> beam.io.Write(beam.
          io.WriteToText(
                file_path_prefix=os.path.join(OUTPUT_DIR,
                'data', 'eval'),
                file_name_suffix=".csv",
                shard_name_template="-SS-of-NN",
                num_shards=1)))

_ = (transform_fn
        | 'WriteTransformFn' >>
        transform_fn_io.WriteTransformFn(os.path.
        join(OUTPUT_DIR, 'metadata')))

preprocess()
```

- The Dataflow pipeline graph is shown in Figure 44-4.

Figure 44-4. *Dataflow pipeline graph*

Training on Cloud MLE

The following code example will train the processed datasets on Google Cloud MLE. At this point, change the Notebook runtime type to Python 3.0.

- Configure GCP project.

```
# configure GCP project - update with your parameters
project_id = 'ekabasandbox'
bucket_name = 'superconductor'
region = 'us-central1'
tf_version = '1.8'

import os
os.environ['bucket_name'] = bucket_name
os.environ['tf_version'] = tf_version
os.environ['project_id'] = project_id
os.environ['region'] = region
```

- Create directory "trainer".

```
# create directory trainer
import os
try:
    os.makedirs('./trainer')
    print('directory created')
except OSError:
    print('could not create directory')
```

- Create file __init__.py.

```
%%writefile trainer/__init__.py
```

- Create the trainer file task.py. Replace the bucket name with your values.

```
%%writefile trainer/task.py
import argparse
import json
import os
```

```python
import tensorflow as tf
from tensorflow.contrib.training.python.training import hparam

import trainer.model as model

def _get_session_config_from_env_var():
    """Returns a tf.ConfigProto instance that has appropriate
    device_filters set.
    """

    tf_config = json.loads(os.environ.get('TF_CONFIG', '{}'))

    if (tf_config and 'task' in tf_config and 'type' in tf_
    config['task'] and
        'index' in tf_config['task']):
        # Master should only communicate with itself and ps
        if tf_config['task']['type'] == 'master':
            return tf.ConfigProto(device_filters=['/job:ps', '/
            job:master'])
        # Worker should only communicate with itself and ps
        elif tf_config['task']['type'] == 'worker':
            return tf.ConfigProto(device_filters=[
                '/job:ps',
                '/job:worker/task:%d' % tf_config['task']['index']
            ])
    return None

def train_and_evaluate(hparams):
    """Run the training and evaluate using the high level API."""

    train_input = lambda: model.input_fn(
        tf.gfile.Glob(hparams.train_files),
        num_epochs=hparams.num_epochs,
        batch_size=hparams.train_batch_size
    )

    # Don't shuffle evaluation data
    eval_input = lambda: model.input_fn(
        tf.gfile.Glob(hparams.eval_files),
```

```python
            batch_size=hparams.eval_batch_size,
            shuffle=False
        )

    train_spec = tf.estimator.TrainSpec(
        train_input, max_steps=hparams.train_steps)

    exporter = tf.estimator.FinalExporter(
        'superconductor', model.SERVING_FUNCTIONS[hparams.export_
        format])
    eval_spec = tf.estimator.EvalSpec(
        eval_input,
        steps=hparams.eval_steps,
        exporters=[exporter],
        name='superconductor-eval')

    run_config = tf.estimator.RunConfig(
        session_config=_get_session_config_from_env_var())
    run_config = run_config.replace(model_dir=hparams.job_dir)
    print('Model dir %s' % run_config.model_dir)
    estimator = model.build_estimator(
        learning_rate=hparams.learning_rate,
        # Construct layers sizes with exponential decay
        hidden_units=[
            max(2, int(hparams.first_layer_size * hparams.scale_
            factor**i))
            for i in range(hparams.num_layers)
        ],
        config=run_config,
        output_dir=hparams.output_dir)

    tf.estimator.train_and_evaluate(estimator, train_spec, eval_spec)

if __name__ == '__main__':
    parser = argparse.ArgumentParser()
    # Input Arguments
    parser.add_argument(
        '--train-files',
```

```python
        help='GCS file or local paths to training data',
        nargs='+',
        # update the bucket name
        default='gs://{}/preproc_csv/data/{}*{}*'.format('super
        conductor', tf.estimator.ModeKeys.TRAIN, 'of'))
parser.add_argument(
        '--eval-files',
        help='GCS file or local paths to evaluation data',
        nargs='+',
        # update the bucket name
        default='gs://{}/preproc_csv/data/{}*{}*'.format('super
        conductor', tf.estimator.ModeKeys.EVAL, 'of'))
parser.add_argument(
        '--job-dir',
        help='GCS location to write checkpoints and export models',
        default='/tmp/superconductor-estimator')
parser.add_argument(
        '--num-epochs',
        help="""\
        Maximum number of training data epochs on which to train.
        If both --max-steps and --num-epochs are specified,
        the training job will run for --max-steps or --num-epochs,
        whichever occurs first. If unspecified will run for
        --max-steps.\
        """,
        type=int)
parser.add_argument(
        '--train-batch-size',
        help='Batch size for training steps',
        type=int,
        default=20)
parser.add_argument(
        '--eval-batch-size',
        help='Batch size for evaluation steps',
        type=int,
        default=20)
```

```
    parser.add_argument(
        '--learning-rate',
        help='The training learning rate',
        default=1e-4,
        type=float)
    parser.add_argument(
        '--first-layer-size',
        help='Number of nodes in the first layer of the DNN',
        default=256,
        type=int)
    parser.add_argument(
        '--num-layers', help='Number of layers in the DNN',
        default=3, type=int)
    parser.add_argument(
        '--scale-factor',
        help='How quickly should the size of the layers in the DNN
        decay',
        default=0.7,
        type=float)
    parser.add_argument(
        '--train-steps',
        help="""\
        Steps to run the training job for. If --num-epochs is not
        specified,
        this must be. Otherwise the training job will run
        indefinitely.\
        """,
        default=100,
        type=int)
    parser.add_argument(
        '--eval-steps',
        help='Number of steps to run evalution for at each
        checkpoint',
        default=100,
        type=int)
```

```python
    parser.add_argument(
        '--export-format',
        help='The input format of the exported SavedModel binary',
        choices=['JSON', 'CSV', 'EXAMPLE'],
        default='CSV')
    parser.add_argument(
        '--output-dir',
        help='Location of the exported model',
        nargs='+')
    parser.add_argument(
        '--verbosity',
        choices=['DEBUG', 'ERROR', 'FATAL', 'INFO', 'WARN'],
        default='INFO')

    args, _ = parser.parse_known_args()

    # Set python level verbosity
    tf.logging.set_verbosity(args.verbosity)
    # Set C++ Graph Execution level verbosity
    os.environ['TF_CPP_MIN_LOG_LEVEL'] = str(
        tf.logging.__dict__[args.verbosity] / 10)

    # Run the training job
    hparams = hparam.HParams(**args.__dict__)
    train_and_evaluate(hparams)
```

- Create the file model.py that contains the model code.

```python
%%writefile trainer/model.py
import six

import tensorflow as tf
from tensorflow.python.estimator.model_fn import ModeKeys as Modes

# Define the format of your input data including unused columns.
CSV_COLUMNS = [
    'number_of_elements', 'mean_atomic_mass', 'entropy_atomic_mass',
    'wtd_entropy_atomic_mass', 'range_atomic_mass',
```

```
        'wtd_range_atomic_mass', 'mean_fie', 'wtd_mean_fie',
        'wtd_entropy_fie', 'range_fie', 'wtd_range_fie',
        'mean_atomic_radius', 'wtd_mean_atomic_radius',
        'range_atomic_radius', 'wtd_range_atomic_radius', 'mean_
        Density',
        'entropy_Density', 'wtd_entropy_Density', 'range_Density',
        'wtd_range_Density', 'mean_ElectronAffinity',
        'wtd_entropy_ElectronAffinity', 'range_ElectronAffinity',
        'wtd_range_ElectronAffinity', 'mean_FusionHeat', 'gmean_
        FusionHeat',
        'entropy_FusionHeat', 'wtd_entropy_FusionHeat', 'range_
        FusionHeat',
        'wtd_range_FusionHeat', 'mean_ThermalConductivity',
        'wtd_mean_ThermalConductivity', 'gmean_ThermalConductivity',
        'entropy_ThermalConductivity', 'wtd_entropy_
        ThermalConductivity',
        'range_ThermalConductivity', 'wtd_range_ThermalConductivity',
        'mean_Valence', 'wtd_mean_Valence', 'range_Valence',
        'wtd_range_Valence', 'wtd_std_Valence', 'critical_temp'
    ]

CSV_COLUMN_DEFAULTS = [[0.0] for i in range(0, len(CSV_COLUMNS))]
LABEL_COLUMN = 'critical_temp'

# Define the initial ingestion of each feature used by your model.
# Additionally, provide metadata about the feature.
INPUT_COLUMNS = [tf.feature_column.numeric_column(i) for i in CSV_
COLUMNS[:-1]]

UNUSED_COLUMNS = set(CSV_COLUMNS) - {col.name for col in INPUT_
COLUMNS} - \
    {LABEL_COLUMN}

def build_estimator(config, output_dir, hidden_units=None,
learning_rate=None):
    """

    Deep NN Regression model.
```

```
Args:
    config: (tf.contrib.learn.RunConfig) defining the runtime
    environment for
      the estimator (including model_dir).
    hidden_units: [int], the layer sizes of the DNN (input
    layer first)
    learning_rate: (int), the learning rate for the optimizer.
Returns:
    A DNNRegressor
"""

(number_of_elements,mean_atomic_mass,entropy_atomic_mass,wtd_
entropy_atomic_mass, \
  range_atomic_mass,wtd_range_atomic_mass,mean_fie,wtd_mean_
  fie,wtd_entropy_fie,range_fie,\
  wtd_range_fie,mean_atomic_radius,wtd_mean_atomic_
  radius,range_atomic_radius,wtd_range_atomic_radius,\
  mean_Density,entropy_Density,wtd_entropy_Density,range_
  Density,wtd_range_Density,mean_ElectronAffinity,\
  wtd_entropy_ElectronAffinity,range_ElectronAffinity,wtd_
  range_ElectronAffinity,mean_FusionHeat,\
  gmean_FusionHeat,entropy_FusionHeat,wtd_entropy_
  FusionHeat,range_FusionHeat,wtd_range_FusionHeat,\
  mean_ThermalConductivity,wtd_mean_ThermalConductivity,gmean_
  ThermalConductivity,entropy_ThermalConductivity,\
  wtd_entropy_ThermalConductivity,range_
  ThermalConductivity,wtd_range_ThermalConductivity,mean_
  Valence,\
  wtd_mean_Valence,range_Valence,wtd_range_Valence,wtd_std_
  Valence) = INPUT_COLUMNS

columns = [number_of_elements,mean_atomic_mass,entropy_atomic_
mass,wtd_entropy atomic_mass, \
  range_atomic_mass,wtd_range_atomic_mass,mean_fie,wtd_mean_
  fie,wtd_entropy_fie,range_fie,\
  wtd_range_fie,mean_atomic_radius,wtd_mean_atomic_
  radius,range_atomic_radius,wtd_range_atomic_radius,\
```

```
        mean_Density,entropy_Density,wtd_entropy_Density,range_
        Density,wtd_range_Density,mean_ElectronAffinity,\
        wtd_entropy_ElectronAffinity,range_ElectronAffinity,wtd_
        range_ElectronAffinity,mean_FusionHeat,\
        gmean_FusionHeat,entropy_FusionHeat,wtd_entropy_FusionHeat,
        range_FusionHeat,wtd_range_FusionHeat,\
        mean_ThermalConductivity,wtd_mean_ThermalConductivity,
        gmean_ThermalConductivity,entropy_ThermalConductivity,\
        wtd_entropy_ThermalConductivity,range_ThermalConductivity,
        wtd_range_ThermalConductivity,mean_Valence,\
        wtd_mean_Valence,range_Valence,wtd_range_Valence,wtd_std_
        Valence]

    estimator = tf.estimator.DNNRegressor(
      model_dir=output_dir,
      config=config,
      feature_columns=columns,
      hidden_units=hidden_units or [256, 128, 64],
      optimizer=tf.train.AdamOptimizer(learning_rate)
    )

    # add extra evaluation metric for hyperparameter tuning
    estimator = tf.contrib.estimator.add_metrics(estimator, add_
    eval_metrics)
    return estimator

def add_eval_metrics(labels, predictions):
    pred_values = predictions['predictions']
    return {
        'rmse': tf.metrics.root_mean_squared_error(labels,
        pred_values)
    }

# [START serving-function]

def csv_serving_input_fn():
    """Build the serving inputs."""
    csv_row = tf.placeholder(shape=[None], dtype=tf.string)
```

```python
    features = _decode_csv(csv_row)
    # Ignore label column
    features.pop(LABEL_COLUMN)
    return tf.estimator.export.ServingInputReceiver(features,
                                        {'csv_row': csv_row})
def example_serving_input_fn():
    """Build the serving inputs."""
    example_bytestring = tf.placeholder(
        shape=[None],
        dtype=tf.string,
    )
    features = tf.parse_example(
        example_bytestring,
        tf.feature_column.make_parse_example_spec(INPUT_COLUMNS))
    return tf.estimator.export.ServingInputReceiver(
        features, {'example_proto': example_bytestring})

def json_serving_input_fn():
    """Build the serving inputs."""
    inputs = {}
    for feat in INPUT_COLUMNS:
        inputs[feat.name] = tf.placeholder(shape=[None],
        dtype=feat.dtype)

    return tf.estimator.export.ServingInputReceiver(inputs, inputs)

# [END serving-function]

SERVING_FUNCTIONS = {
    'JSON': json_serving_input_fn,
    'EXAMPLE': example_serving_input_fn,
    'CSV': csv_serving_input_fn
}

def _decode_csv(line):
    """Takes the string input tensor and returns a dict of rank-2
    tensors."""
```

```
    # Takes a rank-1 tensor and converts it into rank-2 tensor
    row_columns = tf.expand_dims(line, -1)
    columns = tf.decode_csv(row_columns, record_defaults=CSV_
    COLUMN_DEFAULTS)
    features = dict(zip(CSV_COLUMNS, columns))

    # Remove unused columns
    for col in UNUSED_COLUMNS:
        features.pop(col)
    return features

def input_fn(filenames, num_epochs=None, shuffle=True, skip_
header_lines=1, batch_size=200):
    """Generates features and labels for training or evaluation.
    This uses the input pipeline based approach using file name queue
    to read data so that entire data is not loaded in memory.
    Args:
        filenames: [str] A List of CSV file(s) to read data from.
        num_epochs: (int) how many times through to read the data.
        If None will loop through data indefinitely
        shuffle: (bool) whether or not to randomize the order of
        data. Controls randomization of both file order and line
        order within files.
        skip_header_lines: (int) set to non-zero in order to skip
        header lines in CSV files.
        batch_size: (int) First dimension size of the Tensors
        returned by input_fn
    Returns:
        A (features, indices) tuple where features is a dictionary of
            Tensors, and indices is a single Tensor of label indices.
    """
    dataset = tf.data.TextLineDataset(filenames).skip(skip_header_
    lines).map(
        _decode_csv)
```

```
if shuffle:
    dataset = dataset.shuffle(buffer_size=batch_size * 10)
iterator = dataset.repeat(num_epochs).batch(
    batch_size).make_one_shot_iterator()
features = iterator.get_next()
return features, features.pop(LABEL_COLUMN)
```

- Create the hyper-parameter config file.

```
%%writefile hptuning_config.yaml
trainingInput:
  hyperparameters:
    hyperparameterMetricTag: rmse
    goal: MINIMIZE
    maxTrials: 4 #20
    maxParallelTrials: 2 #5
    enableTrialEarlyStopping: True
    algorithm: RANDOM_SEARCH
    params:
      - parameterName: learning-rate
        type: DOUBLE
        minValue: 0.00001
        maxValue: 0.005
        scaleType: UNIT_LOG_SCALE
      - parameterName: first-layer-size
        type: INTEGER
        minValue: 50
        maxValue: 500
        scaleType: UNIT_LINEAR_SCALE
      - parameterName: num-layers
        type: INTEGER
        minValue: 1
        maxValue: 15
        scaleType: UNIT_LINEAR_SCALE
      - parameterName: scale-factor
        type: DOUBLE
```

```
                minValue: 0.1
                maxValue: 1.0
                scaleType: UNIT_REVERSE_LOG_SCALE
```

- The following code executes the training job on Cloud MLE.

```
%%bash
JOB_NAME=superconductor_$(date -u +%y%m%d_%H%M%S)
HPTUNING_CONFIG=hptuning_config.yaml
GCS_JOB_DIR=gs://$bucket_name/jobs/$JOB_NAME

echo $GCS_JOB_DIR

gcloud ai-platform jobs submit training $JOB_NAME \
                                --stream-logs \
                                --runtime-version $tf_version \
                                --job-dir $GCS_JOB_DIR \
                                --module-name trainer.task \
                                --package-path trainer/ \
                                --region us-central1 \
                                --scale-tier=STANDARD_1 \
                                --config $HPTUNING_CONFIG \
                                -- \
                                --train-steps 5000 \
                                --eval-steps 100
```

```
gs://superconductor/jobs/superconductor_181222_040429
endTime: '2018-12-22T04:24:50'
jobId: superconductor_181222_040429
startTime: '2018-12-22T04:04:35'
state: SUCCEEDED
```

- Cloud MLE training output is shown in Figure 44-5.

Figure 44-5. *Cloud MLE training output*

Deploy Trained Model

The best model trial with the lowest **objectiveValue** is deployed for inference on Cloud MLE:

- Display content of selected trained model directory.

```
%%bash
gsutil ls gs://${BUCKET}/jobs/superconductor_181222_040429/4/
export/superconductor/1545452450
```

```
'Output':
gs://superconductor/jobs/superconductor_181222_040429/4/export/
superconductor/1545452450/
gs://superconductor/jobs/superconductor_181222_040429/4/export/
superconductor/1545452450/saved_model.pb
gs://superconductor/jobs/superconductor_181222_040429/4/export/
superconductor/1545452450/variables/
```

- Deploy the model.

```
%%bash
MODEL_NAME="superconductor"
MODEL_VERSION="v1"
MODEL_LOCATION=gs://$bucket_name/jobs/
superconductor_181222_040429/4/export/superconductor/1545452450

echo "Deploying model $MODEL_NAME $MODEL_VERSION"
gcloud ai-platform models create ${MODEL_NAME} --regions us-central1
gcloud ai-platform versions create ${MODEL_VERSION} --model
${MODEL_NAME} --origin ${MODEL_LOCATION} --runtime-version ${tf_
version}
```

Batch Prediction

The following code carries out inference on the deployed model:

- Submit a batch prediction job.

```
%%bash
JOB_NAME=superconductor_prediction
MODEL_NAME=superconductor
MODEL_VERSION=v1
TEST_FILE=gs://$bucket_name/preproc_csv/data/eval-00-of-01.csv
OUTPUT_DIR=gs://$bucket_name/jobs/$JOB_NAME/predictions

echo $OUTPUT_DIR

# submit a batched job
gcloud ai-platform jobs submit prediction $JOB_NAME \
        --model $MODEL_NAME \
        --version $MODEL_VERSION \
        --data-format TEXT \
        --region $region \
        --input-paths $TEST_FILE \
            --output-path $OUTPUT_DIR
```

```
# stream job logs
echo "Job logs..."
gcloud ml-engine jobs stream-logs $JOB_NAME
```

'Output':
```
gs://superconductor/jobs/superconductor_prediction/predictions
Job logs...
INFO    2018-12-22 22:04:22 +0000    service    Validating job
                                                requirements...
INFO    2018-12-22 22:04:22 +0000    service    Job creation
                                                request has been
                                                successfully
                                                validated.
INFO    2018-12-22 22:04:22 +0000    service    Job superconductor_
                                                prediction is
                                                queued.
INFO    2018-12-22 22:09:09 +0000    service    Job completed
                                                successfully.
```

- List the contents of the prediction output directory in GCS.

```
%%bash
gsutil ls gs://superconductor/jobs/superconductor_prediction/
predictions/
```

'Output':
```
gs://superconductor/jobs/superconductor_prediction/predictions/
prediction.errors_stats-00000-of-00001
gs://superconductor/jobs/superconductor_prediction/predictions/
prediction.results-00000-of-00002
gs://superconductor/jobs/superconductor_prediction/predictions/
prediction.results-00001-of-00002
```

- Show predicted RMSE outputs.

```bash
%bash
# read output summary
echo "Job output summary:"
gsutil cat 'gs://superconductor/jobs/superconductor_prediction/
predictions/prediction.results-00000-of-00002'
```

```
'Output':
{"outputs": [0.02159707620739937]}
{"outputs": [0.13300871849060059]}
{"outputs": [0.02054387889802456]}
{"outputs": [0.09370037913322449]}
                    . . .
{"outputs": [0.41005855798721313]}
{"outputs": [0.39907798171043396]}
{"outputs": [0.4040292799472809]}
{"outputs": [0.43743470311164856]}
```

This chapter provided a walk-through of an end-to-end process to model and deploy a machine learning solution on Google Cloud Platform. The next chapter will introduce the concepts of a microservice architecture. It provides an overview of working with Docker containers and their orchestration with Kubernetes on GCP.

PART VIII

Productionalizing Machine Learning Solutions on GCP

## CHAPTER 45

Containers and Google Kubernetes Engine

The microservice architecture is an approach for developing and deploying enterprise cloud-native software applications that involve separating the core business capabilities of the application into decoupled components. Each business capability represents some functionality that the application provides as services to the end user. The idea of microservices is in contrast to the monolithic architecture which involves building applications as a composite of its "individual" capabilities. See an illustration in Figure 45-1.

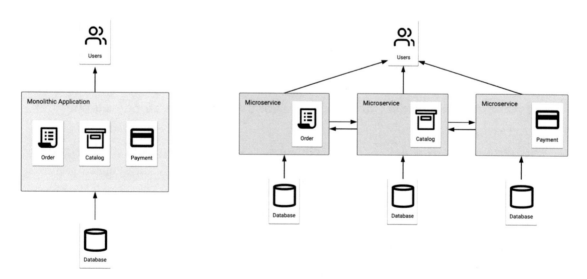

Figure 45-1. *Microservice applications (right) vs. monolithic applications (left)*

© Ekaba Bisong 2019

E. Bisong, *Building Machine Learning and Deep Learning Models on Google Cloud Platform*,
https://doi.org/10.1007/978-1-4842-4470-8_45

Microservices interact with each other using representational state transfer (REST) communications for stateless interoperability. By stateless, we mean that "the server does not store state about the client session." These protocols can be HTTP request/response APIs or an asynchronous messaging queue. This flexibility allows the microservice to easily scale and respond to request even if another microservice fails.

Advantages of Microservices

- Loosely coupled components make the application fault tolerant.

- Ability to scale out making each component highly available.

- The modularity of components makes it easier to extend existing capabilities.

Challenges with Microservices

- The software architecture increases in complexity.

- Overhead in management and orchestration of microservices. We will, however, see in the next sessions how Docker and Kubernetes work to mitigate this challenge.

Docker

Docker is a virtualization application that abstracts applications into isolated environments known as containers. The idea behind a container is to provide a unified platform that includes the software tools and dependencies for developing and deploying an application.

The traditional way of developing applications is where an application is designed and hosted on a single server. This is illustrated in Figure 45-2. This setup is prone to several problems including the famous "it works on my machine but not on yours". Also in this architecture, apps are difficult to scale and to migrate resulting in huge costs and slow deployment.

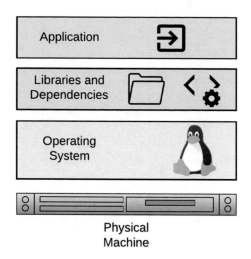

Figure 45-2. *Application running on a single server*

Virtual Machines vs. Containers

Virtual machines (VMs), illustrated in Figure 45-3, emulate the capabilities of a physical machine making it possible to install and run operating systems by using a hypervisor. The hypervisor is a piece of software on the physical machine (the host) that makes it possible to carry out virtualization where multiple guest machines are managed by the host machine.

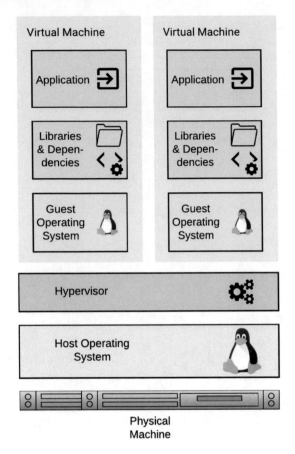

Figure 45-3. *Virtual machines*

Containers on the other hand isolate the environment for hosting an application with its own libraries and software dependencies; however, as opposed to a VM, containers on a machine all share the same operating system kernel. Docker is an example of a container. This is illustrated in Figure 45-4.

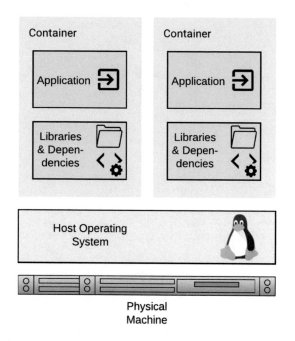

Figure 45-4. *Containers*

Working with Docker

Google Cloud Shell comes pre-configured with Docker.

Key concepts to note are

- Dockerfile: A Dockerfile is a text file that specifies how an image will be created.

- Docker images: Images are created by building a Dockerfile.

- Docker containers: Docker containers are the running instance of an image.

The diagram in Figure 45-5 highlights the process to build an image and run a Docker container.

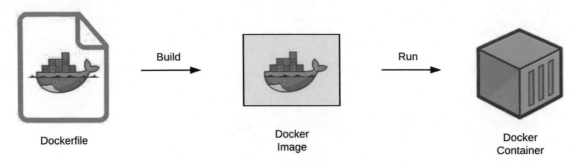

Figure 45-5. *Steps to deploying a Docker container*

Table 45-1 shows key commands when creating a Dockerfile.

Table 45-1. *Commands for Creating Dockerfiles*

Command	Description
FROM	The base Docker image for the Dockerfile.
LABEL	Key-value pair for specifying image metadata.
RUN	It executes commands on top of the current image as new layers.
COPY	Copies files from the local machine to the container file system.
EXPOSE	Exposes runtime ports for the Docker container.
CMD	Specifies the command to execute when running the container. This command is overridden if another command is specified at runtime.
ENTRYPOINT	Specifies the command to execute when running the container. Entrypoint commands are not overridden by a command specified at runtime.
WORKDIR	Set working directory of the container.
VOLUME	Mount a volume from the local machine file system to the Docker container.
ARG	Set Environment variable as a key-value pair when building the image.
ENV	Set Environment variable as a key-value pair that will be available in the container after building.

Build and Run a Simple Docker Container

Clone the book repository to run this example in Cloud Shell; we have a bash script titled **date-script.sh** in the chapter folder. The script assigns the current date to a variable and then prints out the date to the console. The Dockerfile will copy the script from the local machine to the docker container file system and execute the shell script when running the container. The Dockerfile to build the container is stored in **docker-intro/hello-world**.

```
# navigate to the folder with images
cd docker-intro/hello-world
```

Let's view the bash script.

```
cat date-script.sh
```

```
#! /bin/sh
DATE="$(date)"
echo "Todays date is $DATE"
```

Let's view the Dockerfile.

```
# view the Dockerfile
cat Dockerfile
```

```
# base image for building container
FROM docker.io/alpine
# add maintainer label
LABEL maintainer="dvdbisong@gmail.com"
# copy script from local machine to container file system
COPY date-script.sh /date-script.sh
# execute script
CMD sh date-script.sh
```

The Docker image will be built off the Alpine Linux package. See `https://hub.docker.com/_/alpine`. The CMD routine executes the script when the container runs.

Build the Image

Run the following command to build the Docker image.

```
# build the image
docker build -t ekababisong.org/first_image .
```

Build output

```
Sending build context to Docker daemon   2.048kB
Step 1/4 : FROM docker.io/alpine
latest: Pulling from library/alpine
6c40cc604d8e: Pull complete
Digest: sha256:b3dbf31b77fd99d9c08f780ce6f5282aba076d70a513a8be859d8d3a4d0c92b8
Status: Downloaded newer image for alpine:latest
 ---> caf27325b298
Step 2/4 : LABEL maintainer="dvdbisong@gmail.com"
 ---> Running in 306600656ab4
Removing intermediate container 306600656ab4
 ---> 33beb1ebcb3c
Step 3/4 : COPY date-script.sh /date-script.sh
 ---> Running in 688dc55c502a
Removing intermediate container 688dc55c502a
 ---> dfd6517a0635
Step 4/4 : CMD sh date-script.sh
 ---> Running in eb80136161fe
Removing intermediate container eb80136161fe
 ---> e97c75dcc5ba
Successfully built e97c75dcc5ba
Successfully tagged ekababisong.org/first_image:latest
```

Run the Container

Execute the following command to run the Docker container.

```
# show the images on the image
docker images
```

```
# run the docker container from the image
docker run ekababisong.org/first_image
```

```
Todays date is Sun Feb 24 04:45:08 UTC 2019
```

Important Docker Commands

In this section, let's review some important Docker commands.

Commands for Managing Images

Table 45-2 contains commands for managing Docker images.

Table 45-2. *Docker Commands for Managing Images*

Command	Description
docker images	List all images on the machine.
docker rmi [IMAGE_NAME]	Remove the image with name IMAGE_NAME on the machine.
docker rmi $(docker images -q)	Remove all images from the machine.

Commands for Managing Containers

Table 45-3 contains commands for managing Docker containers.

Table 45-3. *Docker Commands for Managing Containers*

Command	Description
docker ps	List all containers. Append –a to also list containers not running.
docker stop [CONTAINER_ID]	Gracefully stop the container with [CONTAINER_ID] on the machine.
docker kill CONTAINER_ID]	Forcefully stop the container with [CONTAINER_ID] on the machine.
docker rm [CONTAINER_ID]	Remove the container with [CONTAINER_ID] from the machine.
docker rm $ (docker ps -a -q)	Remove all containers from the machine.

Running a Docker Container

Let's break down the following command for running a Docker container:

```
docker run -d -it --rm --name [CONTAINER_NAME] -p 8081:80 [IMAGE_NAME]
```

where

- `-d` runs the container in detached mode. This mode runs the container in the background.

- `-it` runs in interactive mode, with a terminal session attached.

- `--rm` removes the container when it exits.

- `--name` specifies a name for the container.

- `-p` does port forwarding from host to the container (i.e., host:container).

Kubernetes

When a microservice application is deployed in production, it usually has many running containers that need to be allocated the right amount of resources in response to user demands. Also, there is a need to ensure that the containers are online, are running, and are communicating with one another. The need to efficiently manage and coordinate clusters of containerized applications gave rise to Kubernetes.

Kubernetes is a software system that addresses the concerns of deploying, scaling, and monitoring containers. Hence, it is called a container orchestrator. Examples of other container orchestrators in the wild are Docker Swarm, Mesos Marathon, and HashiCorp Nomad.

Kubernetes was built and released by Google as an open source software, which is now managed by the Cloud Native Computing Foundation (CNCF). Google Cloud Platform offers a managed Kubernetes service called Google Kubernetes Engine (GKE). Amazon Elastic Container Service for Kubernetes (EKS) also provides a managed Kubernetes service.

Features of Kubernetes

The following are some features of Kubernetes:

- Horizontal auto-scaling: Dynamically scales containers based on resource demands

- Self-healing: Re-provisions failed nodes in response to health checks

- Load balancing: Efficiently distributes requests between containers in a pod

- Rollbacks and updates: Easily update or revert to a previous container deployment without causing application downtime

- DNS service discovery: Uses Domain Name System (DNS) to manage container groups as a Kubernetes service

Components of Kubernetes

The main components of the Kubernetes engine are

- Master node(s): Manages the Kubernetes cluster. There may be more than one master node in high availability mode for fault-tolerance purposes. In this case, only one is the master, and the others follow.

- Worker node(s): Machine(s) that runs containerized applications that are scheduled as pod(s).

The illustration in Figure 45-6 provides an overview of the Kubernetes architecture.

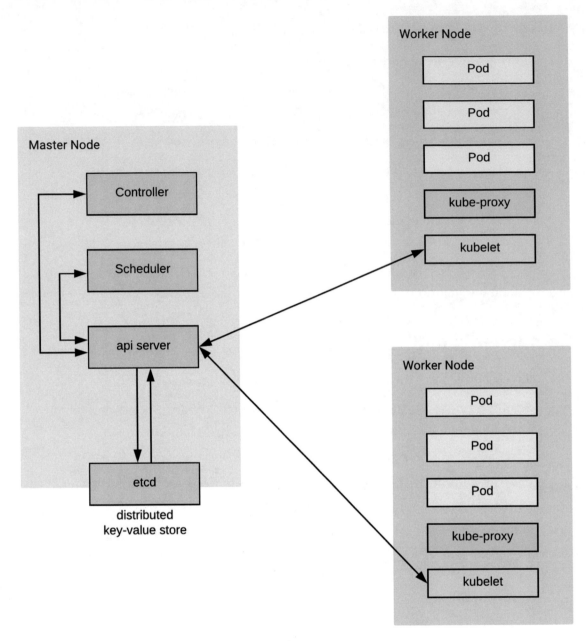

Figure 45-6. *High-level overview of Kubernetes components*

Master Node(s)

The master node consists of

- **etcd (distributed key-store):** It manages the Kubernetes cluster state. This distributed key-store can be a part of the master node or external to it. Nevertheless, all master nodes connect to it.

- **api server:** It manages all administrative tasks. The api server receives commands from the user (kubectl cli, REST or GUI); these commands are executed and the new cluster state is stored in the distributed key-store.

- **scheduler:** It schedules work to worker nodes by allocating pods. It is responsible for resource allocation.

- **controller:** It ensures that the desired state of the Kubernetes cluster is maintained. The desired state is what is contained in a JSON or YAML deployment file.

Worker Node(s)

The worker node(s) consists of

- **kubelet:** The kubelet agent runs on each worker node. It connects the worker node to the api server on the master node and receives instructions from it. It ensures the pods on the node are healthy.

- **kube-proxy:** It is the Kubernetes network proxy that runs on each worker node. It listens to the api server and forwards requests to the appropriate pod. It is important for load balancing.

- **pod(s):** It consists of one or more containers that share network and storage resources as well as container runtime instructions. Pods are the smallest deployable unit in Kubernetes.

Writing a Kubernetes Deployment File

The Kubernetes deployment file defines the desired state for the various Kubernetes objects. Examples of Kubernetes objects are

- **Pods:** It is a collection of one or more containers.

- **ReplicaSets:** It is part of the controller in the master node. It specifies the number of replicas of a pod that should be running at any given time. It ensures that the specified number of pods is maintained in the cluster.

- **Deployments:** It automatically creates ReplicaSets. It is also part of the controller in the master node. It ensures that the cluster's current state matches the desired state.

- **Namespaces:** It partitions the cluster into sub-clusters to organize users into groups.

- **Service:** It is a logical group of pods with a policy to access them.

  - *ServiceTypes:* It specifies the type of service, for example, ClusterIP, NodePort, LoadBalancer, and ExternalName. As an example, LoadBalancer exposes the service externally using a cloud provider's load balancer.

Other important tags in writing a Kubernetes deployment file

- **spec:** It describes the desired state of the cluster

- **metadata:** It contains information of the object

- **labels:** It is used to specify attributes of objects as key-value pairs

- **selector:** It is used to select a subset of objects based on their label values

The deployment file is specified as a yaml file.

Deploying Kubernetes on Google Kubernetes Engine

Google Kubernetes engine (GKE) provides a managed environment for deploying application containers. To create and deploy resources on GCP from the local shell, the Google command-line SDK gcloud will have to be installed and configured. If this is not the case on your machine, follow the instructions at https://cloud.google.com/sdk/gcloud/. Otherwise, a simpler option is to use the Google Cloud Shell which already has gcloud and kubectl (the Kubernetes command-line interface) installed.

Creating a GKE Cluster

Run the following command to create a cluster of containers on GKE. Assign the cluster name.

```
# create a GKE cluster
gcloud container clusters create my-gke-cluster-name
```

A Kubernetes cluster is created on GCP with three nodes (as default). The GKE dashboard on GCP is shown in Figure 45-7.

```
Creating cluster ekaba-gke-cluster in us-central1-a... Cluster is being
deployed...done.
Created [https://container.googleapis.com/v1/projects/oceanic-sky-230504/
zones/us-central1-a/clusters/ekaba-gke-cluster].
To inspect the contents of your cluster, go to: https://console.
cloud.google.com/kubernetes/workload_/gcloud/us-central1-a/ekaba-gke-
cluster?project=oceanic-sky-230504
kubeconfig entry generated for ekaba-gke-cluster.
NAME                 LOCATION        MASTER_VERSION  MASTER_IP      MACHINE_
TYPE    NODE_VERSION  NUM_NODES  STATUS
ekaba-gke-cluster  us-central1-a  1.11.7-gke.4     35.226.72.40  n1-
standard-1  1.11.7-gke.4   3           RUNNING
```

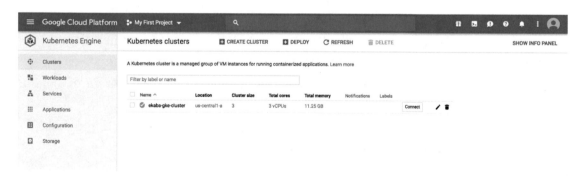

Figure 45-7. *Google Kubernetes Engine dashboard*

To learn more about creating clusters with Google Kubernetes Engine, visit `https://cloud.google.com/kubernetes-engine/docs/how-to/creating-a-cluster`.

Run the following command to display the nodes of the provisioned cluster on GKE.

```
# get the nodes of the kubernetes cluster on GKE
kubectl get nodes
```

NAME	STATUS	ROLES	AGE
VERSION			
gke-ekaba-gke-cluster-default-pool-e28c64e0-8fk1	Ready	<none>	45m
v1.11.7-gke.4			
gke-ekaba-gke-cluster-default-pool-e28c64e0-fmck	Ready	<none>	45m
v1.11.7-gke.4			
gke-ekaba-gke-cluster-default-pool-e28c64e0-zzz1	Ready	<none>	45m
v1.11.7-gke.4			

Delete the Kubernetes Cluster on GKE

Run the following command to delete a cluster on GKE.

```
# delete the kubernetes cluster
gcloud container clusters delete my-gke-cluster-name
```

Note   Always remember to clean up cloud resources when they are no longer needed.

This chapter introduced the concepts of a microservice architecture and provided an overview of working with Docker containers for building applications in isolated environments/sandboxes. In the event that many of such containers are deployed in production, this chapter introduces Kubernetes as a container orchestrator for managing the concerns of deploying, scaling, and monitoring containers.

The next chapter will discuss on Kubeflow and Kubeflow Pipelines for deploying machine learning components into production on Kubernetes.

## CHAPTER 46

Kubeflow and Kubeflow Pipelines

Machine learning is often and rightly viewed as the use of mathematical algorithms to teach the computer to learn tasks that are computationally infeasible to program as a set of specified instructions. However, it turns out that these algorithms constitute only a small fraction of the overall learning pipeline from an engineering perspective. Building high-performant and dynamic learning models includes a number of other critical components. These components actually dominate the space of concerns for delivering an end-to-end machine learning product.

A typical machine learning production pipeline looks like the illustration in Figure 46-1.

© Ekaba Bisong 2019

E. Bisong, *Building Machine Learning and Deep Learning Models on Google Cloud Platform*, https://doi.org/10.1007/978-1-4842-4470-8_46

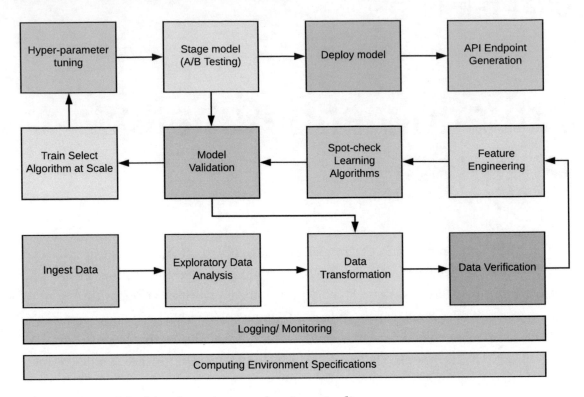

Figure 46-1. *Machine learning production pipeline*

From the preceding diagram, observe that the process flow in the pipeline is iterative. This repetitive pattern is central to machine learning experimentation, design, and deployment.

The Efficiency Challenge

It is easy to recognize that the pipeline requires a significant amount of development operations for the seamless transition from one component to another when building a learning model. This interoperability of parts has given rise to Machine Learning Ops, also known as MLOps. The term is coined as an amalgam of Machine Learning and DevOps.

The conventional way of doing machine learning is to perform all of the Experiment and development work in Jupyter notebooks, and the model is exported and sent off to the software development team for deployment and endpoint generation for integration into downstream software products, while the DevOps team handles the infrastructure and configuration of the machine for model development. This monolithic style of working results in a machine learning process that is not reusable, difficult to scale and maintain, and even tougher to audit and perform model improvement, and it is easily fraught with errors and unnecessary complexities.

However, by incorporating the microservice design pattern to machine learning development, we can address a host of these concerns and really streamline the productionalization process.

Kubeflow

Kubeflow is a platform that is created to enhance and simplify the process of deploying machine learning workflows on Kubernetes. Using Kubeflow, it becomes easier to manage a distributed machine learning deployment by placing components in the deployment pipeline such as the training, serving, monitoring, and logging components into containers on the Kubernetes cluster.

The goal of Kubeflow is to abstract away the technicalities of managing a Kubernetes cluster so that a machine learning practitioner can quickly leverage the power of Kubernetes and the benefits of deploying products within a microservice framework. Kubeflow has its history as an internal Google framework for implementing machine learning pipelines on Kubernetes before being open sourced late 2017.

Table 46-1 is a sample of some of the components that run on Kubeflow.

Table 46-1. *Sample of Kubeflow Components*

Component	Description
 Chainer	**Chainer** is a define-by-run deep learning neural network framework. It also supports multi-node distributed deep learning and deep reinforcement algorithms.
 Jupyter	**Jupyter** provides a platform for the rapid prototyping and easy sharing of reproducible codes, equations, and visualizations.
 ksonnet	**ksonnet** provides a simple way to create and edit Kubernetes configuration files. Kubeflow makes use of ksonnet to help manage deployments.
 Istio	**Istio** eases microservice deployments by providing a uniform way to connect, secure, control, and observe services.
 K a t i b **Katib**	**Katib** is a deep learning framework agnostic hyper-parameter tuning framework. It is inspired by Google Vizier.
 MXNet	**MXNet** is a portable and scalable deep learning library using multiple frontend languages such as Python, Julia, MATLAB, and JavaScript.

(continued)

Table 46-1. (*continued*)

Component	Description
PyTorch	**PyTorch** is a Python deep learning library developed by Facebook based on the Torch library for Lua, a programming language.
 **NVIDIA TensorRT**	**TensorRT** is a platform for high-performance and scalable deployment of deep learning models for inference.
Seldon	**Seldon** is an open source platform for deploying machine learning models on Kubernetes.
TensorFlow	**TensorFlow** provides an ecosystem for the large-scale productionalization of deep learning models. This includes distributed training using TFJob, serving with TF Serving, and other Tensorflow Extended components such as TensorFlow Model Analysis (TFMA) and TensorFlow Transform (TFT).

Working with Kubeflow

1. **Set up a Kubernetes cluster on GKE.**

    ```
    # create a GKE cluster
    gcloud container clusters create ekaba-gke-cluster

    # view the nodes of the kubernetes cluster on GKE
    kubectl get nodes
    ```

2. **Create OAuth client ID to identify Cloud IAP:** Kubeflow uses
    Cloud Identity-Aware Proxy (Cloud IAP) to connect to Jupyter and
    other running web apps securely. Kubeflow uses email addresses
    for authentication. In this section, we'll create an OAuth client ID
    which will be used to identify Cloud IAP when requesting access
    to a user's email account:

- Go to the APIs & Services ➤ Credentials page in GCP Console.

- Go to the OAuth consent screen (see Figure 46-2).

  - Assign an Application name, for example, My-Kubeflow-App.

  - For authorized domains, use [YOUR_PRODJECT_ID].
    cloud.goog.

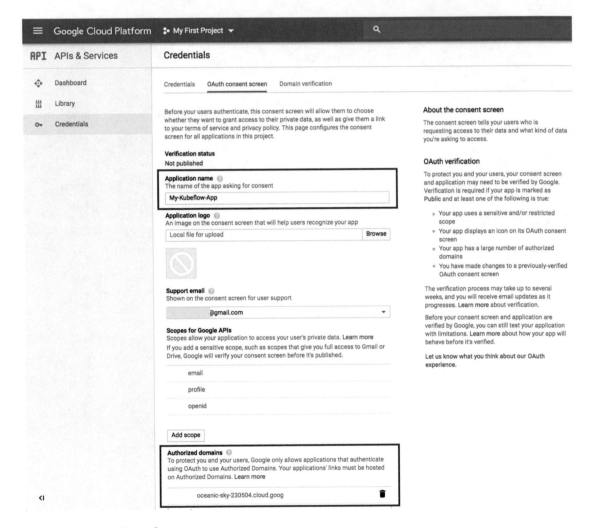

Figure 46-2. *OAuth consent screen*

- Go to the Credentials tab (see Figure 46-3).

    - Click Create credentials, and then click OAuth client ID.

    - Under Application type, select Web application.

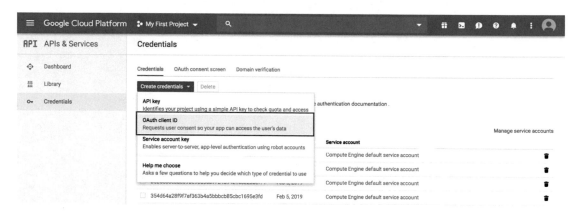

Figure 46-3. *GCP Credentials tab*

- Choose a **Name** to identify the OAuth client ID (see Figure 46-4).

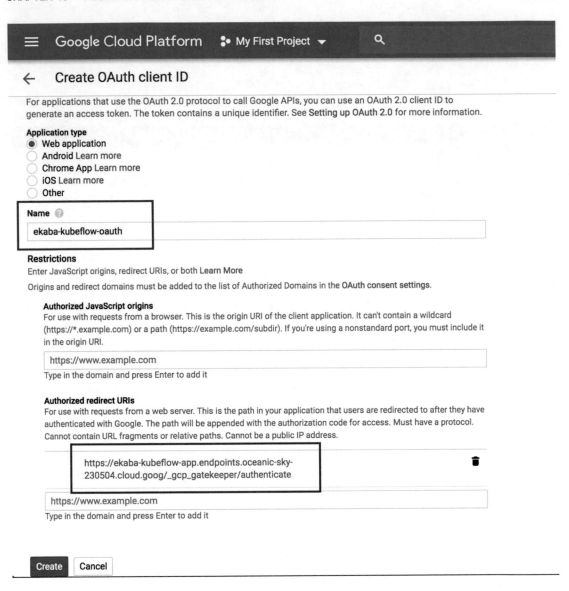

Figure 46-4. Create OAuth client ID

- In the Authorized redirect URIs box, enter the following:

  ```
  https://<deployment_name>.endpoints.<project>.cloud.
  goog/_gcp_gatekeeper/authenticate
  ```

- <deployment_name> must be the name of the Kubeflow deployment.

- <project> is the GCP project ID.

- In this case, it will be

  ```
  https://ekaba-kubeflow-app.endpoints.oceanic-
  sky-230504.cloud.goog/_gcp_gatekeeper/authenticate
  ```

- Take note of the client ID and client secret that appear in the OAuth client window. This is needed to enable Cloud IAP.

  ```
  # Create environment variables from the OAuth client ID and
  secret earlier obtained.
  export CLIENT_ID=506126439013-drbrjo36hihvdolgki6lflovm4bjb6c1.
  apps.googleusercontent.com
  export CLIENT_SECRET=bACWJuojIVm7PIMphzTOYz9D
  export PROJECT=oceanic-sky-230504
  ```

Download kfctl.sh

The file kfctl.sh is the Kubeflow installation shell script. As at this time of writing, the latest Kubeflow tag is 0.5.0.

```
# create a folder on the local machine
mkdir kubeflow

# move to created folder
cd kubeflow

# save folder path as a variable
export KUBEFLOW_SRC=$(pwd)

# download kubeflow `kfctl.sh`
export KUBEFLOW_TAG=v0.5.0

curl https://raw.githubusercontent.com/kubeflow/kubeflow/${KUBEFLOW_TAG}/
scripts/download.sh | bash

# list directory elements
ls -la
drwxr-xr-x    6 ekababisong   staff    204 17 Mar 04:15 .
drwxr-xr-x   25 ekababisong   staff    850 17 Mar 04:09 ..
drwxr-xr-x    4 ekababisong   staff    136 17 Mar 04:18 deployment
```

```
drwxr-xr-x  36 ekababisong  staff  1224 17 Mar 04:14 kubeflow
drwxr-xr-x  16 ekababisong  staff   544 17 Mar 04:14 scripts
```

Deploy Kubeflow

Run the following code block to deploy Kubeflow.

```
# assign the name for the Kubeflow deployment
# The ksonnet app is created in the directory ${KFAPP}/ks_app
export KFAPP=ekaba-kubeflow-app

# run setup script
${KUBEFLOW_SRC}/scripts/kfctl.sh init ${KFAPP} --platform gcp --project
${PROJECT}

# navigate to the deployment directory
cd ${KFAPP}

# creates config files defining the various resources for gcp
${KUBEFLOW_SRC}/scripts/kfctl.sh generate platform

# creates or updates gcp resources
${KUBEFLOW_SRC}/scripts/kfctl.sh apply platform

# creates config files defining the various resources for gke
${KUBEFLOW_SRC}/scripts/kfctl.sh generate k8s

# creates or updates gke resources
${KUBEFLOW_SRC}/scripts/kfctl.sh apply k8s

# view resources deployed in namespace kubeflow
kubectl -n kubeflow get  all
```

Kubeflow is available at a URL that will be unique for your deployment. In this case, Kubeflow is available to me at `https://ekaba-kubeflow-app.endpoints.` `oceanic-sky-230504.cloud.goog/` (see Figure 46-5). Again, this URL is unique for your deployment.

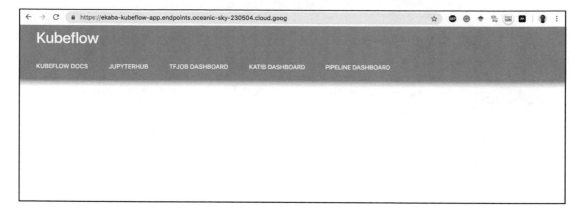

Figure 46-5. *The Kubeflow homescreen*

> **Note**   It can take 10–15 minutes for the URI to become available. Kubeflow needs to provision a signed SSL certificate and register a DNS name.

Kubeflow Pipelines – Kubeflow for Poets

Kubeflow Pipelines is a simple platform for building and deploying containerized machine learning workflows on Kubernetes. Kubeflow pipelines make it easy to implement production-grade machine learning pipelines without bothering on the low-level details of managing a Kubernetes cluster.

Kubeflow Pipelines is a core component of Kubeflow and is also deployed when Kubeflow is deployed. The Pipelines dashboard is shown in Figure 46-6.

Figure 46-6. *Kubeflow Pipelines dashboard*

Components of Kubeflow Pipelines

A Pipeline describes a machine learning workflow, where each component of the pipeline is a self-contained set of codes that are packaged as Docker images. Each pipeline can be uploaded individually and shared on the Kubeflow Pipelines user interface (UI). A pipeline takes inputs (parameters) required to run the pipeline and the inputs and outputs of each component.

The Kubeflow Pipelines platform consists of

- A user interface (UI) for managing and tracking Experiments, jobs, and runs

- An engine for scheduling multi-step ML workflows

- An SDK for defining and manipulating pipelines and components

- Notebooks for interacting with the system using the SDK (taken from `www.kubeflow.org/docs/pipelines/pipelines-overview/`)

Executing a Sample Pipeline

1. Click the name **[Sample] Basic - Condition** (see Figure 46-7).

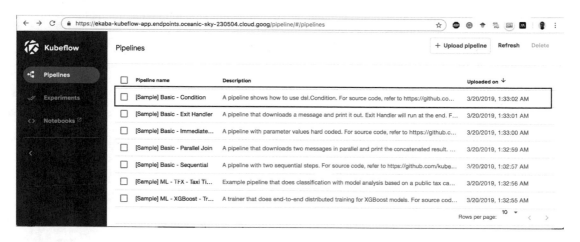

Figure 46-7. *Select a Pipeline*

2. Click **Start an Experiment** (see Figure 46-8).

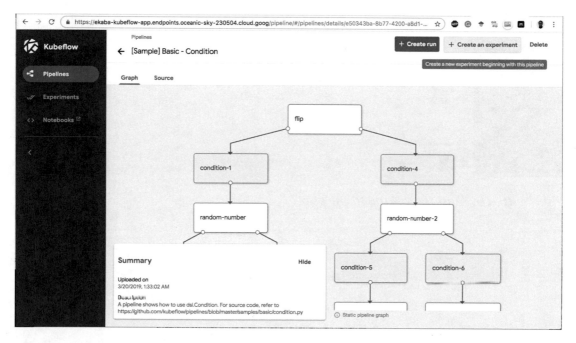

Figure 46-8. *Create a new Experiment*

3.   Give the Experiment a name (see Figure 46-9).

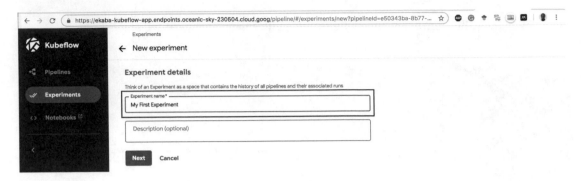

Figure 46-9. *Assign a name to the Experiment*

4.   Give the run a name (see Figure 46-10).

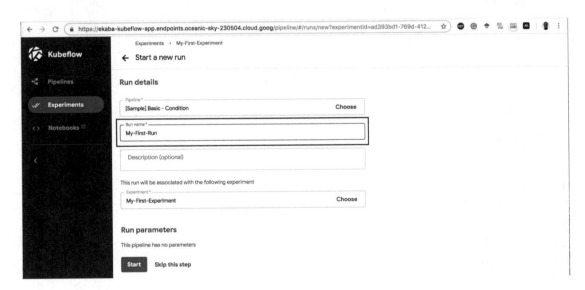

Figure 46-10. *Assign a name to the run*

5. Click the **Run Name** to start the run (see Figure 46-11).

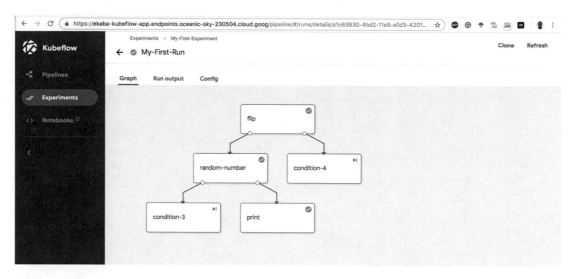

Figure 46-11. *Run the Pipeline*

Note   Always remember to clean up cloud resources when they are no longer needed.

This chapter covered setting up Kubeflow on Kubernetes and introduced working with Kubeflow Pipelines to manage containerized machine learning workflows. The next chapter will deploy an end-to-end machine learning solution with Kubeflow Pipelines.

CHAPTER 47

Deploying an End-to-End Machine Learning Solution on Kubeflow Pipelines

A Kubeflow pipeline component is an implementation of a pipeline task. A component is a step in the workflow. Each task takes one or more artifacts as input and may produce one or more artifacts as output.

Each component usually includes two parts:

- Client code: The code that talks to endpoints to submit jobs, for example, code to connect with the Google Cloud Machine Learning Engine.

- Runtime code: The code that does the actual job and usually runs in the cluster, for example, the code that prepares the model for training on Cloud MLE.

A component consists of an interface (inputs/outputs), the implementation (a Docker container image and command-line arguments), and metadata (name, description).

© Ekaba Bisong 2019

E. Bisong, *Building Machine Learning and Deep Learning Models on Google Cloud Platform*, https://doi.org/10.1007/978-1-4842-4470-8_47

Overview of a Simple End-to-End Solution Pipeline

In this simple example, we will implement a deep neural regressor network to predict the closing prices of Bitcoin crypto-currency. The machine learning code itself is pretty basic as it is not the focus of this article. The goal here is to orchestrate a machine learning engineering solution using microservice architectures on Kubernetes with Kubeflow Pipelines. The code for this chapter is in the book code repository. Clone the repository from the GCP Cloud Shell.

The pipeline consists of the following components:

1. Move raw data hosted on GitHub to a storage bucket.

2. Transform the dataset using Google Dataflow.

3. Carry out hyper-parameter training on Cloud Machine Learning Engine.

4. Train the model with the optimized hyper-parameters.

5. Deploy the model for serving on Cloud MLE.

Create a Container Image for Each Component

First, we'll package the client and runtime code into a Docker image. This image also contains the secure service account key to authenticate against GCP. For example, the component to transform the dataset using Dataflow has the following files built into its image:

- __ Dockerfile: Dockerfile to build the Docker image.

- __ build.sh: Script to initiate the container build and upload to Google Container Registry.

- __ dataflow_transform.py: Code to run the beam pipeline on Cloud Dataflow.

- __ service_account.json: Secure key to authenticate container on GCP.

- __ local_test.sh: Script to run the image pipeline component locally.

Build Containers Before Uploading to Kubeflow Pipelines

Before uploading the pipeline to Kubeflow Pipelines, be sure to build the component containers so that the latest version of the code is packaged and uploaded as images to the container registry. The code provides a handy bash script to build all containers.

Compile the Pipeline Using the Kubeflow Pipelines DSL Language

The pipeline code contains a specification on how the components interact with one another. Each component has an output that serves as an input to the next component in the pipeline. The Kubeflow pipeline DSL language dsl-compile from the Kubeflow Pipelines SDK is used to compile the pipeline code in Python for upload to Kubeflow Pipelines.

Ensure the Kubeflow Pipelines SDK is installed on the local machine by running

```
# install kubeflow pipeline sdk
pip install https://storage.googleapis.com/ml-pipeline/release/0.1.12/kfp.
tar.gz --upgrade

# verify the install
which dsl-compile
```

Compile the pipeline by running

```
# compile the pipeline
python3 [path/to/python/file.py] [path/to/output/tar.gz]
```

For the sample code, we used

```
python3 crypto_pipeline.py crypto_pipeline.tar.gz
```

Upload and Execute the Pipeline to Kubeflow Pipelines

The following steps upload and execute the compiled pipeline on Kubeflow Pipelines:

1. Upload the pipeline to Kubeflow Pipelines (Figure 47-1).

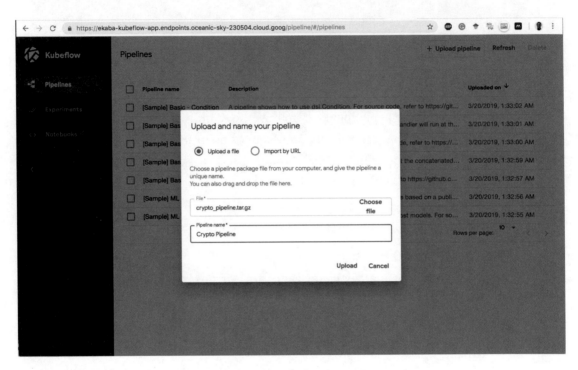

Figure 47-1.   *Upload the compiled pipeline to Kubeflow Pipelines*

2. Click the pipeline to see the static graph of the flow (Figure 47-2).

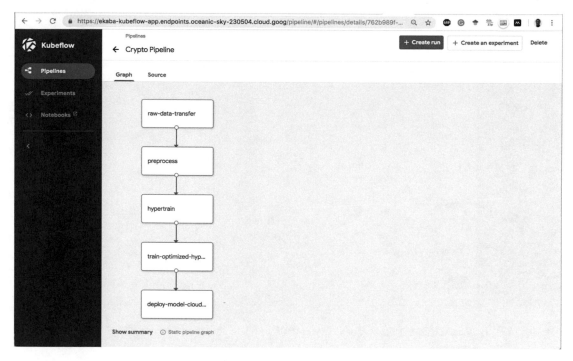

Figure 47-2. *Pipeline summary graph*

3. Create an Experiment and run to execute the pipeline
    (Figure 47-3).

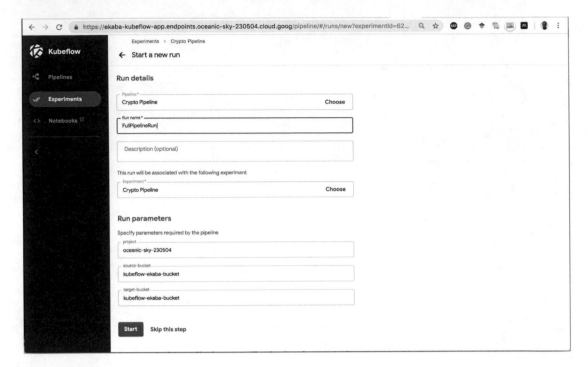

Figure 47-3. *Create and run the Experiment*

4.   Completed Pipeline run (Figure 47-4).

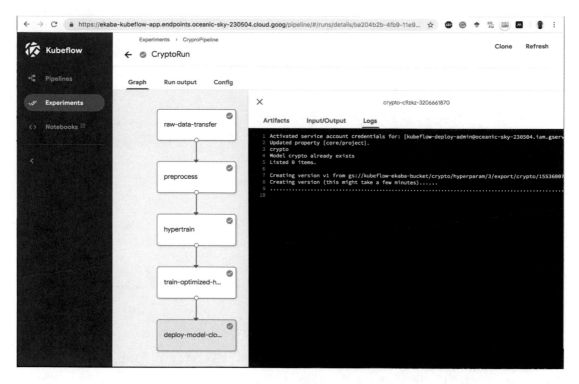

Figure 47-4.   *Completed Pipeline run*

Completed Dataflow Pipeline: The completed run of the second component of the Pipeline, which is to transform the dataset with Cloud Dataflow, is illustrated in Figure 47-5.

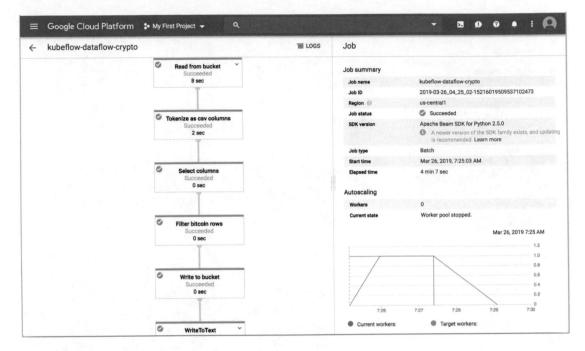

Figure 47-5. *Completed Dataflow run*

Deployed model on Cloud MLE: The deployed model on Cloud MLE, which is the fifth component of the Pipeline, is illustrated in Figure 47-6.

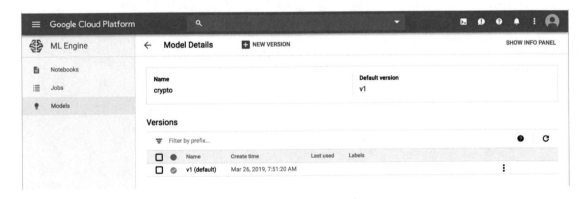

Figure 47-6. *Deployed model on Cloud MLE*

Note   Always remember to clean up cloud resources when they are no longer needed.

Delete Kubeflow: Run the script to delete the deployment.

```
# navigate to kubeflow app
cd ${KFAPP}
```

```
# run script to delete the deployment
${KUBEFLOW_SRC}/scripts/kfctl.sh delete all
```

Delete the Kubernetes cluster: Replace name with your own cluster name.

```
# delete the kubernetes cluster
gcloud container clusters delete ekaba-gke-cluster
```

This chapter covered building an end-to-end machine learning product as a containerized application on Kubernetes with Kubeflow and Kubeflow pipelines. Again, the code for this chapter may be accessed by cloning the book repository to the Cloud Shell.

This concludes this book.

Index

A

Accuracy, 181, 294
Activation functions
    hyperbolic tangent (tanh), 341
    Leaky ReLU, 342, 343
    Maxout, 343
    non-linear function, 339
    ReLU, 342
    sigmoid, 340
Adaptive learning rates, 413
Alpine Linux package, 661
append() method, 76, 128, 129
Area under the receiver operating curve
    (AUC-ROC), 183–184, 294
argparse.ArgumentParser()
    method, 558
Artificial neural network
    (ANN), 329, 331, 332
assign method, 127
Autoencoder
    architecture, 476
    defined, 475
    denoising, 481, 482
    undercomplete, 475
AutoML NLP
    custom language classification
        model (*see* Custom language
        classification model)
    dataset, training, 602–604
    GCP, 599–601

B

Backpropagation
    algorithm, 337, 338
Backpropagation through time
    (BPTT), 453, 454
Bar plot, 154, 155
Batch gradient descent
    algorithm, 205
Batch learning, 199–200
Batch normalization, 408–410
Beam programming
    data processing pipeline
        build/run, 541, 542
        creation, 540
        preprocessing, 543
    pipeline transformation
        I/O transforms, 539
        Pcollection, 538
        Ptransform, 538
Bias *vs.* variance trade-Off
    hidden layers, 402–403
    high bias, 178
    high variance, 178
    machine learning, 177
    quality/performance, 178
BigQuery
    defined, 485
    first query
        census_bureau_international, 490
        Query editor, 491

© Ekaba Bisong 2019
E. Bisong, *Building Machine Learning and Deep Learning Models on Google Cloud Platform*,
https://doi.org/10.1007/978-1-4842-4470-8

R

Printed in the United States
By Bookmasters